Molekülverbindungen und Koordinationsverbindungen
in Einzeldarstellungen

Herausgegeben von
G. Briegleb · F. Cramer · H. Hartmann

Chr. Klixbüll Jørgensen

Oxidation Numbers and Oxidation States

Springer-Verlag New York Inc. 1969

Dr. Chr. Klixbüll Jørgensen

Cyanamid European Research Institute
Cologny, Geneva

Title No. 6388

Preface

The correlation of spectroscopic and chemical investigations in recent years has been highly beneficial of many reasons. Around 1950, no valid explanation was available of the colours of compounds of the five transition groups. Later, it was possible to identify the excited levels with those expected for an electron configuration with a definite number of electrons in the partly filled shell. It is not generally recognized that this is equivalent to determining spectroscopic oxidation states related to the preponderant electron configuration and not to estimates of the fractional atomic charges. This brings in an entirely different type of description than the formal oxidation numbers used for characterizing compounds and reaction schemes. However, it must be realized that collectively oxidized ligands, formation of cluster-complexes and catenation may prevent the oxidation state from being well-defined.

The writer would like to express his gratitude to many, but first of all to DR. CLAUS SCHÄFFER, University of Copenhagen, who is the most efficient group-theoretical engineer known to the writer; his comments and discussions have been highly valuable. The writer's colleague, Professor FAUSTO CALDERAZZO (now going to the University of Pisa) has been most helpful in metallo-organic questions. Thanks are also due to Professors E. RANCKE-MADSEN and K. A. JENSEN for correspondence and conversations about formal oxidation numbers. It was finally of great importance for this book that Professor JANNIK BJERRUM (of whom we both were students) in 1952 required Miss LENE RASMUSSEN, now at the University of Aarhus, to write a dissertation about unusual oxidation states in the first transition group. The subsequent discussions are not finished yet; chemistry is complicated though fascinating.

Geneva, October 1968   CHR. KLIXBÜLL JØRGENSEN

Table of Contents

Oxidation Numbers
and Oxidation States

1. Introduction

Chemistry is very young, as a science, and the most plausible and coherent hypotheses about the nature of the chemical bonding have changed very rapidly. However, the waves of new discoveries have deposited a sediment of theory, which is not so thixotropic as sand though it is not as firm as rock.

During the Nineteenth Century, the two main components of modern theory were suggested by a variety of chemists. The electrovalent bonding caused by negative and positive charges on the atoms participating in chemical compounds and the covalent bonding caused by electron pairs being bound to two adjacent atoms more firmly than to the isolated atoms, are both necessary. There exist compounds, such as crystalline NaCl, which are nearly exclusively electrovalent, whereas homonuclear molecules such as H_2, N_2, P_4, S_8 and crystalline diamond, Si, and Te are exclusively covalent.

BERZELIUS suggested residual fractional charges on the atoms in molecules, somewhat related to our concept of electronegativities. For instance in oxides such as CaO and SO_3, oxygen carries a fractional negative charge and calcium and sulphur positive charges. According to BERZELIUS' dualistic theory, there exists an attraction between these two molecules, forming $CaO \cdot SO_3$ under considerable evolution of heat. Since CaO and SO_3 manifestly are electrically neutral materials, as are O_2 and calcium metal, for that matter, this residual attraction in the "sulphate of lime" must be due to some kind of dipole or multipole interactions, or, what is more plausible to us, some kind of redistribution of electric charges during the chemical reaction. We are going to discuss this question of fractional atomic charges extensively.

However, FARADAY started another line of thought involving integral atomic charges. It is indeed just that 96488 coulomb, the electric charge of a mole of protons, is now called a faraday. There is not the slightest doubt that by electrolysis, elemental bromine forms Br^-, silver Ag^+ and cadmium Cd^{++} in aqueous solution. It took a long time before chemists accepted the idea of ions not being formed specifically only under the circumstances of electrolysis, but having a quite stable existence in water and other solvents. ARRHENIUS had the courage to emit the hypothesis of ionic dissociation of electrolytes, and quantitative treatment on the basis of the mass-action law succeeded fairly well, though, as N.

BJERRUM pointed out, it is doubtful whether strong solutions of Na$^+$ and Cl$^-$ contain any undissociated NaCl at all. When BRAGG initiated X-ray crystallography, it turned out that no NaCl molecules exist in crystals; each Na$^+$ is surrounded by six equivalent Cl$^-$ in a regular octahedron, and each Cl$^-$ surrounded by six equivalent Na$^+$. However, it is the atomic positions which are determined; it needs extremely refined measurements to get convinced about complete ionicity, i.e. that the sodium nucleus is surrounded by ten electrons only, and the chlorine nucleus by eighteen.

As we shall see, life is not so simple as to allow a sharp distinction between electrovalent and covalent compounds. In this sense, we are going to define formal oxidation numbers. It is completely legitimate when many chemists consider oxidation numbers as a purely formal tool for the purpose of book-accounting of redox (i.e. reduction-oxidation) reactions. However, the progress in the understanding of chemical bonding and in the spectroscopy of transition group complexes which has taken place since 1950 allows a more significant definition of various forms of oxidation states in many, but not all, compounds. Cases exist, where either the formal oxidation number or the oxidation state cannot be defined reasonably, but where the other concept is quite clear-cut. In the cases where both can be defined, they do not need to coincide, as we shall see.

It is common for the formal oxidation number and the some four different forms of oxidation states that they are all based on some perfect or approximate separation of an integral number of electrons from all the other electrons in the rest of the universe. From the point of view of quantum chemistry, this is a question of distinct volumes containing an integral number of electrons. The actual SCHRÖDINGER-HEISENBERG formulation emphasizes that the electrons are indistinguishable, and that *a* number of electrons remains in such a volume. This technicality may cover more profound changes to be expected in future versions of quantum mechanics. As first discussed by DIRAC and by LÖWDIN (cf. JØRGENSEN, 1962), if only two-particle interactions occur in electronic systems, any observable quantity is connected with the electronic density in our three-dimensional space and with the correlated electronic second-order density-matrix (or second-order density-operator, according to KUTZELNIGG) considering the positions of two electrons at a time and hence having six spatial and two spin variables. Hence, the number of electrons in a system has not exactly the same connotations as the number of pedestrians in a street, but is rather a quantum number, as we shall discuss further at a later point.

In dilute gases, monatomic entities and molecules are sufficiently separated in this sense and are either neutral or carry charges which are a positive or negative integer multiplied by the protonic charge. We reserve *Arabic numerals* for these, sharply defined, charges and write Li^+, Mg^{++} or Mg^{+2}, Cr^{+3}, H_2^+ and KrH^+ for such species. The IUPAC nomenclature commission recommends at present to write Mg^{2+}, Cr^{3+}, ... but the writer cannot help feeling that there is in no sense a positively charged entity the multiple of which should be indicated as a number preceding the plus sign. In the gaseous state, we have a more serious problem of nomenclature. There is a long tradition among atomic spectroscopists to name the neutral chromium atom Cr I (which we write Cr or Cr^0), to name Cr^+ Cr II, Cr^{+2} Cr III, etc. This is not as aberrant as it might seem to a chemist. The first spectrum of an element M is that of the neutral atom, the second spectrum is that of M^+, and so on. Until recently, the gaseous monatomic anions such as Cl^- had not been studied by atomic spectroscopists, and even now (cf. BERRY et al., 1962) there is no evidence for discrete, excited levels of gaseous Cl^- below the two ionization limits. Hence, we leave Roman numerals on line without parentheses to atomic spectroscopists, and we use them nowhere in this book.

In solution or in solids, polyatomic *complex ions* such as $N(CH_3)_4^+$, $Co(NH_3)_6^{+3}$, BF_4^-, SO_4^{-2}, $PtCl_6^{-2}$ and FeF_6^{-3} retain a considerable individuality, and we write Roman numerals though this is a less perfect approximation than for gaseous species. It is a less clear-cut question what to do with hydrated ions. We write Cl^- and I^- in aqueous solution, because the evidence available suggests approximately localized charge distributions. On the other hand, spectroscopic measurements show that Ni^{+2} and Cr^{+3} do not carry two or three full positive charges in aqueous solution, and that defined aqua ions $Ni(H_2O)_6^{+2}$ and $Cr(H_2O)_6^{+3}$ are formed, retaining (at room temperature) the individuality for roughly 10^{-4} sec. and one day, respectively. There is no doubt that Mg^{+2} is not as good a description as $Mg(H_2O)_6^{+2}$ or Mg(II) aqua ions, but the case is doubtful for Li(I) and Ca(II) aqua ions. On the other hand, we can write K^+, Cs^+ and Ba^{+2} in aqueous solution, or Ca^{+2} in CaF_2 crystals, with the arguments about good charge separation.

It is clear that individual monatomic species in aqueous solution meriting Arabic numeral charges are a far less numerous class than species for which oxidation states written with Roman numerals can be defined. In this book, we are going to discuss five concepts:

formal oxidation numbers which we propose to write as superscript small Roman numerals: Ca^{II}, Cr^{III}, Ni^0, Cl^{-I}. We commit the anachronism to introduce zero as a Roman numeral and to introduce the minus sign for negative numbers.

(spectroscopic) oxidation states derived directly or by analogy from the excited levels observed. The Roman numerals are written on line in

parentheses Cr(III), S(VI), Cl(−I) and occur in the description of *chromophores* such as $Cr(III)O_6$, $S(VI)F_6$, $Ni(II)Cl_6$ etc.

conditional oxidation states derived either from parentage configurations of atomic spectroscopy considering only the partly filled shell having the smallest average radius (e.g. the electron configuration $[Ar]3d^2$ of Ti^{+2}, $[Ar]3d^24s$ of Ti^+ and $[Ar]3d^24s^2$ of Ti^0 are all Ti[II] whereas $[Ar]3d^34s$ of Ti^0 is Ti[I]) or derived from magnetic measurement on black or metallic compounds, alloys and elements (e.g. the presence of the half-filled shell $4f^7$ corresponding to Eu[II], Gd[III] and Tb[IV]). The concept of *preponderant configuration* classifying correctly the groundstate and the low-lying excited levels is of fundamental importance both for spectroscopic and conditional oxidation states. In certain cases, one can define *collectively oxidized sets of ligands* producing fractional oxidation states. As an exception, we use also Arabic numerals for this purpose and talk about the electron transfer bands of $IrCl_6^{-2}$ involving transitions from the groundstate (containing the well-defined oxidation states Ir(IV) and Cl(−I)) to states having preponderant configurations corresponding to Ir(III) and Cl_6^{-5}. Also clusters of central atoms can be collectively oxidized.

Quanticule oxidation states are connected with a description where electron pairs or larger groups of electrons connect either atomic cores isoelectronic with the noble gases (e.g. C{IV}, N{V}) or atoms containing lone-pairs (e.g. N{III}, O{II}). Thus, methane can be described C{IV} $((e_2)H\{I\})_4$, ethane $(e_2)(C\{IV\}((e_2)H\{I\})_3)_2$, benzene $(e_6)((e_2)C\{IV\}$ $(e_2)H\{I\})_6$, ammonia $N\{III\}((e_2)H\{I\})_3$ and the nitrogen molecule $N\{III\}(e_6)N\{III\}$ or $N\{V\}(e_{10})N\{V\}$. This description is derived from FAJANS' theory of quanticules and needs further comments in chapter 11.

Distributed quanticule oxidation states share the electrons in the quanticules equally between the adjacent atoms bonded, and one obtains $C\langle O\rangle$ and $H\langle O\rangle$ in CH_4 and C_2H_6, $N\langle I\rangle$ and $H\langle O\rangle$ in NH_4^+, $N\langle O\rangle$ in NH_3 and N_2 etc. This concept is closely related to *formal charges* in organic chemistry.

The fundamental reason why one cannot define an universal oxidation number or oxidation state in a way satisfactory for all compounds is the mixture of electrovalent and covalent bonding actually occurring. If inorganic chemists had no knowledge of organic chemistry, they would probably restrict themselves to the three first of the five concepts. However, the catenation of carbon in C_2H_6, C_2H_4, C_6H_6 and cyclo-hexane C_6H_{12} and the near homopolar character of most carbon-hydrogen bonds disturb this simple picture. Further on, from the point of view of spectroscopic oxidation states, hydrogen is in a particularly unfavourable position because it is nearly impossible to make a distinction between H(−I) and H(I).

This book is not intended to be a historical survey, but rather a presentation of present status of oxidation numbers and states. However, PALMER's excellent books can be recommended for their treatment of valency theory from 1800 to 1930.

Bibliography

BERRY, R. S., C. W. REIMANN, and G. N. SPOKES: J. Chem. Phys. **37**, 2278 (1962).
JØRGENSEN, C. K.: Orbitals in Atoms and Molecules. London: Academic Press 1962.
PALMER, W. G.: Valency, 2. Ed. Cambridge: University Press 1959.
— A History of the Concept of Valency to 1930. Cambridge: University Press 1965.

2. Formal Oxidation Numbers

Many authors introduce oxidation numbers in their text-books of inorganic chemistry. Perhaps one of the clearest accounts is given by RANCKE-MADSEN (1940). This book, as well as a more recent book (1963) exist only in Danish. After fruitful discussions with Professor RANCKE-MADSEN the writer attempts a somewhat more axiomatic representation. The fundamental axiom for all definitions of oxidation numbers is:

Axiom 1. The sum of the oxidation numbers of the atoms in a monatomic or polyatomic entity is the electric charge of the entity in protonic units.

There may exist compounds where the oxidation numbers cannot be defined; but in any case where they can be defined, axiom 1 is valid. There may be formulated two other axioms which are far more open to discussion than axiom 1:

Axiom 2. If there is no serious reasons to do otherwise, identical oxidation numbers are ascribed to atoms of the same element in a given compound.

Thus, the atoms in Cl_2 have the oxidation number 0 and not, for instance, $-I$ and $+I$. Chemical evidence can constitute "serious reasons"; one may either choose to ascribe the common oxidation number II to the two sulphur atoms in thiosulphate $S_2O_3^{-2}$ or VI to the central atom and $-II$ to the terminal atom. Crystal structures and similar evidence can be even stronger; though $GaCl_2$ definitely can be described as Ga^{II}, the structure suggests $Ga^I(Ga^{III}Cl_4)$.

The practical utility of the concept of formal oxidation numbers is to a large extent based on an axiom of the kind:

Axiom 3. Oxidation numbers can be ascribed by analogy (substitution of similar atoms or groups) to compounds where the oxidation numbers are determined from the specific rules.

One may also include an axiom which to a certain extent is a tautologic definition of what we understand with (BRØNSTED) acid-base reactions not accompanied by *redox reactions* (i.e. reactions where some of the atoms change oxidation number) viz.

Axiom 4. Reactions of an entity with the characteristic acid or base of a solvent (hydrated H_3O^+ and OH^- in aqueous solution) does not modify the oxidation numbers of the individual atoms.

As we shall see later, one *cannot* maintain axiom 4 in an absolute sense. A clear-cut counter-example is the reaction

$$Cl_2 + 2OH^- = ClO^- + H_2O + Cl^- \tag{2.1}$$

where Cl^I and Cl^{-I} are formed by *disproportionation* of Cl^0 (cf. axiom 2).

Besides these four axioms, of which only the first is absolute, we need specific rules for assigning numerical values. It is quite clear that this system is fundamentally arbitrary; we are just trying to make it consistent. RANCKE-MADSEN and several other authors tend to use hydrogen as a test element. H is assumed to have the oxidation number I except in H_2 (axiom 2), salt-like hydrides (of the type Li^IH^{-I} and $Ca^{II}(H^{-I})_2$) and in metallic alloys (Pd_2H) and a few other compounds (e.g. UH_3). As seen in PALMER's books, it was already a problem in classical valency theory that hydrogen does not form compounds with all elements, and that the compounds formed are not always very typical. MENDELEEV considered rather oxygen as the best test element, and the Periodic Table was mainly based on the highest oxide formed by each element. However, for our purposes, fluorine is the most suitable reference, and we suggest

Rule 1. With the exception of elemental F_2 (zero according to axiom 2), all fluorine atoms in chemical materials have the oxidation number $-I$.

The words "chemical materials" is a protection against F^+, F^{+2}, ... studied by atomic spectroscopists and species found by mass spectroscopy in violent electric discharges, such as F_2^+ and NeF^+. Combined with axiom 3, rule 1 already allows the assignment of oxidation states in great many compounds, because all elements with the exception of helium, neon and argon are known to form fluorides.

The second test element can be chosen as oxygen:

Rule 2. All oxygen atoms in chemical materials have the oxidation number $-II$ with the following exceptions:

$$O^{II}(F^{-I})_2 \text{ and } (O^I)_2(F^{-I})_2 \text{ (cf. rule 1)}.$$

Salts such as $O_2^+BF_4^-$, $O_2^+PtF_6^-$ with the oxidation number $+1/2$ (we do not write fractions with Roman numerals).

O_2 and O_3 zero (cf. axiom 2).

O_2^- and a variety of superoxo complexes $-1/2$.

O_2^{-2} and a variety of peroxo complexes $-I$.

It is quite clear that in order to be consistent, rule 3, 4, ... for a third, fourth, ... test element are going to be far more complicated than rule 2, because one each time has to compile all the exceptions involving the previous test elements (OF_2 and O_2F_2 in rule 2, whereas the four last examples rather are related to axiom 2). Whereas rule 1 could be kept invariant for all time, rules 2, 3, . . . are subject to additions each time new compounds are discovered necessitating such revisions. It might be tempting to include hydrogen as the third test element. We shall see in chapter 9 that good reasons can be given for considering H^I only in hydrides of F, Cl, Br, I, O, S, Se, Te and N, making all other hydrides H^{-I}.

It is quite clear that further development of the concept of formal oxidation numbers is to a great extent done by chemical reasoning rather than by choosing more than two (F, O) or three (F, O, H) test elements. After all, we are interested more in chemistry than in mathematics.

The authors of text-books such as RANCKE-MADSEN use formal oxidation numbers as a tool for rapid solution of the following type of problem:

We have a characteristic solvent. In the following, we assume it being water. We have a characteristic acid (hydrated H_3O^+) and a characteristic base (OH^-). Reactants undergo simultaneously acid-base and redox (i.e. change of formal oxidation number) reactions. What is the stoichiometric equation for the process?

The universal scheme for solving such problems is the following:

1. Indicate the reactants (differing from H_2O, H_3O^+ and OH^-).

2. Determine coefficients to the reactants in such a way that the sum of oxidation numbers of the elements participating in redox reactions is the same on both sides of the equality sign.

3. Determine coefficients for reactants not participating in redox reactions in such a way that the number of atoms of each element (excluding H and O) is the same on both sides of the equality sign.

4. Make the sum of electrical charges invariant by adding H_3O^+ (or H^+) on one of the two sides, if acid solutions are considered; by adding OH^- on one of the two sides, if alkaline solutions are considered; and by adding H_3O^+ (or H^+) or OH^- on the right-hand side, if an originally neutral solution is considered. However, this rule may be attenuated by our additional knowledge of simultaneous equilibria taking place. Thus, it would be unreasonable to produce both OH^- and NH_4^+ in a neutral solution, because it is known that the main part of such a mixture rearranges to H_2O and NH_3. On the other hand, this question cannot be settled in many other cases, and one may provisionally speak about Ti^{+4} and Ce^{+4} in acid solution though there is little doubt that these oxidation states nearly exclusively are represented by hydroxo or anion complexes.

5. Make the sum of hydrogen atoms and the sum of oxygen atoms invariant by appropriate addition of a number of water molecules.

This scheme may be illustrated by the reaction in acid solution between $Mn^{VII}O_4^-$ and $(Br^{-I})^-$ forming $(Mn^{II})^{+2}$ and $(Br^0)_2$:

$$1: \quad Br^- + \quad MnO_4^- \qquad = Br_2 + Mn^{+2}$$
$$2: 10\ Br^- + 2\ MnO_4^- \qquad = 5\ Br_2 + 2\ Mn^{+2}$$
$$3: \text{(no further information)} \qquad\qquad (2.2)$$
$$4: 10\ Br^- + 2\ MnO_4^- + 16\ H^+ = 5\ Br_2 + 2\ Mn^{+2}$$
$$5: 10\ Br^- + 2\ MnO_4^- + 16\ H^+ = 5\ Br_2 + 2\ Mn^{+2} + 8\ H_2O\ .$$

It is customary to neglect hydration (in the example of H^+ and Mn^{+2}) in order to simplify the operations. This is reasonable in sofar the water concentration in the solutions is normally fairly large, This would no longer be true for reactions in concentrated or fuming sulphuric acid, where production of water significantly may contribute to make a reaction progressing.

Precipitates are not given in ionic form; only the stoichiometry of the solid reactants count. A somewhat complicated example is the precipitation of (mixed crystals) of $BaMn^{VI}O_4$ and $BaS^{VI}O_4$ in an (alkaline) $Ba(OH)_2$ solution formed by the reaction between stoichiometric amounts of $Mn^{VII}O_4^-$ and $S^{IV}O_3^{-2}$:

$$1: SO_3^{-2} + \quad MnO_4^- \qquad\qquad = SO_4^{-2} + \quad MnO_4^{-2}$$
$$2: SO_3^{-2} + 2\ MnO_4^- \qquad\qquad = SO_4^{-2} + 2\ MnO_4^{-2}$$
$$3: SO_3^{-2} + 2\ MnO_4^- + 3\ Ba^{+2} \qquad = BaSO_4 + 2\ BaMnO_4 \qquad (2.3)$$
$$4: SO_3^{-2} + 2\ MnO_4^- + 3\ Ba^{+2} + 2\ OH^- = BaSO_4 + 2\ BaMnO_4$$
$$5: SO_3^{-2} + 2\ MnO_4^- + 3\ Ba^{+2} + 2\ OH^- = BaSO_4 + 2\ BaMnO_4 + H_2O\ .$$

An instructive example is the dissolution of arsenopyrite FeAsS in concentrated nitric acid:

$$1: FeAsS + \quad HNO_3 \qquad = Fe^{+3} + H_3AsO_4 + HSO_4^- + NO_2$$
$$2: FeAsS + 14\ HNO_3 \qquad = Fe^{+3} + H_3AsO_4 + HSO_4^- + 14\ NO_2$$
$$3: \text{(no further information)} \qquad\qquad (2.4)$$
$$4: FeAsS + 14\ HNO_3 + 2\ H^+ = Fe^{+3} + H_3AsO_4 + HSO_4^- + 14\ NO_2$$
$$5: FeAsS + 14\ HNO_3 + 2\ H^+ = Fe^{+3} + H_3AsO_4 + HSO_4^- + 14\ NO_2 + 6\ H_2O\ .$$

In such cases, it is recommended to control the resulting equation by counting the number of oxygen atoms, here 42, on both sides. In view of the comment made above on step 4, it is worth noting that it is the oxidation numbers of the products Fe^{III} (and not Fe^{II}), As^V (and not

As^{III}), S^{VI} (and not S^0) and N^{IV} (and not N^{II}) which are important for the redox reaction rather than the composition of the species which might as well have involved H_2SO_4 and SO_4^{-2} (known to be present only in minor concentrations under the circumstances described) and the complex $FeSO_4^+$ with only insignificant consequences for the final equation. This is a general problem of possible simultaneous non-redox reactions accompanying redox reactions.

An even more important aspect of eq. (2.4) is the fact that the individual oxidation states of the atoms in the solid compound FeAsS are not needed. The most rapid calculation is based on the assumption of all three elements having the oxidation number zero. However, it would not change the final result if two of the three elements already had the same oxidation number in the solid as in the resultant solution, say As^V and S^{VI}, producing the chemically meaningless Fe^{-XI}; we only need to know that taken together, the oxidation number sum of all three elements increase fourteen units.

Once the crystal structure and physical properties of FeAsS are known, it becomes clear that it is a semi-conducting (and not metallic) compound containing di-anions AsS^{-3} in close analogy to marcasite and pyrite, two modifications of FeS_2 which both can be considered as a di-sulphide of Fe^{II}. We shall discuss in Chapter 8 the reasons to attribute the oxidation state Fe^{III} to iron in FeAsS and Co^{IV} to the analogous non-metallic $CoAs_2$ in a sense containing (strongly deformed) As_2^{-4}. The paradoxical situation is that the addition of further elemental arsenic to form skutterudite $CoAs_3$ is a reduction to Co^{III} surrounded by an almost regular octahedron of six arsenic atoms. However, these considerations of oxidation states to be further defined have hardly any connection with oxidation numbers required for redox equations; and from the point of view of writing satisfactory stoichiometric equations, pyrite might equally well have been Fe^{IV} sulphide as frequently suggested before the crystal structure was known. Metallic alloys sometimes have properties as if one of the constituents were an anion. Thus, the fact that hydrogen cannot reduce aluminium oxide but that the reaction

$$6 \, Pt + Al_2O_3 + 3 \, H_2 = 2 \, Pt_3Al + 3 \, H_2O \qquad (2.5)$$

readily takes place might suggest some meaning of the oxidation states Pt^{-I} and Al^{III}. Semi-conducting CsAu is almost certainly closed-shell systems Au^{-I} and Cs^I. However, in all of these cases, it is more practical to define the oxidation numbers all to be zero.

The formal oxidation numbers can be useful also for reactions in which no characteristic solvent and corresponding acids or bases participate. However, some of the problems of scientific interest are closely related to the definition of acids and bases. In BRØNSTED's time, electrons and protons were considered to be the only elementary particles. Though the behaviour of elementary particles constitutes one of the least elementary parts of physics, it is even true today that electrons and protons are among the few particles which are not radioactive, or at least decay with so long a half-life that the reaction has not been observed. Hence, it was tempting for BRØNSTED to idealize two large classes of chemical reactions, redox reactions where electrons are exchanged between the reactants, and acid-base reactions where protons are exchanged.

Of course, one can generalize acid-base reactions in several other ways, as done by LEWIS. JANDER (1949) points out that solvents not containing any hydrogen atoms, such as liquid SO_2 or ICl still shows many analogies to water, containing a characteristic LEWIS acid and LEWIS base. J. BJERRUM suggested in 1951 to call LEWIS acids *antibases* since they neutralize bases, and he argued that nearly all LEWIS bases are also BRØNSTED bases. A logical reason for this asymmetry is that antibases not containing hydrogen atoms (such as SO_3) cannot conceivably loose protons, whereas one can always, hypothetically, imagine the adduct of a base with a proton. As a matter of fact, very few typical LEWIS bases are known not to be typical BRØNSTED bases. CO certainly is a LEWIS base, whereas the hypothetical proton adduct HCO^+ must be a very strong BRØNSTED acid. An interesting class of LEWIS bases are the free metals. It is certainly true in one sense that acids can be neutralized by magnesium metal. J. BJERRUM pointed out to the writer that one may ascribe the LEWIS basicity to the free electrons (we are going in chapter 8 to discuss in what sense Mg can be considered as a Mg[II] electronide) and to write the reaction

$$2e^- + 2\,H^+ = H_2 \tag{2.6}$$

which is the inevitable link joining redox and acid-base reactions, together with the somewhat similar reaction

$$H^- + H^+ = H_2 \,. \tag{2.7}$$

In other words, axiom 4 is not valid for H^{-I}. We see in chapter 9 that we have good reasons to consider CH_4 as containing C^{IV} and H^{-I} though we cannot guarantee the spectroscopic oxidation states C(IV) and H(−I) (what we can in CCl_4 having the choice between C(IV) and Cl(−I) or the chemically unacceptable C(−IV) and Cl(I)). There are many advantages of such a rule; the properties of CF_xH_{4-x} or CCl_xH_{4-x} are much easier to understand, and the oxidation of alkanes to alcohols or aldehydes to carboxylic acids are all examples of the general reaction

$$H^{-I} + O^0 = H^I + O^{-II} \tag{2.8}$$

avoiding C^{-IV} in CH_4 and C^{-II} in CH_3OH. However, there are also a few disadvantages; solid Al_4C_3 reacting with aqueous acid oxidizes from C^{-IV} to C^{IV} (the redox reaction involving simultaneous formation of strongly bound hydride ligands in CH_4).

Similar problems are involved in the rule that $P-H$ bonds contain H^{-I}. This makes hypophosphite $H_2PO_2^{-2}$ a di-hydrido-di-oxo complex of P^V which is quite reasonable, and agrees with the comparable stereo-chemistry of PH_3 and PCl_3. However, a minor disavantage is that

$$P^{III}(H^{-I})_3 + (H^I)^+ = P^V(H^{-I})_4^+ \qquad (2.9)$$

involves oxidation from P^{III} to P^V. A similar equation for NH_3 and NH_4^+ would be rather intolerable, and this is the reason why nitrogen was added to the group of halogens and chalcogens forming H^I hydrides.

This discussion takes care only of one of the two main paradoxes produced by organic chemistry, i.e. the homopolar nature of C—H bonds. The other paradox is connected with the catenation.

It is not easy to accept that C_2H_6 does not contain C^{IV} essentially in the same way as CH_4. But then, an electron pair is left over. In chapter 10, we are going to discuss whether such a description can be made consistent.

The inorganic chemists have never argued that it is possible to define formal oxidation numbers for the atoms in *all* compounds. N_4S_4 and B_2H_6 are not easy to handle in any satisfactory manner. Many of the problems are related to the C—H paradox. The methyl radical CH_3 or the phenyl radical C_6H_5 combines indifferently with halogens or with metal-loid and metallic elements, forming for instance $Hg(CH_3)_2$, $N(CH_3)_3$, $B(C_6H_5)_4^-$, $Sn(CH_3)_4$, $P(CH_3)_4^+$ and $P(C_6H_5)_5$. Metallo-organic chemis-try involving M—C bonds discriminates against formal oxidation numbers and prefers the type of treatment explaining the C—C bond in ethane. Surprisingly enough, these difficulties are more pronounced for single σ-bonds M—C, whereas we shall see in chapter 7 that ferrocene $Fe(C_5H_5)_2$ containing ten equivalent carbon atoms of the two cyclo-pentadienide ligands allow the oxidation state of the central atom to be determined Fe(II).

In sofar formal oxidation numbers go, it is permissible to have a kind of double truth, considering CH_3^- as a halide-similar constituent of the Hg^{II} complex $Hg(CH_3)_2$ and CH_3^+ as a constituent of CH_3F. In the cases where spectroscopic oxidation states can be defined, this is no longer an arbitrary decision by human beings, but is, in one sense, far more objec-tive.

Bibliography

BJERRUM, J.: Naturwiss. **38**, 461 (1951).

JANDER, G.: Die Chemie in wasserähnlichen Lösungsmitteln. Berlin: Springer 1949.

RANCKE-MADSEN, E.: Analytisk Kemi. Copenhagen: G. E. C. Gad 1940.

— Lærebog i Kemi, 7. Ed. Copenhagen: G. E. C. Gad 1963.

3. Configurations in Atomic Spectroscopy

The formal oxidation numbers described in chapter 2 can be assigned to the atoms in many, but not all, compounds. Physical measurements allow an extension in quite unexpected directions. One can apply group-theoretical arguments to the spectroscopic data now available. A *group-theoretical engineering* may even be said to exist in the sense of a reasonable estimate of the importance of the local symmetry, the influence of atoms not being nearest neighbours frequently being negligible. A variety of books on group theory for chemists have been published recently (COTTON, 1963; JAFFÉ and ORCHIN, 1965; SCHONLAND, 1965) but it is worth remembering the problem of relevant and irrelevant symmetry components (JØRGENSEN, 1962). Because of the interelectronic repulsion, electron configurations may be less appropriate when next-nearest neighbour atoms, or generally, atoms at larger distance, are taken into account.

The isolated atom or monatomic ion adapts to the conditions of spherical symmetry. One of the most interesting consequences (JØRGENSEN, 1965b; BERRY, 1966) of this very high symmetry is that the *orbitals*, the one-electron functions, can be separated in a product

$$\psi_{nl} = A_l \cdot R_{nl}/(2\sqrt{\pi}\, r) \tag{3.1}$$

of an angular function A_l which is hydrogenic in the sense of being the same as for one-electron systems, and of a radial function R_{nl} which is different in the various more-electron systems but showing the similarity with the hydrogen atom that it is zero (passing from positive to negative values) for $(n-l-1)$ finite values of r (in the interval $0 < r < \infty$). The normalization by the denominator $(2\sqrt{\pi}r)$ in eq. (3.1) assures the values of A_l^2 being on the average one on a given spherical surface (r constant) and the integral

$$\int_0^\infty R_{nl}^2 \, dr = 1 \,. \tag{3.2}$$

The angular functions can be written as linear combinations of homogeneous polynomials of l'th degree in the Cartesian coordinates:

$$A_l = \Sigma \, N_p x^a y^b z^c / r^l \tag{3.3}$$

with the normalization factors N_p and the condition on the non-negative, integral coefficients

$$a+b+c=l. \qquad (3.4)$$

It is possible to choose the angular function for the one s orbital ($l=0$), three p orbitals ($l=1$) and five d orbitals ($l=2$):

$$
\begin{aligned}
l=0: \text{ s} \quad & 1 \\
l=1: \text{ p}\sigma \quad & \sqrt{3}\, z/r \\
\text{p}\pi c \quad & \sqrt{3}\, x/r \\
\text{p}\pi s \quad & \sqrt{3}\, y/r \\
l=2: \text{ d}\sigma \quad & \sqrt{5}\left(z^2 - \tfrac{1}{2}x^2 - \tfrac{1}{2}y^2\right)/r^2 = \tfrac{3}{2}\sqrt{5}\left(\tfrac{z^2}{r^2} - \tfrac{1}{3}\right) \qquad (3.5) \\
\text{d}\pi c \quad & \sqrt{15}\, xz/r^2 \\
\text{d}\pi s \quad & \sqrt{15}\, yz/r^2 \\
\text{d}\delta c \quad & \sqrt{15}\,(x^2-y^2)/2r^2 \\
\text{d}\delta s \quad & \sqrt{15}\, xy/r^2
\end{aligned}
$$

adapted to linear symmetry. We shall later define the quantum number $\lambda=0,1,2,\ldots$ having the trivial names $\sigma,\pi,\delta,\ldots$ in analogy to the l-values, and the distinction between c and s for positive λ. The orbitals (3.5) are adapted to cubic symmetry as well, though the d-shell would fall in two sub-shells, one consisting of dσ and dδc having the group theoretical quantum number e_g in MULLIKEN's and γ_{3g} in BETHE's notation, and the other sub-shell consisting of the three orbitals dπc, dπs and dδs (t_{2g} or γ_{5g}). The quantum number ç is c (cosine) for the orbital having the maximum in direction of the x-axis, and s (sine) for the corresponding orbital turned $90°/\lambda$ in the xy-plane.

It is a question of minor interest to us that the orbitals adapted to linear or to cubic symmetry can be different for higher l-values. Thus, the seven f orbitals adapted to linear symmetry can be written:

$$
\begin{aligned}
l=3: \text{ f}\sigma \quad & \tfrac{1}{2}\sqrt{7}\ \ z(5z^2-3r^2)/r^3 \\
\text{f}\pi c \quad & \tfrac{1}{4}\sqrt{42}\ \ x(5z^2-r^2)/r^3 \\
\text{f}\pi s \quad & \tfrac{1}{4}\sqrt{42}\ \ y(5z^2-r^2)/r^3 \\
\text{f}\delta c \quad & \tfrac{1}{2}\sqrt{105}\ \ z(x^2-y^2)/r^3 \qquad (3.6) \\
\text{f}\delta s \quad & \sqrt{105}\ \ xyz/r^3 \\
\text{f}\varphi c \quad & \tfrac{1}{4}\sqrt{70}\ \ x(3y^2-x^2)/r^3 \\
\text{f}\varphi s \quad & \tfrac{1}{4}\sqrt{70}\ \ y(3x^2-y^2)/r^3 \,.
\end{aligned}
$$

It may be remarked that if the Cartesian coordinates x and y alone are considered, the polynomials are homogeneous of λ'th degree, as is dσ written on the equivalent form $\frac{3}{2}\sqrt{5}\ (z^2-\frac{1}{3}r^2)/r^2$. The f orbitals adapted to cubic symmetry fall in three sub-shells and can be written:

$$a_{2u} \qquad \sqrt{105}\ xyz/r^3$$

$$t_{1u} \begin{cases} \frac{5}{2}\sqrt{7}\ (z^3-\frac{3}{5}zr^2)/r^3 \\ \frac{5}{2}\sqrt{7}\ (x^3-\frac{3}{5}xr^2)/r^3 \\ \frac{5}{2}\sqrt{7}\ (y^3-\frac{3}{5}yr^2)/r^3 \end{cases}$$

$$t_{2u} \begin{cases} \frac{1}{2}\sqrt{105}\ z(x^2-y^2)/r^3 \\ \frac{1}{2}\sqrt{105}\ x(z^2-y^2)/r^3 \\ \frac{1}{2}\sqrt{105}\ y(z^2-x^2)/r^3\ . \end{cases}$$

(3.7)

It may be remarked that many f orbitals look like normalized products of a p and a d orbital. What is more important for our purposes is the number, $(2l+1)$, of orbitals in each nl-shell. One might have expected six independent polynomials of second degree in the Cartesian coordinates. However, $(x^2+y^2+z^2)/r^2$ is another way of writing 1, and one actually obtains five d and one s orbital. In eq. (3.5), the four last d orbitals have the simplest polynomials, and actually are equivalent when the coordinate system is appropriately rotated. It is only dσ which has been manifestly orthogonalized on the s orbital. It may be noted that this peculiarity has been selected in eq. (3.5) to occur along the z axis; it is equally valid to select the x axis:

$$-\frac{1}{2}\left(d\sigma\right)+\frac{\sqrt{3}}{2}\left(d\delta c\right) = \sqrt{5}\left(x^2-\frac{1}{2}y^2-\frac{1}{2}z^2\right)/r^2$$

$$\frac{\sqrt{3}}{2}\left(d\sigma\right)+\frac{1}{2}\left(d\delta c\right) = \sqrt{15}\left(z^2-y^2\right)/2r^2\ .$$

(3.8)

By the same token, one might have expected ten independent polynomials of third degree, but three are removed by the orthogonalization on the three p orbitals, leaving only seven f orbitals. This process can be seen to operate on fσ, fπc and fπs in eq. (3.6) and on the t_{1u} sub-shell in eq. (3.7). The reader familiar with group theory realizes that σ and π are the quantum numbers characterizing the p shell in linear symmetry and t_{1u} in cubic symmetry.

Quite generally, among the $(l+1)\ (l+2)/2$ independent polynomials of l'th degree, $l(l-1)/2$ disappear because of the need of orthogonalization on the shells having lower values of l of the same parity, even or odd, and only $(2l+1)$ orbitals remain in the l-shell.

This is a very important result for the Periodic Table, because each orbital can accommodate two electrons with opposite spin-direction, and hence, each nl-shell contains at most $(4l+2)$ electrons. However, the order of energy of these shells depends on the detailed structure of the system considered, i.e. the atomic number Z and the ionic charge z. In one-electron systems with $z=Z-1$, the energy of all l-values for a given n is identical, when we neglect relativistic effects. In more-electron systems, one can show that the higher l-values for a given value of n have the lower ionization energy. However, comparing shells having different n is not a clear-cut question. The "Aufbau principle" of the Periodic Table is that to a first approximation, the order of energy of the shells is

$$1s \ll 2s < 2p \ll 3s < 3p \ll 4s \sim 3d < 4p \ll 5s \sim 4d < 5p$$
$$\ll 6s \sim 4f \sim 5d < 6p \ll 7s \sim 5f \sim 6d < \ldots \tag{3.9}$$

for the neutral atoms where the shells start getting filled in the lowest electron configuration. The double inequality signs indicate the positions of the noble gases containing 2, 10, 18, 36, 54 and 86 electrons. The approximation signs indicate the behaviour of the transition group where the relative energy is very strongly dependent on the ionic charge z and other circumstances. There is a strong tendency for the average radii to fulfill the conditions (like in one-electron systems)

$$3d < 4s \qquad 4f < 5d < 6s$$
$$4d < 5s \qquad 5f < 6d < 7s \tag{3.10}$$
$$5d < 6s$$

and hence, the electron affinity and the ionization energy of the partly filled d or f shell in a given situation has a much larger difference than for the s orbital (Jørgensen, 1962 and 1969).

When the ionic charge z increases for a given atomic number Z, the series (3.9) tends to rearrange to the hydrogenic order according to n alone. This is also true for the excitations of inner shells of nearly neutral atoms studied by X-ray spectroscopy of compounds. For chemistry, a very important consequence is that nearly all transition group compounds correspond to the series characterizing ionic charges $z=+2$, $+3$ and $+4$ (with the exception of La^{+2}, Ac^{+2} and Th^{+2} having $(n+1)d$ below nf.):

$$1s \ll 2s < 2p \ll 3s < 3p \ll 3d < 4s < 4p \ll 4d < 5s < 5p$$
$$\ll 4f < 5d < 6s < 6p \ll 5f < 6d < 7s < \ldots \tag{3.11}$$

which is much simpler than (3.9). The possibility of defining spectroscopic oxidation states for most compounds is to a large extent based on the observations condensed into eq. (3.11).

Returning to atomic spectroscopy, the energy levels of monatomic entities are characterized by the spin quantum number S and by the quantum number of orbital angular momentum L. However, these two quantum numbers defined in Russell-Saunders' coupling scheme are not strict, and only the parity (even or odd) and the total angular momentum quantum number J are completely sharply defined. We are not going here to explain how the multiplet terms characterized by definite values of S and L and the J values for the individual energy levels can be classified according to definite electron configurations (CONDON and SHORTLEY, 1953; JØRGENSEN, 1962 and 1969).

We note that this classification is highly successful in predicting the low-lying energy levels, but that serious doubts can be had as to whether the total wavefunctions Ψ do indeed correspond to *well-defined electron configurations*. HUND (1927) describes the reasons for assigning a definite number of electrons to each nl-shell for a given energy level.

The study of regularities in atomic spectra is considerably older than N. BOHR's postulates for the hydrogen atom in 1913. RYDBERG investigated alkali metal atoms and found striking similarities. The trivial names s, p, d, f, . . . are derived from the series observed in e.g. sodium atoms:

$$
\begin{array}{lll}
\text{principal series:} & 3\text{s} \rightarrow n\text{p} & n = 3, 4, 5, \ldots \\
\text{sharp series:} & 3\text{p} \rightarrow n\text{s} & n = 4, 5, 6, \ldots \\
\text{diffuse series:} & 3\text{p} \rightarrow n\text{d} & n = 3, 4, 5, \ldots \\
\text{fundamental series:} & 3\text{d} \rightarrow n\text{f} & n = 4, 5, 6, \ldots
\end{array}
\tag{3.12}
$$

In all of these series, the energy levels correspond to a closed-shell atomic core (in the case of Na^+ $1s^2 2s^2 2p^6$; traditionally, the number of electrons in each shell is given as a right-hand superscript) to which one nl-electron has been added. Since the groundstate contains one 3s electron, only the principal series can be observed in absorption in vapour which is not very hot and it can be obtained in emission already in gas flames. The difference between the sharp and the diffuse series may originate in the possibility of distant sodium atoms in the vapour separating the energies of the five degenerate nd orbitals a little, whereas the ns orbitals cannot be thus influenced. The fundamental series got its name from the fact that both the 3d and the nf orbitals are sufficiently outside the atomic core that the energy levels are essentially those of a hydrogen atom. Hence, such emission lines can be observed in the near infra-red of both Li, Na, K and Rb atoms, whereas in Cs, the 4f orbital is somewhat more stable than for a hydrogen atom, as a precursor of the pronounced stability in the lanthanides.

Monopositive ions of the alkaline earths, Be^+, Mg^+, Ca^+, Sr^+, Ba^+ and Ra^+ have essentially the same relation to the one-electron system He^+ as the alkali metal atoms have to H. Since the isoelectronic series starting with hydrogen H, He^+, Li^{+2}, . . . have energy differences between 1s and the other orbitals proportional to $(z+1)^2$ one understands why the series of the "spark spectra" (of M^+, as contrasted to "arc spectra" of M^0) generally move from the visible out in the ultraviolet. A valuable extension of the concept of series spectra was made by PASCHEN in 1919 studying neon, MEISSNER in 1926 studying argon, HUMPHREYS and MEGGERS studying krypton and xenon since 1929, and RASMUSSEN studying emanation in 1932. In all of these cases, series were observed very similar to eq. (3.12) where five of the six np electrons of the groundstate remain as an invariant background for the excited electron, e.g.

$$np^6 \rightarrow np^5(n+1)s \text{ (in the far ultraviolet)}$$
$$np^5(n+1)s \rightarrow np^5(n+1)p \text{ and higher } n'p \qquad (3.13)$$
$$np^5(n+1)p \rightarrow np^5n'd$$

the two latter series in emission being the main origin of the bright colours of electric discharges in neon signs. However, each electron configuration of eq. (3.13) (except the closed-shell np^6) correspond to several J-levels, and the determination of all these energy levels showed perfect agreement with HUND's classification in terms of well-behaved electron configurations.

The noble gases (heavier than helium) represent intermediate cases between the typical "series" and "multiplet" spectra. The latter occur mainly in the transition groups characterized by a partly filled d shell. CATALÁN was the first, in 1922, to resolve such complications in the neutral manganese atom. Actually, the three volumes of CHARLOTTE MOORE-SITTERLY's "Atomic Energy Levels" are a monument over the classificatory power of electron configurations. Hardly any low-lying energy level is known to resist this classification, and frequently, all the J-levels predicted for a complicated configuration have been identified. The fourth volume related to partly filled 4f and 5f shells has not yet been completed. Certain problems are much more difficult in this case. As DIEKE and CROSSWHITE (1963) pointed out, the higher ionic charges such as M^{+2} and M^{+3} have relatively simpler spectra than the neutral M^0 or M^+. In the former ions, the configurations $4f^q$ have considerably lower energy than $4f^{q-1}5d$ and $4f^{q-1}6s$, whereas in the latter case, the configurations $4f^q6s^2$ (to which the groundstate usually belongs), $4f^{q-1}5d6s^2$, and $4f^{q-1}5d^26s$ tend to overlap strongly. The same is true for the d groups, where nd^q has much lower energy than $nd^{q-1}(n+1)s$ for M^{+2} and higher ionic charges, whereas these two configurations as well as $nd^{q-2}(n+1)s^2$ tend to coincide in M^0 and M^+. One expects roughly the number of observed spectral lines to be proportional to the square of the number of energy levels contributing, and this explains why spectra of neutral d group atoms have hundreds of lines in a spectral region where an alkali metal atom has one or two. The f group atoms litterally have tens of thousands of lines in a similar region, and it is clear that one has to measure the wavenumbers extremely accurately in order to obtain meaningful results. However, even when the large work of identifying the corresponding energy levels can be performed by electronic computers, there always remains many accidental coincidences distorting the interpretation. Experience of atomic spectroscopists shows characteristic variations of emission line intensity as a function of the quantum numbers and preponderant configuration of the two energy levels involved, and the strongest line groups can usually be identified at first. Thus, DIEKE and CROSSWHITE (1963) compared the ultraviolet spectra of the series of 4f group M^{++} and pointed out that the most intense lines are due to transitions from 4f^{q}6s to 4f^{q}6p. The interesting fact is that these configurations look as if they have a much smaller spreading of energy for a given ion than they are known to have from the width of 4f^q alone. The explanation is that the intense transitions take place nearly exclusively between levels having 4f^q in the same state, and the energy difference between the 4f^q contributions of the upper and the lower level hence nearly cancels.

This is a characteristic case of *parentage* of energy levels involving partly filled shells. Originally, this was introduced empirically by atomic spectroscopists, and a look at "Atomic Energy levels" clearly shows why. There is a technical sense of parentage for a single partly filled shell l^q. RACAH introduced the seniority number v as a measure of separating out electron pairs $l^v(s^2)^{(q-v)/2}$ of the total wavefunction (cf. JØRGENSEN, 1962). This concept can be further generalized to the separation of other l' values $l^{q-2}(l')^2$ (cf. also JØRGENSEN, 1966). In the classification of f^q energy levels, the coefficients of fractional parentage buildung l^q wavefunction from l^{q-2} wavefunctions have great importance (JUDD, 1963; WYBOURNE, 1965) but they will not be further discussed here.

In the original sense used by atomic spectroscopists, the parentage refers to various coupling schemes. This pictorial expression did not survive unchanged in quantum mechanics, and we have to remember that the energy differences between L, S-multiplet terms are mainly due to different values of the average interelectronic repulsion $\langle e^2/r_{12} \rangle$ and in the RUSSELL-SAUNDERS coupling, the smaller energy differences between J-levels of the same multiplet are caused by relativistic effects though they are traditionally called spin-orbit coupling. If the atom has a fairly low atomic number Z, or if the l value of the *external electron* (Leuchtelektron in German) is higher than 1, the relativistic effects are usually rather small. If the external electron has a very small *"squared overlap"* $\int \psi_1^2 \psi_2^2 \, d\tau$ (cf. JØRGENSEN, 1962) with partly filled orbitals at smaller distance from the nucleus, the K-integrals of interelectronic repulsion (between the charge distribution $\psi_1\psi_2$ and itself) become rather negligible, and we observe the typical case of one external electron coupled to a definite state of the other electrons. It becomes a good approximation to ascribe effective values of the quantum numbers S, L and J to the central part of the atom without external electron. Experimentally, this is observed in series spectra having various low-lying levels of the ionized species M^{+Z+1} as limits for different spectral series of M^{+Z}. Actually, in the case of the transitions (3.13) for heavier noble gases such as Kr, Xe and Em, the two levels having $J=\frac{3}{2}$ and $J=\frac{1}{2}$ belonging to the ground configuration np^5 of M^+ can be seen as originators of individual sets of energy levels (cf. JØRGENSEN, 1967 b).

In the cases where one partly filled l-shell has much smaller average radius than one or more external electrons, we talk about spectroscopic oxidation states. In order to keep in touch with the chemical part of the present book, we apply the sharp brackets Ne[I] for the excited states involving the preponderant configuration $1s^2 2s^2 2p^5$ and one external electron. Though the ionic charge is zero for such an excited neon atom, and not one, as for Ne^+, we consider both situations as Ne[I]. This decision is not as arbitrary as it may look at first, and its connections with chemistry will be discussed in chapter 8.

It may be worth analyzing the concept "external electron". The fundament is RYDBERG's formula for energy levels of alkali-metal-like atoms, counted negative with the closed-shell M^+ groundstate as zero-point:

$$E = -\frac{Z_0^2\, ry}{(n-\delta)^2}. \tag{3.14}$$

The RYDBERG constant ry is 109.7373 kK for atoms of usual mass, but is slightly smaller for very light atoms; e.g. 109.6788 for normal hydrogen and 109.7086 for deuterium. We use the energy unit kK = kilokayser = 1000 cm^{-1} which is related to other units of the energy in the following way:

$$1 \text{ hartree} = 2 \text{ ry} = 219.4746 \text{ kK}$$
$$1 \text{ eV} = 8.0678 \text{ kK}$$
$$1 \text{ kcal/mole} = 0.351 \text{ kK}$$
$$1 \text{ k Joule/mole} = 0.0834 \text{ kK}.$$

Z_0 in (3.14) indicates the ionic charge plus one, i.e. the effective charge seen by the external electron at large distances; n is the principal quantum number and δ the quantum defect. There is little doubt that δ is generally positive for genuine cases of one external electron. It is true that the ^{1}P levels of the configurations 1s np of the helium atom apparently have negative δ; but this is an effect of interelectronic repulsion between the two electrons. If one either considers the baricenter of the configuration $((^1\mathrm{P}) + 3\ (^3\mathrm{P}))/4$ representing the average energy of the twelve states, or the mean of the energy levels $((^1\mathrm{P}) + (^3\mathrm{P}))/2$ representing the J-integral of interelectronic repulsion (cf. JØRGENSEN, 1962) between the charge distributions ψ^2_{1s} and ψ^2_{np}, δ is again positive. At the time around 1920, it was not completely clear how to define n in heavy atoms, and it was doubted whether δ could be so much larger than one as is now accepted.

δ depends very strongly on l and is very close to zero for sufficiently large l. Thus, for atoms having Z below 18, it nearly vanishes for $l \geq 2$ and for atoms having Z below 54 for $l \geq 3$. In the heaviest known atoms, δ for g-electrons (i.e. $l = 4$) is still very small, thus 0.019 for 5g of Ra${}^+$. This regularity can be explained by the knowledge of HARTREE-FOCK radial functions; when very little of the density ψ^2 of the external electron is inside the atomic core, δ is nearly zero. The interesting point about RYDBERG's equation (3.14) is that δ for a given M^{+Z} and a given l-value is so nearly invariant; and this fact has not yet been completely explained by modern quantum mechanics (cf. JASTROW, 1948). Actually, the accurate determinations of energy levels with some six significant figures allow the detection of many minor variations of δ as a function of n for a given l. This may either be a monotonic variation; or a mild oscillation; or a definite singularity between two values of n. VAN VLECK and WHITE-LAW (1933) demonstrated that such perturbed series are caused by the presence of a foreign level having the same symmetry type L, S, J. However, these small variations should not make us forget the surprising constancy of δ for a given series. If the lowest n-value has a large δ for a given l, the subsequent n-values have δ of the same order of magnitude. Thus, Ra${}^+$ has 5g at -17.691 kK (relative to the groundstate of Ra^{+2}) only 0.133 kK below the hydrogenic value (for $Z_0=2$ and $\delta=0$) whereas both 5f, 6f and 7f are some 80 percent more negative than the hydrogenic values.

In order to obtain a quantitative impression of the behaviour of an external electron, Table 3.1 gives the energy (relative to the ionization limit) and the deviation ΔE from the hydrogenic value for the isoelectronic series of three-electron atoms, and Table 3.2 the same quantities for the nineteen-electron series. In the case of s and p electrons, ΔE has been evaluated in the latter series both for the principal quantum number n and for an effective value two units smaller.

In most treatises (cf. also JØRGENSEN, 1969) the RYDBERG formula (3.14) is invoked in the description of deivations from hydrogenic behaviour. It is certainly true that for sufficiently external orbitals, (3.14) is a far better function of n than the formula for internal orbitals

$$E = - \frac{(Z-Z_\sigma)^2 \, \text{ry}}{n^2} \tag{3.15}$$

which is applied to the excitation of inner shells as observed in X-ray spectroscopy, Z_0 being the screening constant. The asymptotic behaviour of the internal and the external cases can be derived from the HARTREE potential $U(r)$. The SCHRÖDINGER equation for the stationary states for one electron can be written

$$\mathfrak{P}(x,y,z) + U(x,y,z) = E \tag{3.16}$$

where $\mathfrak{P}$ is the kinetic pseudopotential (in atomic units, the unit of energy being 1 hartree $= 2$ry and the unit of length 1 bohr $= 0{,}5292$ Å):

$$\mathfrak{P} = - \frac{1}{2} \frac{\left[\frac{\delta^2 \psi}{\delta x^2} + \frac{\delta^2 \psi}{\delta y^2} + \frac{\delta^2 \psi}{\delta z^2} \right]}{\psi}. \tag{3.17}$$

U the normal (electrostatic) potential and E the (negative) eigenvalue. The virial theorem assures that the total kinetic energy

$$T - \int \mathfrak{P}\psi^2 d\tau = -E \tag{3.18}$$

where $d\tau$ symbolizes integration over all the coordinates, is related to the total potential energy by the ratio -2:

$$\int U\psi^2 d\tau = 2E. \tag{3.19}$$

As RUEDENBERG pointed out to the writer (cf. JØRGENSEN, 1967a and 1968) it is possible to define another local contribution to the kinetic energy

$$\mathfrak{R} = \frac{1}{2} \left[\left(\frac{\delta\psi}{\delta x} \right)^2 + \left(\frac{\delta\psi}{\delta y} \right)^2 + \left(\frac{\delta\psi}{\delta z} \right)^2 \right] \tag{3.20}$$

never being negative in contrast to (3.17) and having the same integrated result

$$\int \mathfrak{R} d\tau = T = -E \tag{3.21}$$

though it does not allow a conservative potential such as (3.16). It is seen that the local contributions to the kinetic energy $\mathfrak{P}\psi^2$ or $\mathfrak{R}$ can be

Table 3.1. *Energy levels E in kK = 1000 cm⁻¹ of systems containing one external electron. The penetration energy ΔE is evaluated by comparison with the hydrogenic situation. δ is Rydberg's quantum defect defined in eq. (3.14)*

		Li	Be⁺	B⁺²	C⁺³	N⁺⁴	O⁺⁵	F⁺⁶
2s	E	−43.49	−146.88	−305.93	−520.18	−789.53	−1114.00	−1493.66
	ΔE	−16.06	− 37.14	− 59.02	− 81.23	−103.67	−126.36	− 149.38
	δ	0.412	0.271	0.203	0.162	0.136	0.117	0.103
2p	E	−28.59	−114.95	−257.55	−455.63	−708.90	−1017.27	−1380.76
	ΔE	− 1.16	− 5.21	− 10.64	− 16.68	− 23.04	− 29.63	− 36.48
	δ	0.034	0.046	0.042	0.038	0.033	0.029	0.026
3s	E	−16.28	− 58.65	−125.73	−217.33	−333.40	− 473.96	− 639.04
	ΔE	− 4.09	− 9.88	− 15.99	− 22.24	− 28.57	− 35.01	− 41.62
	δ	0.40	0.26	0.20	0.16	0.13	0.113	0.099
3p	E	−12.56	− 50.38	−112.97	−200.11	−311.71	− 447.78	− 608.34
	ΔE	− 0.37	− 1.61	− 3.23	− 5.02	− 6.88	− 8.83	− 10.92
	δ	0.04	0.05	0.04	0.04	0.04	0.030	0.027
3d	E	−12.21	− 48.83	−109.86	−195.29	−305.12	− 439.35	− 597.98
	ΔE	− 0.02	− 0.06	− 0.12	− 0.20	− 0.29	− 0.40	− 0.56
	δ	0.001	0.002	0.002	0.002	0.002	0.001	0.001

divided in three contributions according to the directions of the three Cartesian axes. Actually, it is possible to define another operator $\mathfrak{D}$ having the x-component

$$\mathfrak{D}_x = 2\psi \frac{\delta\psi}{\delta x} \tag{3.22}$$

being the gradient of the electronic density ψ^2 in the direction x (we assume for convenience that all the ψ are real, not complex, which is no restriction for stationary states in the absence of an external magnetic field). The definition of the y- and z-components is analogous. It is then possible to differentiate once more in a definite direction

$$\frac{\delta\mathfrak{D}_x}{\delta x} = 2\psi \frac{\delta^2\psi}{\delta x^2} + 2\left(\frac{\delta\psi}{\delta x}\right)^2 = -4\mathfrak{P}\psi^2 + 4\mathfrak{R}. \tag{3.23}$$

It has been discussed elsewhere (JØRGENSEN, 1968) whether the operator

$$\mathfrak{Q} = \mathfrak{P}\psi^2 - \mathfrak{R} \tag{3.24}$$

might have an unexpected connection with the nature of kinetic energy in quantum mechanics, but this is not of importance for our argumentation here.

Table 3.2. The 19-isoelectronic series. Notation as in Table 3.1. For certain orbitals, ΔE is evaluated both by comparison with n and $(n-2)$ for the hydrogenic case

	K	Ca$^+$	Sc^{+2}	Ti^{+3}	V^{+4}	Cr^{+5}	Mn^{+6}	Fe^{+7}
3d E	−13.48	−82.07	−199.58	−348.59	−525.6	−730.3	−961.19	−1218.24
ΔE	− 1.29	−33.30	− 89.84	−153.50	−220.8	−291.3	−363.57	− 437.89
δ	0.146	0.686	0.776	0.756	0.72	0.67	0.635	0.599
4s E	−35.01	−95.75	−174.16	−268.44	−377.9	−503.1	−643.27	—
ΔE ($n=4$)	−28.15	−68.32	−112.43	−158.70	−206.4	−256.2	−307.20	—
($n=2$)	− 7.58	+13.99	+ 72.75	+170.51	+312	+484.5	+701.01	—
δ	2.23	1.86	1.62	1.44	1.30	1.20	1.11	—
4p E	−21.99	−70.41	−137.28	−220.36	−318.7	−433.2	−562.71	− 707.04
ΔE ($n=4$)	−15.13	−42.98	− 75.55	−110.62	−147.3	−186.3	−226.64	− 268.09
($n=2$)	+ 4.44	+39.33	+109.62	+218.59	+367.2	+554.4	+781.57	+1048.76
δ	1.766	1.50	1.32	1.18	1.06	0.98	0.91	0.85
4d E	− 7.61	−38.90	− 87.42	−151.98	—	—	—	—
ΔE	− 0.75	−11.47	− 25.69	− 42.24	—	—	—	—
δ	0.20	0.64	0.64	0.60	—	—	—	—
5s E	−13.98	−43.58	− 84.84	−136.42	−197.8	—	−348.07	− 435.69
ΔE ($n=5$)	− 9.59	−26.02	− 45.33	− 66.19	− 88.1	—	−132.99	− 155.03
($n=3$)	− 1.79	+ 5.19	+ 24.90	+ 59.32	+107	—	+249.35	+ 344.39
δ	2.20	1.82	1.59	1.41	1.28	—	1.07	0.99
5p E	−10.30	−35.17	—	−118.01	—	—	—	—
ΔE ($n=5$)	− 5.91	−17.61	—	− 47.78	—	—	—	—
($n=3$)	+ 1.89	+13.60	—	+ 77.73	—	—	—	—
δ	1.74	1.47	—	1.14	—	—	—	—

If the potential U only depends on the distance r from the nucleus given by

$$r^2 = x^2 + y^2 + z^2 \qquad (3.25)$$

all text-books in quantum mechanics indicate why ψ can be separated in the product of an angular and a radial part (3.1) where the angular part A_l (3.3) is a linear combination of homogeneous polynomials of l'th degree in the Cartesian coordinates. The kinetic pseudo-potential $\mathfrak{P}$ (3.17) can be separated in the sum $\mathfrak{P}_{rad} + \mathfrak{P}_{ang}$ of a radial part

$$\mathfrak{P}_{rad} = -\frac{1}{2\,R_{nl}}\frac{\delta^2 R_{nl}}{\delta r^2} \qquad (3.26)$$

and an angular part

$$\mathfrak{P}_{ang} = \frac{l(l+1)}{2r^2} \qquad (3.27)$$

showing that for spherically symmetric ψ, i.e. $l=0$, the deviation from standing waves in a constant potential U consists in $\mathfrak{P}$ no longer being constant but adding (-1) times the Coulomb potential, determining a definite radial function R_{nl}, whereas for positive l, an apparent repulsion (3.27) can be added to the potential contribution of (3.16) explaining why ψ vanishes at the nucleus and why the R_{nl} close to the nucleus is proportional to r^{l+1}.

In many-electron systems, the Hartree potential U varies between the limits $(-Z_0/r)$ for large r and $(-Z/r)$ for small r, though it is not becoming as negative because of the repulsion from other electrons (external screening). Said in other words, the quantity rU is expected to increase smoothly from a value somewhat above $-Z$ for r=0 to $-Z_0$ for large r, Z_0 being the ionic charge plus one. The screening constant Z_0 of eq. (3.15) can be considered as the average value of rU which for the inner orbital is $-(Z-Z_\sigma)$. There is a very profound difference between monatomic anions on one side and monatomic entities being neutral or carrying positive charge on the other. In the former case, there may exist none, one, more, or an infinite number of discrete, stationary states, whereas for positive Z_0, the energies of the hydrogenic nl-levels $(-Z_0^2/2n^2)$ hartree are upper limits for an infinite number of discrete states, explaining why δ of eq. (3.14) cannot be negative. However, this argument neglects the (very remote) possibility in many-electron systems that the correlation energy (cf. JØRGENSEN, 1962) might be larger in the remaining core M^{+Z_0} than in the state of M^{+Z_0-1} considered.

It is possible (cf. JØRGENSEN, 1962, p. 12) to solve the SCHRÖDINGER equation (3.16) for the potential

$$U = -\frac{Z_0}{r} - \frac{k}{r^2} \tag{3.28}$$

occurring as the first-order approximation to the quantity

$$rU = -Z_0 - k/r \tag{3.29}$$

having as eigenvalues (in hartrees) for positive l:

$$E = -\frac{Z_0^2}{2\,(l_* + 1)^2} \tag{3.30}$$

where l_* is implicitly given by

$$2\,k = l(l+1) - l_*(l_*+1) \tag{3.31}$$

because of (3.27). The lowest wavefunction has the normalized (by N) radial function

$$R_{nl} = N r^{l_*+1} e^{-\mu r} \tag{3.32}$$

with the exponential coefficient

$$\mu = Z_0/(l_* + 1)\,. \tag{3.33}$$

It is tempting to identify $(n - \delta)$ of eq. (3.14) with (l_*+1) of eq. (3.30), but there are various difficulties. Atomic spectroscopy shows that δ behaves in much the same way for s-electrons as for positive l values, and since eq. (3.27) vanishes for $l=0$, it appears that the two first terms of eq. (3.29) are not sufficient for an appropriate description of rU. Further on, it is not yet clearly understood why δ is so nearly invariant for a given l when n increases. JASTROW (1948) discussed the general question why the RYDBERG formula is so good an approximation. TIETZ (1964, 1965) proposed a slightly more flexible approximation than (3.32) to the radial functions (cf. also COOK and MURRELL, 1965 and PARSON and WEISSKOPF, 1967). The effect on core orbitals can be represented by a HELLMAN pseudopotential (SZASZ and MCGINN, 1967; IAFRATE, 1967).

We are going to describe the intermediate cases between external (3.14) and internal (3.15) orbitals as a *penetration energy* ΔE indicating how much lower the ionization energy of effective one-electron systems is compared to the hydrogenic value $-Z_0^2/2n^2$ hartree. It is seen from

Table 3.1 that for 2s-electrons, ΔE is roughly a linear function of the atomic number Z (and actually roughly proportional to $(Z-2)$) whereas for 2p-electrons, ΔE is much smaller and approximately proportional to $(Z-2)^2$. If the two core 1s-electrons coincided with the nucleus, the 2p orbitals would be hydrogenic, and $\Delta E = 0$. However, the actual state of affairs is different for the 2s orbital. Because of the previously filled 1s orbital, the main condition for the 2s function is to be orthogonal on the 1s function, and this can be obtained by a suitable node in the radial function in such a way that the contributions of $\psi_{1s}\psi_{2s}$ in regions were this quantity is positive, can be integrated and exactly cancel the contributions from regions where $\psi_{1s}\psi_{2s}$ is negative. The end result of this adaptation is that the 2s orbital penetrates in a region where rU no longer is $-Z_0$ but considerably more negative. However, the virial theorem does not apply directly to the ionization energy I of one electron in a many-electron system; and in particular, the kinetic energy tends to be much higher than I. It is not easy at present to formulate useful rules for the penetration energy; it is noted that the very large ΔE-values for 2s-electrons in Table 3.1 correspond to a situation where a hydrogenic 2s-orbital has acquired some 2 to 3 percent 1s character by the orthogonalization. On the other hand, it is quite clear that 3d orbitals are external in the three-electron systems.

Atomic spectroscopists early divided the external electrons (Leuchtelektronen in German) in more or less penetrating ones. Quite generally, the values of $(n-\delta)$ in alkali metal atoms tend to be close to 2 (actually between 1,58 for Li to 1.87 for Cs) for the lowest external s-orbital, and 1.96 to 2.35 for the lowest external p-orbital, and spectroscopic nomenclature for a long time remained uncertain about the n-values. In the caesium atom, the degree of penetration goes the long way $6s > 6p > 5d > 4f > 5g$ between strongly penetrating s-orbitals and hydrogenic g-orbitals. However, Table 3.2 rather suggests that the penetration of external orbitals can develop along two different lines. The 4s, 4p, 5s and 5p orbitals of the potassium isoelectronic series are 1.4 to 5.1 times more stable than the corresponding hydrogenic nl-orbitals, but this factor decreases monotonically as a function of increasing ionic charge, like it does in Table 3.1. On the other hand, the 3d-orbitals are nearly hydrogenic both in the neutral potassium atom and as an asymptotic behaviour for very large ionic charges. The ratio of stabilization $(\Delta E/E)$ goes through a maximum 0.45 for Sc^{++}. This maximum is very asymmetric, because the stabilization ratio is 0.096 for K, suddenly raises to 0.406 for Ca^+, but only decreases very slightly, roughly 0.02 per unit of atomic number above the maximum. By the same token, the stabilization ratio $(\Delta E/E)$ for 4d orbitals is 0.10 for K, 0.295 for both Ca^+ and Sc^{++}, and then slowly decreases.

It is as if there is an absolute sense in which 3d orbitals become strongly penetrating from Sc^{++} on, and that Ca^+ nearly made it too. This situation is rather universal in all the transition groups, and for the monatomic species, it is actually the most prominent characteristic of the beginning of such a group.

In order to understand the characteristic behaviour of transition groups it is necessary to consider the effects of interelectronic repulsion in systems containing more than one electron outside the closed shells of the core. In two-electron systems, we propose the general expression for the *baricenter* (the average energy of the $(2l+1)\,(2l'+1)$ states) of the configuration $(nl)^1(n'l')^1$

$$E = -I_\varepsilon(nl) - I_\varepsilon(n'l') + A*(nl,\,n'l')\,. \tag{3.34}$$

The energies I_ε are the ionization energies of the corresponding one-electron systems. For instance, the values of E given in Table 3.1 for 2s- and 2p-electrons would be used as $-I_\varepsilon$ in the description of the four-electron beryllium isoelectronic series. When the configuration $(nl)\,(n'l')$ contains several energy levels, their average is obtained after weighting with the number $(2J+1)$ of states in each energy level. The multiplet terms in RUSSELL-SAUNDERS coupling contain each $(2S+1)\,(2L+1)$ states. We emphasize that the parameter $A*(nl,\,n'l')$ of eq.(3.34) is obtained per definition from the experimentally known quantities E and the two I_ε, though it is represented in the theory of interelectronic repulsion first elaborated by SLATER (cf. CONDON and SHORTLEY, 1953) by definite parameters relevant for well-defined configurations. Thus, if nl is different from $n'l'$,

$$A*(nl,\,n'l') = J_{av}(nl,\,n'l') - \tfrac{1}{2}\,K_{av}(nl,\,n'l') \tag{3.35}$$

where the average values for the J- and K-integrals (cf. JØRGENSEN, 1962) are obtained by considering the interaction between the filled shells containing $(4l+2)$ and $(4l'+2)$ electrons, respectively. In this case, J_{av} is the average value of the $(4l+2)\,(4l'+2)$ J-integrals, whereas K_{av} is the average value of the $(4l+2)\,(2l'+1)$ K-integrals. If at least one of the two l-values is zero, there is only one definite J and one definite K-integral, and no average value needs to be evaluated.

If the two electrons are situated in the same nl-shell, the situation is slightly more complicated. If the shell is completely filled by $(4l+2)$ electrons, we define an average value J'_{av} of the $(2l+1)$ J-integrals having the two electrons in the same orbital, another average value J_{av} of the $(2l+1)4l$ J-integrals between two electrons in two different orbitals of the shell, and $(2l+1)2l$ K-integrals between such electrons averaged to K_{av}. With this definition

$$A_*(nl, nl) = \frac{1}{4l+1} J'_{av} + \frac{4l}{4l+1} J_{av} - \frac{2l}{4l+1} K_{av} \ . \qquad (3.36)$$

Table 3.3. *Average parameters of interelectronic repulsion evaluated without correction for interaction between the configurations* $2s^2 2p^q$ *and* $2p^{q+2}$

	$A_*(2s, 2s)$	$A_*(2s, 2p)$	$A_*(2p, 2p)$
Be	71.7	66.9	~ 68
B+	103.0	100.9	~ 107
B	100.1	96.8	97.9
C++	134.0	134.4	148.2
C+	133.2	129.9	139.4
N+3	164.6	167.1	187.4
N++	165.5	162.8	179.3
O+4	195.3	200.0	226.6
O+3	197.9	195.2	218.5
O++	197.5	191.2	210.3
F+5	226.1	232.7	265.6
F+4	229.0	228.4	257.6
F+3	~ 224	~ 220	~ 245
F++	227	216.8	238.7

Table 3.4 $K(2s, 2p)$ *evaluated from energy differences between terms belonging to the configurations* $2s\,2p^q$

Be 13.6	B 14.6	C 18.0, 16.9, 22.3	
B+ 23.2	C+ 22.5	N+ 27.0, 26.0, 28.8	O+ 30.9
C++ 32.5	N++ 29.5	O++ 34.2, 33.5, 34.0	F++ 38.2
N+3 41.6	O+3 36.4	F+3 41.0, 40.5, 41.1	
O+4 50.7	F+4 42.9		
F+5 59.8			

If two partly filled shells occur in the configuration $(nl)^a\,(n'l')^b$ we write its baricenter energy relative to the previous closed shell with all $(a+b)$ electrons removed

$$E = -aI_\varepsilon(nl) - bI_\varepsilon(n'l') + \frac{a(a-1)}{2} A_*(nl, nl) + ab\,A_*(nl.n'l')$$

$$+ \frac{b(b-1)}{2} A_*(n'l', n'l') \ . \qquad (3.37)$$

This equation can readily be generalized to a larger number of partly filled shells containing a, b, c, . . . electrons:

$$E = -\sum x \, I_\varepsilon(nl) + \sum \frac{x(x-1)}{2} \, A_*(nl, nl) + \sum_{nl \neq n'l'} xy \, A_*(nl, n'l') . \quad (3.38)$$

Table 3.3 gives the values of A_* obtained from the baricenters of the observed $2s^a 2p^b$ configurations in CHARLOTTE MOORE-SITTERLY's "Atomic Energy Levels". It is seen that these parameters increase strongly with the atomic number. For a given element, $A_*(2s, 2p)$ and $A_*(2p, 2p)$ decrease moderately when the number of electrons $(a+b)$ is increased, i.e. the ionic charge lowered, whereas $A_*(2s, 2s)$ seems to stay roughly invariant as a function of ionic charge. When $(a+b)$ is large, the parameters tend to differ $A_*(2p, 2p) > A_*(2s, 2s) > A_*(2s, 2p)$. The hydrogenic values of the three parameters are in hartrees

$$A_*(2s, 2s) = F_0(2s, 2s) = \frac{77}{512} Z_0$$

$$A_*(2s, 2p) = F_0(2s, 2p) - \frac{1}{2} G_1(2s, 2p) = \left(\frac{83}{512} - \frac{1}{2} \cdot \frac{15}{512} \right) Z_0 = \frac{75.5}{512} Z_0 \quad (3.39)$$

$$A_*(2p, 2p) = F_0(2p, 2p) - 2 F_2(2p, 2p) = \left(\frac{93}{512} - 2 \cdot \frac{9}{2560} \right) Z_0 = \frac{79.4}{512} Z_0$$

according to LINDERBERG and SHULL (1960). It is worth noting that even for hydrogenic radial functions having identically $I_\varepsilon(2s) = I_\varepsilon(2p) = (Z_0/8)$ hartrees, the three A_* parameters are not equal.

However, if we want a satisfactory analysis of the energy levels belonging to the configurations $1s^2 2s^a 2p^b$, we rather correct the results of Table 3.3 for certain inherent errors. In order to recognize why these corrections are sensible, we rather at first consider the other parameters of interelectronic repulsion determining the term distances. The simplest case is $2s2p$ (we omit again the core electrons $1s^2$) having the energy levels

$$\begin{aligned}
{}^3P &- I_\varepsilon(2s) - I_\varepsilon(2p) + A_*(2s, 2p) - \tfrac{1}{2} K(2s, 2p) \\
{}^1P &- I_\varepsilon(2s) - I_\varepsilon(2p) + A_*(2s, 2p) + \tfrac{3}{2} K(2s, 2p)
\end{aligned} \qquad (3.40)$$

giving the values for $K(2s, 2p)$ in Table 3.4. As discussed by CONDON and SHORTLEY (1953) the distances between ${}^4P$ and ${}^2P$ in either the configuration $2s2p^2$ or $2s2p^4$ are $3K(2s, 2p)$, between ${}^5S$ and ${}^3S$ of $2s2p^3$ $4K(2s, 2p)$, and between ${}^3D$ and ${}^1D$ or between ${}^3P$ and ${}^1P$ of this configuration $2K(2s, 2p)$. It is seen from Table 3.4 that $K(2s, 2p)$ is roughly a linear function of Z in a given isoelectronic series though the difference quotient $\sim 9\,$kK is somewhat larger than the value 6.4 kK derived from LINDERBERG und SHULL's hydrogenic formula $K(2s,2p) = G_1(2s,2p) = (15\,Z_0/512)$ hartree. It has been recognized for a long time (cf. CONDON

Table 3.5. $K_{av}\,(2p,2p)$ *evaluated from term differences in a given configuration after correction for configuration interaction between* $2s^2 2p^q$ *and* $2p^{q+2}$. *The figures in parentheses are* $-E$ *in kK of the baricenters relative to the* $1s^2$-*systems*

q=		$2s^2 2p^q$	$2s\,2p^{q+1}$	$2p^{q+2}$
0	B+	— (494.6)	— (462.5)	1.7, 5.6 (407.8)
	C++	— (887.2)	— (841.5)	4.2, 5.8 (764.5)
	N+3	— (1389.9)	— (1331.3)	6.7, 10.5 (1232.2)
	O+4	— (2002.9)	— (1931.3)	8.9, 8.8 (1810.0)
	F+5	— (2727)	— (2641.8)	11.4, 7.3 (2499.0)
1	B	— (571.5)	2.2, 5.2 (529.5)	—
	C+	— (1096.9)	4.7, 7.2 (1032.4)	2.8, 6.0 (950.7)
	N++	— (1789.0)	7.1, 10.0 (1702.5)	5.4, 9.7 (1591.2)
	O+3	— (2647.4)	9.6, 12.5 (2539.6)	8.0, 7.2 (2399.3)
	F+4	— (3671.3)	11.8, 14.9 (3541.0)	10.2, 14.6 (3673.0)
2	C	5.1, 5.7 (1187.0)	4.0, 6.9 (1119.5)	—
	N+	7.6, 9.1 (2024.5)	5.8, 9.2 (1936.9)	—
	O++	10.0, 11.7 (3085.1)	8.4, 11.3 (2961.4)	7.2, 10.9 (2812.1)
	F+3	12.4, 10.7 (4395.0)	10.8, 24.0 (4242.7)	— ($\sim$4060)
3	N	6.4, 6.4 (2133.8)	— ($\sim$2033)	—
	O+	8.9, 9.8 (3355.7)	7.6, 9.9 (3124.1)	—
	F++	11.4, 13.0 (4883.1)	10.0, 12.7 (4717.3)	— (4525.2)
4	O	7.9, 7.6 (3484.9)	— (3350.7)	—
	F+	10.4, 10.6 (5189.7)	— (5016.7)	—

and SHORTLEY, 1953) that the first-order expressions for term distances in configurations $2s^2 2p^q$ do not agree with the experimental values:

$$
\begin{aligned}
2p^2 \text{ and } 2p^4 : {}^3P \quad & E_0 - K_{av}(2p,2p) \\
{}^1D \quad & E_0 + K_{av}(2p,2p) \\
{}^1S \quad & E_0 + 4\,K_{av}(2p,2p) \\
2p^3 : {}^4S \quad & E_0 - 3\,K_{av}(2p,2p) \\
{}^2D \quad & E_0 \cdot \\
{}^2P \quad & E_0 + 2\,K_{av}(2p,2p)
\end{aligned}
\tag{3.41}
$$

where E_0 are the baricenters evaluated from eq. (3.37). LINDERBERG and SHULL (1960) have shown that the main reason for the discrepancies is the *configuration interaction* between the terms having the same symmetry type S, L of the configurations $2s^2 2p^q$ and $2p^{q+2}$ which have the non-diagonal elements of interelectronic repulsion

$$
\begin{aligned}
q = 0 \text{ and } 4: \quad &{}^1S \quad \sqrt{3}\, K(2s, 2p) \\
q = 1 \text{ and } 3: \quad &{}^2P \quad \sqrt{2}\, K(2s, 2p) \\
q = 2: \qquad\qquad &{}^3P \quad K(2s, 2p) \\
&{}^1D \quad K(2s, 2p) \\
&{}^1S \quad 2\,K(2s, 2p)
\end{aligned}
\tag{3.42}
$$

which can be evaluated from the known values in Table 3.4 (cf. also CLEMENTI and VEILLARD, 1966). If the observed energies of the interacting terms are E_a and E_b, the correction x can be expressed by second-order perturbation theory as an implicit function of the non-diagonal element H_{ab}:

$$
x = H_{ab}^2 / (E_a - E_b + x) .
\tag{3.43}
$$

Table 3.5 gives the values for $K_{av}(2p, 2p)$ obtained for the corrected energy levels. In parentheses are given $-E$ for the corrected baricenters having values some two orders of magnitude larger than the K_{av} parameters. The latter show many unexplained discrepancies. However, in a given isoelectronic series, their variation per unit of atomic number, 2.4 kK, is roughly what is expected from the hydrogenic value $K_{av} = 3F_2 = (27\, Z_0/2560)$ hartree, i.e 2.31 kK. It is an interesting question to what extent the configuration interaction between $2s^2 2p^q$ and $2p^{q+2}$

Table 3.6. *Average parameters of interelectronic repulsion obtained after correction of the configuration baricenters for interaction between $2s^2\, 2p^q$ and $2p^{q+2}$ (cf. Table 3.3)*

	$A*(2s, 2s)$	$A*(2s, 2p)$	$A*(2p, 2p)$
Be	80.5	66.9	~ 68
B$^+$	117.2	100.9	107.2
B	106.4	95.7	100
C^{++}	153.2	134.4	146.9
C$^+$	138.8	130.2	138.8
N^{+3}	189.1	167.1	185.6
N^{++}	172.6	163.2	178.5
O^{+4}	225.1	200.0	224.6
O^{+3}	206.4	195.8	217.5
O^{++}	201.0	192.0	209.5
F^{+5}	260	232.7	262.6
F^{+4}	239.1	228.9	256.5
F^{+3}	$\sim$227	$\sim$221	$\sim$247
F^{++}	224.7	218.1	237.9

described by eq. (3.42) takes care of all the deviations observed. Though the configurations $2s2p^{q+1}$ have no possibility of changing their number of 2s electrons by two, they show many of the same irregularities as the corresponding $2p^{q+2}$ before the correlation for configuration intermixing.

    Table 3.6 gives the A_* parameters obtained from the baricenters corrected for intermixing of $2s^22p^q$ and $2p^{q+2}$. They show a much more regular evolution than Table 3.3. In particular, $A_*(2s,2s)$ have changed the most and has become an increasing function of the ionic charge for a given element. It is possible to approximate the parameters of eq. (3.37) by simple polynomials:

$$I_\varepsilon(2s) = -3.5 + 20.0(Z-2) + 27.7(Z-2)^2$$
$$I_\varepsilon(2p) = -3.0 + 3.8(Z-2) + 27.7(Z-2)^2 \tag{3.44}$$

whereas the results of Table 3.6 are slightly more complicated as a function of $(a+b)=$

$$2 \quad A_*(2s,2s) = 9 + 36(Z-2)$$
$$A_*(2s,2p) = 2 + 33(Z-2)$$
$$A_*(2p,2p) = -10 + 39(Z-2)$$
$$3 \quad A_*(2s,2s) = 7 + 33(Z-2) \tag{3.45}$$
$$A_*(2s,2p) = -2 + 33(Z-2)$$
$$A_*(2p,2p) = -17 + 39(Z-2)$$
$$5 \quad A_*(2s,2s) \sim -6 + 33(Z-2).$$

If eq. (3.45) had been an exact linear function of $(a+b)$ and $(Z-2)$, eq. (3.37) would have been a rather simple polynomial of third degree with many constraints on the higher terms; e.g. $-27.7\ (a+b)\ (Z-2)^2$ would be the only contribution of second degree in $(Z-2)$, and if the first line of (3.45) had contained the coefficient $+33$ rather than $+36$, $(Z-2)$ would have had the combined coefficients

$$-20.0a - 3.8b + \left[\frac{a(a-1)}{2} + ab\right]33 + \frac{b(b-1)}{2}39$$
$$= -36.5a - 23.3b + 16.5a^2 + 33ab + 19.5b^2. \tag{3.46}$$

However, it is quite clear that for $(a+b)$ higher than four, eq. (3.37) is very difficult to extrapolate because of the differences between enormous quantities. For instance, the experimental groundstate $^3P_0$ of the neutral oxygen atom is -3492.9 kK and the corrected baricenter of $2s^22p^4$ -3484.9 kK relative to O^{+6} and six electrons at very large distance, whereas $-2I_\varepsilon(2s) - 4I_\varepsilon(2p)$ is -6297.2 kK according to Table 3.1 and -6295.4 kK according to eq. (3.44). One would extrapolate $A_*(2s,2p) = 185$ kK and $A_*(2p,2p) = 195$ kK from Table 3.6, whereas $A_*(2s,2s)$ is much more uncertain ~190 kK though, fortunately, its coefficient in eq. (3.37) is only 1. Hence, the baricenter is extrapolated to $-6297 + 2840 = -3457$ kK. It is not surprising that this expression is 28 kK off.

We talk about a *phenomenological baricenter polynomial* Φ as a Taylor series terminating at an early stage giving approximate values of the configuration baricenters as a definite function of the occupation numbers a, b, c, ... of the individual shells and of the atomic number Z. This concept represents a remarkably accurate model of the behaviour observed for monatomic entities. We then introduce *spin-pairing energy* as a small effect superposed on Φ. In the case of one partly filled shell nl containing q electrons, it has been discussed (JØRGENSEN, 1962 and 1965 a) how the baricenter of all states having a given value of the total spin quantum number S of such a configuration deviates from the baricenter of eq. (3.38) to the extent

$$[\langle S(S+1)\rangle - S(S+1)] D \tag{3.47}$$

where the average value of $S(S+1)$ for the whole configuration l^q is

$$\langle S(S+1)\rangle = \frac{3}{4} q \left[1 - \frac{q-1}{4l+1}\right] \tag{3.48}$$

which is symmetric for q and $(4l+2-q)$, i.e. q "holes". The spin-pairing energy parameter D is given by

$$D = \frac{2l+1}{2l+2} K_{av} + \frac{1}{2l+2} (J'_{av} - J_{av}) \tag{3.49}$$

or, since it seems to be universally valid for the parameters of eq. (3.36)

$$J'_{av} = J_{av} + 2 K_{av} \tag{3.50}$$

it is also true

$$D = \frac{2l+3}{2l+2} K_{av} \tag{3.51}$$

which in the case of the 2p-shell can be obtained from Table 3.5. Though this is a technicality without importance for the rest of our discussion, it may be mentioned that for terms having maximum seniority number v, $D=K_{av}$ and the additional factor $(2l+3)/(2l+2)$ may be thought of as excitation due to decreased v, rather than as genuine spin-pairing.

The average parameters of interelectronic repulsion in eq. (3.49) are defined from their contribution to the energy of the filled shell l^{4l+2}

$$(2l+1) J'_{av} + (8l^2+4l) J_{av} - (4l^2+2l) K_{av} \tag{3.52}$$

which is identically $(2l+1)(4l+1)A*$ according to eq. (3.36).

3*

It is important not to confuse the quantum numbers M_S and S, as done in the treatment of spin-pairing energy in the books of ORGEL and of SCHLÄFER and GLIEMANN. Actually, one can define a parameter of energy decrease for increasing M_S^2 equal to $(2l+2)D/(2l+1)$ or $(2l+3)$ $K_{av}/(2l+1)$ since the baricenter of all states of l^q having a definite value of M_S deviates from the common baricenter for l^q to the extent

$$[\langle M_S^2\rangle - M_S^2]\,\frac{2l+3}{2l+1}\,K_{av} \tag{3.53}$$

where

$$\langle M_S^2\rangle = [-q^2 + (4l+2)q]/(16l+4) . \tag{3.54}$$

It may be noted that for non-equivalent electrons not restricted by the PAULI exclusion principle, $\langle M_S^2\rangle = q/4 = \langle S(S+1)\rangle/3$.

Many of these results were previously given (JØRGENSEN, 1962, 1965a and 1966) but recently, SLATER (1968) gave a more general proof. The reason why chemists tend to make a confusion between M_s and S is that M_S being the arithmetic sum of $m_s = +\frac{1}{2}$ or $-\frac{1}{2}$ for each electron can be pictured as the sum of opposite arrow directions in boxes each representing an orbital. Though it is true that the extreme values $\pm M_S$ of a given configuration also correspond to the maximum value of $S=|M_S|$, it must be realized that $(2S+1)$ such states having all intermediate values for $M_S = S-1, S-2, \ldots$ must be included too. A state having an equal number of electrons with opposite m_s has $M_S = 0$; but it does not necessarily correspond to $S=0$ unless it is the only state of the configuration (i.e. a closed shell system).

JUDD (1967) has raised the interesting question whether M_S may be a better classification and approximate quantum number than the conventional symmetry types for certain purposes for high l-values. The reader is referred to the original paper.

In the case of l^qs configurations, one may think of $K(s,l)$ such as shown in eq. (3.40) as an inter-shell spin-pairing energy parameter. As seen below, one may define significantly the effective S' for a given shell, when more than one shell is partly filled. This may be strictly valid; e.g. all terms of l^2s belong to a definite l^2 term because each L-value is represented only once; it is also true for $2s2p^3$ considered above, but not for all terms of $3d^34s$ because e.g. ^3P can originate in ^4P or ^2P terms of $3d^3$. However, the conditions are rather complicated when two or more partly filled shells have positive l. In this connection, it may be noted that $\frac{3}{4}q$ of (3.48) does not have the subsequent parenthesis in the case of non-equivalent electrons (either n or l or both being different). Though the spin-pairing energy is the main contribution to be superposed on the phenomenological baricenter polynomial Φ, there exist other effects dependent on the symmetry type as we shall see below.

One may consider this discussion from rather different points of view. LINDERBERG and SHULL (1960) argue that the exact non-relativistic energy levels can be expanded in a Taylor series in the atomic number:

$$E = a_2 Z^2 + a_1 Z + a_0 + a_{-1} Z^{-1} + \ldots \qquad (3.55)$$

which technically is the product of Z^2 and a Taylor series in Z^{-1}. There is much evidence that the three first coefficients are much more important than the following ones. However, unfortunately, when more than three electrons are present, the residual effects fluctuate, and there is no tendency for a_{-k} to decrease rapidly with increasing k. The interesting point is that the hydrogenic value obtains

$$a_2 = -\sum \frac{1}{2n^2} \text{ hartree} \qquad (3.56)$$

and that the coefficient a_1 also has quite definite values. The effects of *near-orbital degeneracy*, i.e. intermixing of configurations where two electrons differ by having the same n but different l, such as given in eq. (3.42) contribute to a_1 but other effects of configuration interaction only contribute to a_0 and later coefficients. (The near-orbital degeneracy was first discussed by LAYZER; cf. also SUREAU and BERTHIER, 1967). It is indeed striking that the two coefficients to Z^2 in eq. (3.44), 27.7 kK, both are so close to a quarter rydberg, and the many regularities studied by GLOCKLER, LISITZIN and BAUGHAN (cf. JØRGENSEN, 1962) indicate similar results. However, the relativistic effects are a considerable nuisance from this point of view, because they start earlier than eq. (3.55) with contributions such as $Z^4/137^2$. CLEMENTI (1964) emphasizes that the chemical relativistic effects are not entirely negligible when involving atomic numbers higher than 10. Since the sulphur atom in H_2S, SO_3 and SF_6 undoubtedly carries different fractional charges, an interpolation between the values evaluated for S^-, S, S^+, ... shows an effect on the groundstate energy perhaps a tenth of typical chemical bond energies. However, since we are far from being able to calculate bonding energy in such molecules a priori, this reminder is of restricted practical importance.

One may consider the experimental data of atomic spectroscopy as a kind of intermediate material for chemical calculations, not as universal as the SCHRÖDINGER equation but more fundamental than semi-empirical parameters directly connected with chemical bonding. In this sense, the observed ionization energies can be corrected for spin-pairing energy and other deviations from the configuration baricenters and determine phenomenological baricenter polynomials of which the differential quotient with respect to the occupation number of the partly filled shell is the *differential ionization energy* (JØRGENSEN, 1962). This concept is of great importance in the description of transition group compounds and is further discussed in chapter 4.

Bibliography

BERRY, R. S.: J. Chem. Educ. **43**, 283 (1966).
CLEMENTI, E.: J. Mol. Spectr. **12**, 18 (1964).
—, and A. VEILLARD: J. Chem. Phys. **44**, 3050 (1966).
CONDON, E. U., and G. H. SHORTLEY: Theory of Atomic Spectra, 2. Ed. Cambridge: University Press 1953.
COOK, D. B., and J. N. MURRELL: Mol. Phys. **9**, 417 (1965).
COTTON, F. A.: Chemical Applications of Group Theory. New York: Interscience (John Wiley) 1963.
DIEKE, G. H., and H. M. CROSSWHITE: Applied Optics **2**, 675 (1963).
HUND, F.: Linienspektren und Periodisches System der Elemente. Berlin: Springer 1927.
IAFRATE, G. J.: J. Chem. Phys. **46**, *728* (1967).
JAFFE, H. H., and M. ORCHIN: Symmetry in Chemistry. New York: John Wiley 1965.
JASTROW, R.: Phys. Rev. **73**, 60 (1948).
JØRGENSEN, C. K.: Orbitals in Atoms and Molecules. London: Academic Press 1962.
— Chimica Teorica. VIII Corso Estivo di Chimica Milano 1963, Fondazione Donegani, Accademia Nazionale dei Lincei, Rome, p. 63 (1965a).
— J. Phys. **26**, 825 (1965b).
— Quantum Theory of Atoms, Molecules and the Solid State; a Tribute to John C. Slater, p. 307 (Ed. P. O. LÖWDIN). New York: Academic Press 1966.
— Chem Phys. Letters **1**, 11 (1967a).
— Halogen Chemistry **1**, 265 (1967b) (Ed. V. GUTMANN). London: Academic Press.
— Int. J. Quantum Chem. **2**, 49 (1968).
— Lanthanides and 5f Elements. London: Academic Press 1969.
JUDD, B. R.: Operator Techniques in Atomic Spectroscopy. New York: McGraw-Hill 1963.
— Phys. Rev. **162**, 28 (1967).
LINDERBERG, J., and H. SHULL: J. Mol. Spectr. **5**, 1 (1960).
MOORE, C. E.: Atomic Energy Levels. National Bureau of Standard Circular No. 467. Vol. I (H to V) Washington, 1949; Vol. II (Cr to Nb) Washington, 1952; Vol. III (Mo to Ac except Ce to Lu) Washington, 1958.
PARSON, R. G., and V. F. WEISSKOPF: Z. Physik **202**, 492 (1967).
SCHONLAND, D. S.: Molecular Symmetry. London: Van Nostrand 1965.
SLATER, J. C.: Phys. Rev. **165**, 655 (1968).
SUREAU, A., and G. BERTHIER: Theoret. Chim. Acta **7**, 41 (1967).
SZASZ, L., and G. McGINN: J. Chem. Phys. **47**, 3495 (1967).
TIETZ, T.: J. Chem. Phys. **40**, 2066 (1964).
— Can. J. Phys. **43**, 250 (1965).
VAN VLECK, J. H., and N. G. WHITELAW: Phys. Rev. **44**, 551 (1933).
WYBOURNE, B. G.: Spectroscopic Properties of Rare Earths. New York: Interscience (John Wiley) 1965.

4. Characteristics of Transition Group Ions

The compounds of elements of the five transition groups can contain a partly filled shell, 3d, 4d, 5d, 4f or 5f. This fact is the closest link existing between atomic spectroscopy and inorganic chemistry; and at the same time, it illustrates how surprising chemistry is to the atomic spectroscopist. This surprise has various facets which it may be worth analyzing separately.

1. The k'th ionization energy of an atom going from the ionic charge $z=k-1$ to $z=k$ tends to be roughly proportional to k. Hence, the total energy of M^{+k} relative to the neutral atom M^0 tends to be a parabolic function $\frac{1}{2}(k^2+k)$. On the other hand, chemists are familiar with many non-transition group elements, such as beryllium, aluminium, silicon, thorium, . . . occurring in only one oxidation state in all or nearly all compounds, such as Be(II), Al(III), Si(IV) and Th(IV). Obviously, the chemical bonding (whatever electrostatic Madelung potentials or covalent effects being the most important) is able to over-compensate the cost of energy of the high oxidation state, whereas the corresponding gaseous ions Be^{+2}, Al^{+3}, Si^{+4} and Th^{+4} represent enormously energetic species even compared with Be^+, Al^{+2}, Si^{+3} and Th^{+3}. In the transition groups, a variety of different oxidation states is usually accessible by choosing the ligands connected with the central atom and by varying the reducing or oxidizing character of the ambient medium.

2. The partly filled shell must constitute the loosest bound electrons of the compound since it would otherwise receive further electrons from the ligands. In view of the large difference between the high *ionization energy* and the low *electron affinity* of a partly filled shell having a small average radius, it is more accurate to say that the electron affinity of the partly filled shell is necessarily smaller than the ionization energy of the loosest bound orbitals anywhere else in the compound.

3. From the point of view of quantum mechanics, it is rather surprising that in most transition group complexes, the number of electrons in the partly filled shell is an *integer* allowing also a spectroscopic oxidation state to be defined. Since the partly filled shell as well as all the filled *molecular orbitals* (M.O.) in the molecule are delocalized to some extent over adjacent atoms, this remark should not be construed to mean that the *fractional charge* of the central atom is as high as the oxidation state,

though spectroscopic observations suggest that in most cases, the fractional charge is above $+1$.

Though the total wavefunction Ψ of many-electron systems definitely are not well-defined configurations in the sense of being anti-symmetrized Slater determinants, the electron configurations have a strong classificatory ability, and the *taxological* aspects of correct predictions of the symmetry types (the total spin quantum number S; and L, Λ and Γ_n in spherical, linear and other symmetries) and approximate order of low-lying energy levels are very important. Hence, the parameters of inter-electronic repulsion are to be considered as semi-empirical quantities in the sense that they may include correlation effects not taken into account in well-defined configurations (JØRGENSEN, 1962a and 1962d). With this restriction, it is worth contemplating eq. (3.37) which we may simplify to the case of one partly filled shell containing q electrons:

$$E = E_0 - qI_\varepsilon + \frac{q(q-1)}{2} A_* \tag{4.1}$$

E_0 being the energy of the system having lost all q electrons from the partly filled shell, I_ε being the ionization of one such electron when the $(q-1)$ other electrons already are lost, and A_* being the average value of the repulsion energy between two electrons in the partly filled shell. Eq. (4.1) is an approximation of another reason than the correlation energy, namely because of the small variation of I_ε and A_* from one degree of ionization to another. In a semi-empirical description, such a variation tends to add a term to be called $(-q^3a_2/3)$ below. If the cruder approximation (4.1) is considered, the ionization energy going from q to $(q-1)$ electrons in the partly filled shell is

$$I = I_\varepsilon - (q-1) A_* \tag{4.2}$$

showing that the difference between the ionization energy I and the electron affinity A is essentially A_*. Like all other parameters of inter-electronic repulsion, this quantity is expected to be inversely proportional to the average radius $\langle r \rangle$ of the partly filled shell; or more precisely, to be proportional to $\langle r^{-1} \rangle$. The difference between consecutive ionization energies for gaseous $3d^q$ ions is some 10 eV or 80 kK, and there is good evidence that it is above 100 kK for gaseous $4f^q$ ions.

From a chemical point of view, it is worth emphasizing that the origin of the constant oxidation number $+3$ of aluminium in compounds is another than for lanthanides. The ionization energy of Al^{+2} is 229 kK which obviously is less than the chemical stabilization increasing from Al(II) to Al(III) in aqueous solution. On the other hand, the closed-shell ion Al^{+3} isoelectronic with neon has the ionization energy 965 kK which cannot be paid back by any conceivable chemical stabilization of Al(IV).

It is interesting to compare with Fe^{+2} having the ionization energy 247 kK (LOTZ, 1967) where both $Fe(H_2O)_6^{+2}$ and $Fe(H_2O)_6^{+3}$ are well known; and with Zn^{+2} 320 kK (LOTZ, 1967) where only Zn(II) is known from chemistry. Quite generally (JØRGENSEN, 1955) M(I) are oxidized by water forming H_2 if gaseous M^+ has a lower ionization energy than 20 eV or 160 kK, and M(II) are oxidized if M^{++} has the ionization energy below 30 eV or 240 kK. At this point, it is necessary to make some remarks about standard oxidation potentials measured relative to the normal hydrogen electrode. Classical physico-chemists show a great reluctance to admit single-ion thermodynamic properties; and in partic- ular, it is the general opinion that the absolute potential of the hydrogen electrode cannot be defined. However, the heats of hydration of gaseous ions seem to have a fairly well-defined meaning (ROSSEINSKY, 1965; SHARPE, 1967; MORRIS, 1968) and is for H^+ 261 ±3 kcal/mole. Another way of stating this proposition is that an approximate value for the ioni- zation energy of $\frac{1}{2}H_2$ to give hydrated H_3O^+ and an electron in vacuo is $17,7 + 109,7 - 91.6 = 35,8$ kK, where 17,7 kK is the change of enthalpy for formation of one hydrogen atom, and 109,7 kK the ionization energy forming a proton in gaseous state. In Table 24, p. 236 (JØRGENSEN, 1962b) the heats of hydration were discussed on the basis of 40 kK for this value (cf. also SANDERSON, 1966).

It is well known today (JØRGENSEN, 1963) that many aqua ions are quite definite chemical species, such as $Be(H_2O)_4^{++}$, $Mg(H_2O)_6^{++}$ and $Al(H_2O)_6^{+3}$ though the evidence is far less complete for non-transition group aqua ions than for octahedral $Cr(H_2O)_6^{+3}$, $Rh(H_2O)_6^{+3}$ (both exchanging water ligands with a half-life of several days) and rapidly reacting $Mn(H_2O)_6^{++}$, $Ni(H_2O)_6^{++}$ and quadratic $Pd(H_2O)_4^{++}$ all con- taining partly filled d-shells and having correspondingly characteristic absorption spectra. Hence, it does not seem very appropriate to consider the solvent for such cations as a homogeneous dielectric though other cases exist, such as Na^+, Ag^+, Cs^+ and Ba^{++}, where we have no certain evidence for the fixation of a definite number of aqua ligands. Any dielectric theory involves the self-energy of a sphere with radius r and charge z and the dielectric constant ε of the solvent:

$$\frac{z^2}{2r}\left(1 - \frac{1}{\varepsilon}\right). \tag{4.3}$$

For water, the reciprocal dielectric constant $\varepsilon^{-1} = 0.013$ is so close to zero that it can be neglected in eq. (4.3) and the numerical value is $(57 \text{ kK})z^2/r$ when r is measured in Å. It may be noted that N. BJERRUMs concept of a local dielectric constant ε in the interval between 10 and 80 which is useful for the discussion of the influence of separated charges on complex formation constants (SCHWARZENBACH and SCHNEIDER,

1955) would have a relatively small influence on eq. (4.3). Actually, if one accept GOLDSCHMIDT's ionic radii, eq. (4.3) is a good approximation for halide anions but a bad approximation for cations except Ag^+ since the heat of hydration is about half the expected value for alkaline and alkaline-earth metal ions and a third of the expected value for Al^{+3} and Ga^{+3}. Any explanation invoking local dielectric constants would need $\varepsilon \sim 2$ for the cations which is hardly plausible. LATIMER (1955) suggested another approach, to use crystallographic radii for anions but to increase the value of r by 0.82 Å for cations before inserting in eq. (4.3). This rather curious argumentation undoubtedly works in most cases if one wants to predict the heat of hydration within some 20% accuracy, but it does not explain the intriguing fact that Ag^+, and to a somewhat smaller extent, Cu^+ and Hg^{++}, have heats of hydration some 30 to 20 kK higher than M^+ and M^{++} isoelectronic with the noble gases. ORGEL (1958) argued that besides the ligand field stabilization (shortly discussed in Chapter 5) occurring for partly filled d-shells, the d^{10}-central atoms having low oxidation states are stabilized by mixing of empty s- and filled d-orbitals (cf. also JØRGENSEN, 1957). However, these central atoms are even much more stabilized by iodide and by phosphorus- and sulphur-containing ligands (AHRLAND, 1967; JØRGENSEN, 1966 and 1967 b).

Another modification of eq. (4.3) might occur if the conventional ionic radii were changed. There is a considerable degree of freedom at this point, and SLATER (1964) actually went so far as to study BRAGG's old suggestion of roughly constant *atomic* radii in all solid compounds, which would decrease the conventional anion radii by approximately 0.8 Å and increase the cation radii by 0.8 Å. However, this cannot be the whole truth; in typically electrovalent compounds, the ionic radii must be decreased in an absolute sense, and the fact that the shortest Cs-Cs distance in CsF is shorter than in caesium metal does not confer any properties connected with inter-metallic bonding to the fluoride. MORRIS (1968) argues that precise measurements (by x-ray diffraction) of elec- tronic densities in fairly ionic compounds suggest an increase (relative to the conventional values) of the cation radii and a decrease of the anion radii to the extent of some 0.3 Å. According to MORRIS, these changed values also rationalize the heats of hydration somewhat more satisfac- torily than the conventional ones. Quite generally, single-ion properties tend to involve zero-points with rather large experimental uncertainties. This is less true for ionic conductivities and ionic radii, which are at least positive, than for the thermodynamic quantities which, for the purist refraining from transport of charged species across phase borders, always involve the contribution zC, where C is an undetermined constant. However, in the case of Stokes' law for conductivities and in the case of dielectric effects of the type (4.3), one has the advantage that an extra-polation as a function of the reciprocal ionic radius is possible to a reasonable degree of certainty.

Ionic solids present another type of difficulties than aqua ions at great dilution in aqueous solution. As described by RABINOWITCH and THILO (1930) it was the general opinion in the decade 1920—30 that compounds are *either* covalent *or* electrovalent. In the latter case, it is a reasonable approximation to calculate the heat of formation of the solid compounds from gaseous ions by the formula

$$-\alpha z^2 / r_{MX} \qquad (4.4)$$

where the *Madelung constant* α can be evaluated for lattices (z being the smallest of the two charges of the cation M and of the anion X, and r_{MX} being their internuclear distance) by an ingenious summation giving physical significance to an apparently conditionally convergent series:

$$
\begin{array}{llll}
NaCl & 1.748 & CaF_2 & 5.039 \\
CsCl & 1.763 & TiO_2 & 4.816 \\
wurtzite & 1.641 & Cu_2O & 4.116 \\
zincblende & 1.638 \, .
\end{array}
\qquad (4.5)
$$

Though most people talk about interatomic Coulomb interactions in small molecules, the writer believes that there is a close analogy between the Madelung potential in a crystal and in a heteronuclear molecule, and he defines the Madelung constant α (for which explicit expressions can be given):

$$
\begin{array}{lll}
MX & 1 & \\
linear & MX_2 & 3.5 \\
trigonal & MX_3 & 9 - \sqrt{3} = 7.268 \\
tetrahedral & MX_4 & 16 - \tfrac{3}{2}\sqrt{6} = 12.326 \\
quadratic & MX_4 & 15 - 2\sqrt{2} = 12.172 \\
octahedral & MX_6 & 34.5 - 6\sqrt{2} = 26.015 \, .
\end{array}
\qquad (4.6)
$$

Though the internuclear distance r_{MX} generally is slightly shorter in isolated molecules than in crystalline compounds, the total Madelung energy tends to be some 50% larger for eq. (4.5) because of the larger α. It may be noted that in a molecule MX_N where N goes toward infinity, and where the ligands X are evenly distributed on a spherical surface, α goes toward the asymptotic value $N^2/2$.

The original argument for compounds either being covalent or electrovalent was based on the observation that certain chlorides (such as HCl, BCl$_3$, CCl$_4$, NCl$_3$, SiCl$_4$, PCl$_5$, Cl$_2$, TiCl$_4$, VCl$_4$,...) melt and boil below 200° C and do not conduct electricity in their anhydrous state, and other (such as LiCl, NaCl, $MgCl_2$, KCl, $CaCl_2$, $CrCl_3$,...) melt above 600° C and once molten show ionic conductance. However, MAGNUS was the first to point out that this distinction may not be caused by intrinsically different chemical bonding but rather by the relative ionic radii. Thus, SiF_4, PF_5 and SF_6 have sufficiently small central atoms in such a way that the coordination number N of the fluoride ligands is the same as the oxidation number and hence forms low-boiling, monomeric molecules. The electrostatic interaction between such neutral molecules are far smaller than in ionic salts even if the charge separation in each unit MX_N is considerable, or even, as an extreme case, completely electrovalent. Actually, the higher boiling point of WCl_6 than of WF_6 and UF_6 rather suggests that Van der Waals interactions (of the same type as those producing the relatively high boiling points of Xe and CBr_4) rather than electric charge distributions are important. The crystalline salts NaF, MgF_2 and AlF_3 on the other hand have $N = 6$ and hence have to polymerize to structures where each fluoride is connected to six sodium, three magnesium and two aluminium respectively. A comparable example is the isolated molecule CO_2 and the polymerized modifications of SiO_2 containing $Si(IV)O_4$ and $O(-II)Si_2$ groupings.

However, there remained the feeling that because of the z^2 dependence of eq. (4.4) one would not normally find partly ionic compounds, because if e.g. the NaCl-lattice BaO consisted of Ba^+O^- rather than $Ba^{+2}O^{-2}$ only a quarter of the Madelung energy would be obtained. One could show conclusively that CCl_4 cannot be fully ionic because $C^{+4}(Cl^-)_4$ would be unstable relative to neutral, isolated carbon and chlorine atoms. On the other hand, $TiCl_4$ might conceivably be fully ionic, because the energy needed to produce Ti^{+4} (and the minute energy gained by forming gaseous Cl^- fra $Cl + e^-$) can be compensated by the Madelung energy ($\alpha = 12.326$, $r_{MX} = 2.18$ Å giving -649 kK).

The mistaken opinion that compounds are either ionic *or* covalent got some support from magnetic measurements. In most cases, complexes containing a partly filled shell have a groundstate possessing a quite definite value of the total spin quantum number S which can be detected in usual paramagnetic $(S > 0)$ or diamagnetic $(S = 0)$ compounds not showing either exceptionally strong temperature-independent paramagnetism nor strong ferromagnetic or antiferromagnetic coupling between partly filled shells of different atoms. Magnetochemists then argued that *high-spin* behaviour (e.g. most Fe(II) complexes and CoF_6^{-3} having the same value of $S = 2$ as the groundstate of the corresponding gaseous ions Fe^{+2} and Co^{+3}) indicates ionic bonding (Anlagerungskomplexe in German) whereas *low-spin* behaviour (e.g. $Fe(CN)_6^{-4}$ and a few other Fe(II) complexes, nearly all cobalt (III) complexes, NiF_6^{--}, all known ruthenium (II), rhodium (III), palladium (IV), iridium (III) and platinum (IV) complexes all having $S = 0$) was taken as evidence for covalent bonding (Durchdringungskomplexe in German). As we shall see in Chapter 5, there is no reason to conclude from the sharp division in high-spin and low-spin complexes that either ionic or covalent bonding is exclusively present. Actually, spectroscopic evidence is available for the complexes of all five transition groups showing that all known cases are intermediate cases between these two extremes.

The decades 1930—50 saw a complete reversal from the emphasis on ionic bonding with superposed polarisation effects prevailing in the previous decade (RABINOWITCH and THILO, 1930; FAJANS, 1967). Based on magnetochemical and stereochemical arguments, PAULING suggested partly covalent bonding in many more compounds than previously expected. The hybridization theory suffered from the general weakness of the valence-bond description that it was thought necessary to find N orbitals forming suitably directed linear combinations in the chromophore MX_N. Thus, in SF_6, *six* orbitals are needed for the partly covalent bonding, and the only reasonable hybridization in octahedral symmetry is sp^3d^2, whereas one might discuss whether the four orbitals supposedly involved in the covalent bonding of $TiCl_4$ or MnO_4^- were sp^3 or sd^3. By the same token, if ThO_2 is not entirely ionic, the cubic chromophore $Th(IV)O_8$ occurring in the fluorite lattice would need eight orbitals sp^3d^3f. From the point of view of M.O. theory (JØRGENSEN, 1962b) the assumption of an equal importance of N orbitals of the central atom is entirely unnecessary, and SF_6 may have the main part of its covalent bonding due to *four* orbitals, 3s and 3p of sulphur, or the weak covalent bonding in ThO_2 may be produced by the empty 7s or three of the five empty 6d orbitals of Th(IV).

PAULING (1948) finally admitted that the magnetochemical criterion of high-spin and low-spin behaviour cannot be used, and proposed at the same time the *electroneutrality principle*. The fractional charges of all atoms in compounds were supposed to occur in the interval from -1 to $+1$. This idea is fairly reasonable for the anions, since it is trivially true for halides, and since it is difficult to furnish convincing evidence that even the least covalent oxides carry more than one negative charge on the oxygen atoms. In sofar compounds MX alone are considered, the electroneutrality principle looks fairly safe. However, the situation is entirely different for discrete molecules MX_N with high N. The electroneutrality principle would restrict the fractional charge on the anion to the interval from O to -0.25 in the case of CF_4, $TiCl_4$ and RuO_4, and to the interval from O to -0.167 in the case of SF_6, WCl_6 and IrF_6. The first five molecules mentioned are closed-shell systems, and the arguments of chapter 5 do not apply directly to them, as they do to the $5d^3$-system

IrF_6. However, any reasonable extrapolation of the spectroscopic evidence, as well as various other physical properties, indicate that the fractional charge of the central atom most frequently occurs in the interval between $+1$ and the oxidation state.

The electroneutrality principle would only be plausible if the reacting atoms had large internuclear distances. Actually, the highest known electron affinity of a neutral atom, that of Cl, is smaller than the lowest known ionization energy of a neutral atom, that of Cs. Hence, all compounds dissociate to neutral atoms if all internuclear distances are strongly increased. However, the Madelung potential favours charge separation on adjacent atoms at the actual internuclear distances prevailing in compounds. JØRGENSEN et al. (1967) suggested a model combining the *differential ionization energy* previously defined (JØRGENSEN, 1962d) with the Madelung potential, minimizing the total energy as a function of the ionicity ξ which is chosen in such a way that for neutral molecules MX_N and complex anions MX_N^{-q}, the fractional charge on X is $-\xi$ and on M in the two cases $+N\xi$ and $(N\xi - q)$, respectively. The differential ionization energy $I(z)$ for a given atom having a fractional charge z is an analytical function

$$I(z) = a_0 + a_1 z + a_2 z^2 \qquad (4.7)$$

within each interval of z-values characterizing a definite partly filled shell nl (e.g. for Cl between -1 and $+5$). The ionization energy I_n going from M^{+n-1} to M^{+n} is

$$I_n = \int_{n-1}^{n} I(z)dz = \left(a_0 - \frac{a_1}{2} + \frac{a_2}{3}\right) + (a_1 - a_2)n + a_2 n^2$$
$$= a_0 + \left(n - \frac{1}{2}\right)a_1 + \left[n(n-1) + \frac{1}{3}\right]a_2 . \qquad (4.8)$$

If the electron affinity going from M^0 to M^- falls within the same nl-shell, it is I_0 with this definition.

In order to obtain a smooth analytical function $I(z)$ from the energy levels known from atomic spectroscopy, it is necessary to correct for the spin-pairing energy (3.47). In the limit of Russell-Saunders coupling, neglecting relativistic (spin-orbit coupling) effects, the appropriate configuration baricenter occurs at:

$$
\begin{aligned}
&p^2, p^4: && [^3P] + \tfrac{4}{5}D && \sim ([^3P] + [^1D])/2 \\
&p^3: && [^4S] + \tfrac{12}{5}D && \sim [^2D] \\
&d^2, d^8: && [^3F] + \tfrac{2}{3}D + \tfrac{9}{2}B && \sim [^3F] + 9B \\
&d^3, d^7: && [^4F] + 2D + \tfrac{9}{2}B && \sim [^4F] + 18B \\
&d^4, d^6: && [^5D] + 4D && \sim [^5D] + 28B \\
&d^5 && [^6S] + \tfrac{20}{5}D && \sim [^6S] + 46B
\end{aligned}
\qquad (4.9)
$$

where the sharp brackets indicate the observed energies of definite terms. D is the spin-pairing energy parameter and B one of Racah's parameters of interelectronic repulsion to be discussed in chapter 5. As a matter of fact, the energy difference $[{}^3P] - [{}^3F]$ in d^2 and d^8 systems and $[{}^4P] - [{}^4F]$ in d^3 and d^7 all represent $15B$, whereas $[{}^4G] - [{}^6S]$ in d^5 represent $30\,B$.

Table 4.1 gives the parameters of eq. (4.7) for many elements and for such nl shells which are of chemical significance. In the cases where only two (and not at least three) ionization energies can be determined for insertion in eq. (4.8), the most uncertain parameter, a_2, has been obtained by extrapolation and is given in parentheses in Table 4.1. The configuration baricenters used for the evaluation of the parameters were tabulated by JØRGENSEN et al. (1967). They are uncertain to the extent of some 2 kK, and in particular, a_2 in Table 4.1 should not be taken too seriously. The same set of experimental ionization energies can be represented nearly equally well with several polynomials (4.7) and there is little doubt that the non-monotonic behaviour of a_0 going from Mn to Co in the 3d group really represents a weak increase of a_2 which cannot be safely determined as an entirely free parameter. On the other hand, it is not satisfactory to restrict eq. (4.7) to two terms, putting $a_2 = 0$. By the some token as the phenomenological baricenter polynomial such as eq. (3.46) does not contain third-order terms and tends to deviate when highly differing electron configurations are considered, (4.7) is equivalent to a total energy

$$E = -T(z) = -T_0 + a_0 z + \frac{a_1}{2} z^2 + \frac{a_2}{3} z^3 \qquad (4.10)$$

having as zero point $-T_0$ the energy of the neutral atom. It is remembered that such expressions are only valid approximations within the same nl-shell. Further on, expressions such as eq. (4.10) loose their physical significance for z-values below that (z_0) corresponding to the maximum (for which $I(z_0) = 0$ and is decreasing) since it is quite obvious that the atom does not get more stable when negative charge below z_0 is piled up on it. Actually, it is surprising to what extent (4.10) is a good approximation for $z > z_0$.

MULLIKEN (1965) asked the question whether the concept of *electronegativity* x should be relegated to chemical history. Inorganic chemists certainly agree (JØRGENSEN, 1962 d) that x of Mn(VII) certainly must be much higher than for Mn(II) and it is quite true that the valence state may be different for compounds of an element in the same oxidation state (lead (IV) connected with CH_3^- and other alkyl groups definitely is less oxidizing than in $PbCl_6^-$). However, already the success of a concept such as the optical electronegativity x_{opt} to be discussed in

Table 4.1. *Differential ionization energy parameters of eq. (4.7) in the unit 1 kK = 1 000 cm^{-1}*

Element	Shell	a_0	a_1	a_2	Element	Shell	a_0	a_1	a_2
H	1s	58	104	(0)	Ga	4p	21	50	(6)
Li	2s	21	35	(12)	Ge	4p	30	56	(6)
Be	2s	48	47	(12)	As	4p	41	62	6
B	2p	27	70	(14)	Se	4p	52	66	6
C	2p	40	83	(14)	Br	4p	60	73	6
N	2p	55	96	14	Kr	4p	76	76	6
O	2p	69	110	14	Rb	5s	19	25	(6)
F	2p	83	127	14	Sr	5s	28	31	(6)
Na	3s	19	41	(6)	Zr	4d	9	46	9
Mg	3s	36	47	(6)	Nb	4d	11	52	9
Al	3p	21	50	(6)	Mo	4d	11	60	9
Si	3p	32	57	(6)	Ru	4d	21	66	9
P	3p	45	64	6	Rh	4d	23	72	9
S	3p	55	74	6	Pd	4d	28	75	9
Cl	3p	67	82	6	Ag	5s	35	46	(6)
Ar	3p	81	90	6	Ag	4d	42	75	9
K	4s	19	28	(6)	Cd	5s	44	52	(6)
Ca	4s	30	34	(6)	In	5p	26	38	(6)
Ti	3d	−12	60	12	Sn	5p	32	44	(6)
V	3d	−12	60	15	Sb	5p	42	48	6
Cr	3d	0	60	15	Te	5p	48	56	6
Mn	3d	6	62	15	I	5p	54	68	6
Fe	3d	1	70	15	Xe	5p	58	80	6
Co	3d	−1	76	15	Cs	6s	19	20	(6)
Ni	3d	1	80	15	Ba	6s	27	26	(6)
Cu	4s	35	51	(6)	Pt	5d	36	64	(9)
Cu	3d	4	84	15	Au	6s	49	49	(6)
Zn	4s	45	57	(6)	Hg	6s	54	55	(6)

chapter 7 seems to indicate that x sometimes has an observable background. MULLIKEN's suggestion 1934 of $x = (I_0 + I_1)/2$ makes $x = a_0$ if a_2 is neglected. The arguments conducting to eq. (4.2) identifies a_1 with A_*, a parameter of interelectronic repulsion expected to be proportional to the reciprocal average radius $\langle r_{12}^{-1} \rangle$ of the partly filled shell. In Table 4.1, there is a certain trend in a_0 and a_1 increasing regularly with increasing atomic number in a given nl-group, but there is no tendency of a_0 and a_1 to be universally related. In particular, the 3d group has extraordinarily large a_1 and rather small a_0, in fact smaller than for the alkali metals. This is another way of expressing the situation that a neutral atom with nothing but 3d electrons outside the argon closed-shells such as Ti $3d^4$ is extraordinarily easy to ionize, whereas the configurations $3d^34s$ and $3d^24s^2$ are more stable. In chapter 5, we are returning to our reasons to believe that, under equal circumstances, the average radii of 4d and 5d shells are some 60 to 70% larger than of the 3d shell, and consequently, a_1 is smaller for 4d and 5d group atoms. Again, the 2p group has much smaller radii and higher a_1 than all the subsequent p groups. It turns out that it is the first n-value for each given l-value, 1s, 2p, 3d and 4f, which is characterized by this property. This generalization is only remotely connected with the fact that these orbitals have no radial nodes.

If the differential ionization energy was a kind of chemical potential for electrons, and if the Madelung potential is neglected, the equilibrium condition for fractional electron transfer in the neutral molecule MX_N would be

$$I_M(+N\xi) - I_X(-\xi) = 0 . \tag{4.11}$$

The resulting ionicities ξ obey the electroneutrality principle in the sense that $N\xi$ seems always to be below 1. However, they are not plausible at all; e.g. $\xi = 0.39$ for NaCl which is known to be entirely electrovalent or at least very nearly so. JØRGENSEN, HORNER, HATFIELD and TYREE (1967) then suggested to add the Madelung potentials to eq. (4.11). Calling the quantity $e/r_{MX} = \mu$ (corresponding to 115 kK if the internuclear distance r_{MX} is measured in Å) the crystals MX of eq. (4.5) have the Madelung potentials (for electrons) $+\alpha\xi\mu$ at M sites and $-\alpha\xi\mu$ at X sites. The MX_N molecules in eq. (4.6) all have the Madelung potential $+N\xi\mu$ at the central atom, whereas the X sites have

$$
\begin{array}{lll}
\text{linear} & MX_2 & -1.5\,\xi\mu \\
\text{trigonal} & MX_3 & -1.8452\,\xi\mu \\
\text{tetrahedral} & MX_4 & -2.1628\,\xi\mu \\
\text{quadratic} & MX_4 & -2.0858\,\xi\mu \\
\text{octahedral} & MX_6 & -2.6716\,\xi\mu .
\end{array}
\tag{4.12}
$$

The Madelung constant α is $(N/2)$ times the sum of the absolute values of the coefficients at the M site (i.e. $+N$) and the X site (i.e. the numbers in eq. (4.12) with positive sign). Thus, the equilibrium condition for differential ionization energy becomes

$$I_M(+N\xi) - I_X(-\xi) - \frac{2\alpha}{N}\,\xi\mu = 0 . \tag{4.13}$$

Neither diatomic NaCl ($r_{MX} = 2.36$ Å) nor the crystalline salt ($r_{MX} = 2.81$ Å) has a zero-point of eq. (4.13); the two polynomials obtained,

$(-48+25\xi)$ and $(-48-19\xi)$, respectively, have only physical significance for $\xi \leq 1$ and are invariantly negative. In our model, this means that the compound is most stable as a completely ionic entity. As discussed below, there is no doubt that the influence of the Madelung potential has been exaggerated in eq. (4.13). However, though the potentials (4.12) have been evaluated for point-ions, the same results would have been obtained for spherically symmetric, non-overlapping, ions.

It may be worth, as a numerical example, to consider the tetrahedral molecule $TiCl_4$ for which eq. (4.13) reads (cf. Table 4.1):

$$
\begin{aligned}
I_{Ti}(4\xi) &= -12 + 240\xi + 192\xi^2 \\
-I_{Cl}(-\xi) &= -67 + 82\xi - 6\xi^2 \\
-6.1628\xi\mu &= -325\xi \\
\hline
& -79 - 3\xi + 186\xi^2
\end{aligned}
\tag{4.14}
$$

since $r_{MX} = 2.18$ Å corresponds to $\mu = 52.7$ kK. The appropriate root of the polynomial (4.14) is $\xi = 0.66$ giving $z_{Ti} = 2.64$ and $z_{Cl} = -0.66$.

In complex anions MX_N^{-q}, the Madelung potential at the M site is $+N\xi\mu$ like in the neutral molecules, whereas the X sites have added $+q\mu$ to the expressions (4.12). Thus, for the octahedral anion $TiCl_6^{--}$ having $r_{MX} = 2.35$ Å, the polynomial reads

$$
\begin{aligned}
I_{Ti}(-2+6\xi) &= -84 + 72\xi + 432\xi^2 \\
-I_{Cl}(-\xi) &= -67 + 82\xi - 6\xi^2 \\
+2\mu - 8.6716\xi\mu &= +98 - 425\xi \\
\hline
& -53 - 271\xi + 426\xi^2
\end{aligned}
\tag{4.15}
$$

having the appropriate root $\xi = 0.793$ corresponding to the charge distribution $Ti^{+2.76} Cl_6^{-0.793}$.

Table 4.2 gives the ionicities ξ and central atom fractional charges z_M calculated for 46 molecules and complex anions according to polynomials such as (4.14) and (4.15). Many of the compounds are not known, but are included in the table for comparison. It is striking how invariant z_M is for a given element with a given set of ligands as a function of the oxidation state. The q electrons added to MX_N to give MX_N^{-q} go nearly exclusively on the N ligand atoms in this model. The general tendency for z_M is to increase with the series of ligands $Br < Cl < F < O$, which is perhaps a little surprising with respect to oxygen. One reason may be the relatively short internuclear distances observed in the oxo complexes; it is remembered that the μ values are inversely proportional to r_{MX}. Since the coefficients a_0 and a_1 tend towards a smooth increase across a given nl-shell in Table 4.1, it is not surprising that z_M decreases regularly going from Ti to Ni or from Zr to Pd.

Table 4.2. *Ionicities* ξ *and central atom fractional charges* z_M *calculated from the model of differential ionization energies taking the Madelung potential into account. The internuclear distances* r_{MX} *are given in* Å. *(hyp.) indicates hypothetical compounds; values in sharp brackets are given when the model predicts full ionicity. Internuclear distances in brackets are extrapolated from other molecules*

Compound	r_{MX}	ξ	z_M	Compound	r_{MX}	ξ	z_M
SF_6	1.58	0.555	3.33	ZrF_6^{-2}	2.04	[1]	[4.00]
SCl_6 (hyp.)	(2.02)	0.176	1.06	$ZrCl_6^{-2}$	2.45	0.907	3.44
ClO_4^-	1.43	1.265	4.06	$MoCl_6^{-3}$	(2.50)	0.928	2.57
TiF_6^{-3}	(1.99)	[1]	[3.00]	$MoCl_6^{-2}$	(2.42)	0.775	2.65
$TiCl_6^{-3}$	(2.43)	0.962	2.77	$MoCl_6^-$	(2.34)	0.595	2.57
$TiCl_4$	2.18	0.66	2.64	MoF_6	(1.82)	0.592	3.55
TiF_6^{-2}	1.91	0.90	3.40	$MoCl_6$	(2.26)	0.42	2.52
$TiCl_6^{-2}$	2.35	0.793	2.76	RuO_4^{-2}	(1.85)	1.29	3.16
CrO_4^{-2}	1.65	1.365	3.46	RuF_6	(1.80)	0.50	3.00
CrF_6	(1.68)	0.50	3.00	$RuCl_6$ (hyp.)	(2.24)	0.333	2.00
$CrCl_6$ (hyp.)	(2.12)	0.383	2.30	RuO_4^-	1.79	1.023	3.09
$FeCl_4^{-2}$	(2.30)	0.99	1.96	RuO_4	1.70	0.785	3.14
$FeCl_4^-$	2.19	0.767	2.07	PdF_6^{-2}	1.89	0.748	2.47
$FeBr_4^-$	(2.32)	0.705	1.81	$PdCl_6^{-2}$	(2.31)	0.63	1.78
FeF_6^{-3}	2.03	0.887	2.32	PdF_6 (hyp.)	(1.78)	0.397	2.38
$FeCl_6^{-3}$	(2.45)	0.833	2.00	TeF_6	1.84	0.66	3.96
FeO_4^{-2}	(1.60)	1.275	3.10	$TeCl_6$ (hyp.)	(2.28)	0.352	2.11
FeF_6 (hyp.)	(1.66)	0.45	2.70	XeF_2	1.99	0.28	0.56
$FeCl_6$ (hyp.)	(2.10)	0.33	1.98	XeF_4	1.95	0.23	0.92
FeO_4 (hyp.)	(1.55)	0.715	2.86	XeO_4	(1.75)	0.225	0.90
NiF_6^{-4}	2.01	[1]	[2.00]	PtF_6^{-2}	1.91	0.808	2.85
$NiCl_6^{-4}$	2.50	0.935	1.61	$PtCl_6^{-2}$	2.33	0.675	2.05
NiF_6^{-2}	1.70	0.727	2.36	PtF_6	(1.80)	0.467	2.80

Our model obviously would lack physical significance for isolated atoms at large internuclear distances which are known to contain each an integral number of electrons. A closer analysis (JØRGENSEN, 1965) shows that this breakdown of the idea of differential ionization energies is connected with the coefficient q $(q-1)/2$ to A_* in eq. (4.1) whereas a charge distribution in classical electrostatics shows the coefficient $q^2/2$. However, the delocalized molecular orbitals occurring in molecules at normal internuclear distances attenuates this argument to a considerable extent.

Actually, one can compare heats of formation of neutral molecules from neutral, gaseous atoms with the model (JØRGENSEN et al. 1967).

The first thing to remark is that the p^q or d^q configuration baricenters, through which the differential ionization energy parabola of Table 4.1 passes, have higher energy than the lowest energy level of the neutral atoms. This energy difference is for instance (in kK):

$$
\begin{array}{llllllll}
\text{Ti } 43 & \text{V } 47 & \text{Cr } 49 & \text{Mn } 52 & \text{Fe } 39 & \text{Co } 28 & \text{Ni } 15 \\
\text{Zr } 31 & \text{Nb } 29 & \text{Mo } 36 & \text{Tc } — & \text{Ru } 15 & \text{Rh } 4 & \text{Pd } 0
\end{array}
\tag{4.16}
$$

Thus, the stabilization predicted by our model for $TiCl_4$ is

$$
E_{Ti}(d^4) + \int_0^{2.64} I_{Ti}(z)dz + 4\int_0^{-0.66} I_{Cl}(z)dz - 0.66^2 \cdot 649
\tag{4.17}
$$
$$
= +43 + 251 - 108 - 283 = -97 \text{ kK}
$$

The same result is obtained by adding 43 kK to *four* times the integral

$$
\int_0^{0.66} (-79 - 3\xi + 186\xi^2)\, d\xi = -35 \text{ kK}
\tag{4.18}
$$

of the polynomial (4.14). The reason for the multiplication by four is that the differential electron transfer in eq. (4.13) is $Nd\xi$.

The heat of formation of liquid $TiCl_4$ from metallic Ti and gaseous Cl_2 is -179.3 kcal/mole equivalent to -63 kK. The heat of sublimation of Ti is 39 kK, the heat of formation of four chlorine atoms from two chlorine molecules 40 kK, the heat of vapourization for $TiCl_4$ roughly 2 kK, so that the heat of formation of gaseous $TiCl_4$ from isolated Ti and Cl atoms is -140 kK. This is 43 kK more negative than the value (4.17) predicted by our model. The reason of this deviation is clearly the neglect of covalent bonding, since the model does not include any bond energy of homonuclear molecules.

We can define the difference between the heat of formation observed and the heat of formation predicted by our model as E_{cov}. This quantity is known to be positive in typical ionic salts, such as $+6$ kK for NaCl, $+8$ kK for LiCl and $+2$kK for RbI. E_{cov} contains a positive contribution from the repulsion between the atomic cores (increasing tremendously when the internuclear distances are artificially compressed by applying external pressure) and a negative contribution from the formation of bonding M.O. not having their anti-bonding counterparts occupied. LADD and LEE (1964 and 1965) discussed Madelung energies of such crystals. We can define the *experimental Born-Haber energy* for MX_N

$$
E_{B.H.} = \Delta H - E(M^{+N}) - NE(X^-)
\tag{4.19}
$$

as the difference between the heat of formation from neutral atoms ΔH and the energies of the gaseous ions (assuming univalent anions; for chalkogenides, M^{+2N} and X^{-2} are considered). Thus, $E_{B.H.}$ for crystalline NaCl is -65.4 kK whereas the calculated Madelung energy (4.4) is -71.5 kK. The difference between these numbers is E_{cov} in the special case of $\xi = 1$. LADD and LEE discussed the suggestion that the ratio between the calculated Madelung energy and $E_{B.H.}$ is $1 + (1/gr_{MX})$ where g is constant (roughly 3.1Å^{-1}) for alkali metal halides crystallizing in the NaCl lattice. Thus, this ratio is 1.115 for LiCl and 1.046 for RbI. On the other hand, the ratio is 1.030 for AgF, 0.964 for AgCl and 0.943 for AgBr. When this quantity is smaller than 1, there is no doubt that partly covalent bonding occurs.

Many physicists have elaborated models including the attraction between higher-order electric multipole moments of the constituents of ionic lattices, and it is indeed true that negative values of E_{cov} can be explained that way. However, when a cubic crystal such as AgBr has $E_{cov} = -4$ kK (which would be fully ionic in our model) it is rather difficult to invoke multipole interactions, because the high symmetry prevents static dipole, quadrupole and octupole moments; the 16-pole moment is the first (totally symmetric) one permitted. When FAJANS (1967) talks about polarization effects, they rather are spherically symmetric, modifying the radial functions.

In the case of $TiCl_4$, the experimental Born-Haber energy (4.19) is $-140 - 735 + 116 = -759$ kK. The ratio between the calculated Madelung energy -649 kK for the fully ionic molecule and this quantity is 0.855, somewhat smaller than for crystalline AgBr. In most molecules (with exception of cases such as CO having the most stable, triple bond in the situation C^-O^+ isoelectronic with N_2) and crystals (with the analogous exceptions such as Ga^-As^+ isoelectronic with Ge) the partly covalent bonding tends to stabilize the neutral atoms with less pronounced charge separation than predicted from the Madelung potential alone. There is a persistent argument in literature (PEARSON and GRAY, 1963; FERREIRA, 1964) that approximately

$$E_{cov} = -b(1 - \xi^2)^{1/2}$$
$$dE_{cov}/d\xi = b\xi/(1 - \xi^2)^{1/2} \tag{4.20}$$

preventing ionicities close to the extreme limit. We may also consider the simpler approximation $E_{cov} = b(\xi - 1)$ having a larger differential quotient than eq. (4.20) for $\xi < 0.707$. Supposing $dE_{cov}/d\xi = 100$ kK for $TiCl_4$, we add a quarter of this value (cf. 4.18), viz. 25 kK, to the first term of the polynomial (4.14) obtaining

$$-54 - 3\xi + 186\xi^2 \tag{4.21}$$

having the appropriate root $\xi = 0.545$, i.e. $z_{Ti} = 2.18$. Though no assumption was introduced in (4.21) that $E_{cov} = 0$ for $\xi = 1$ (it is probably slightly positive), the corresponding value of $E_{cov} = 100\ (0.545 - 1) = -45.5$ kK is close to the difference evaluated eq. (4.17).

JØRGENSEN et al. (1967) also found $E_{cov} = -54$ kK for $TiBr_4$, -78 kK for RuO_4, -44 kK for TeF_6 and -33 kK for XeF_4. It may be noted that the quadratic symmetry of the last molecule readily can be explained by the angular dependent part of the covalent bonding; the difference between the Madelung energy (4.12) for quadratic and tetrahedral forms is only 0.5 kK for $\xi = 0.24$. As discussed in chapter 7, the values found for E_{cov} are quite compatible with the effects expected in M.O. theory of covalent bonding with the empty 3d orbitals in $TiBr_4$, empty 4d orbitals in RuO_4, empty 5s and 5p orbitals in TeF_6, and partly filled 5p shell (low-spin $5p^2$) in XeF_4. The only difficulty is that E_{cov} is the difference between the actual energy (at an lower, unknown ξ – value) and the energy minimum calculated in our model for the higher ξ – value given in Table 4.2. Most z_M values are probably about one unit too high, as discussed in chapter 5.

The model of differential ionization energies cannot be applied to *hydrides*. Thus, the polynomial for methane ($r_{MX} = 1.094$ Å)

$$
\begin{aligned}
I_C(4\xi) &+ 40 + 332\xi + 224\xi^2 \\
-I_H(-\xi) &- 58 + 104\xi \\
-6.1628\xi\mu\ &\qquad - 647\xi \\
\hline
&- 18 - 211\xi + 224\xi^2
\end{aligned}
\tag{4.22}
$$

suggests full ionicity $C^{+4}H_4^-$. This argument is not affected by the higher $I(z)$ for the 2s shell of carbon discussed below, and actually, the observed heat of formation $- 140$ kK of CH_4 from isolated atoms compares well with the complete Madelung expression $+ 1193 - 25 - 1294 = -126$ kK. This is not the case for CCl_4 ($r_{MX} = 1.766$ Å)

$$
\begin{aligned}
I_C(4\xi) &+ 40 + 332\xi + 224\xi^2 \\
-I_{Cl}(-\xi) &- 67 + 82\xi - 6\xi^2 \\
-6.1628\xi\mu\ &\qquad - 401\xi \\
\hline
&- 27 + 13\xi + 218\xi^2
\end{aligned}
\tag{4.23}
$$

for which the appropriate root is $\xi = 0.32$ and $z_C = 1.28$ (eq. (7.15) on p. 91, JØRGENSEN 1962d, contains an error in μ). CF_4 ($r_{MX} = 1.32$ Å) poses a very interesting problem. The differential ionization energy I_C in (4.23) is only defined for $z \leq 2$, and the polynomial

$$
\begin{array}{l}
I_C(4\xi) + 40 + 332\xi + 224\xi^2 \\
- I_F(-\xi) - 83 + 127\xi - 14\xi^2 \\
- 6.1628\xi\mu \qquad - 536\xi \\
\hline
-43 - 87\xi + 210\xi^2
\end{array}
\tag{4.24}
$$

would have its zero-point above this value. Since only I_3 and I_4 of carbon is available to define $I(z)$ for the 2s shell, we cannot determine all three parameters. However, the function $103 + 98z + 6z^2$ seems to be a reasonable choice. If a_2 was much larger (e.g. comparable to $a_2 = 14$ for the 2p shells in Table 4.1), a_1 would be smaller than 83, the value for 2p, which does not seem quite reasonable. Inserting this function in eq. (4.13), we obtain

$$
\begin{array}{l}
I_{C2s}(4\xi) + 103 + 392\xi + 96\xi^2 \\
- I_F(-\xi) - 83 + 127\xi - 14\xi^2 \\
- 6.1628\xi\mu \qquad - 536\xi \\
\hline
+ 20 - 17\xi + 82\xi^2
\end{array}
\tag{4.25}
$$

not having any real root, and it can be concluded that in our model, the fluorine atoms are able to extract exactly the two 2p electrons and none of the two 2s electrons of the carbon atom; and hence the minimum energy occurs for $\xi = 0.50$ and $z_C = 2.00$.

The writer is grateful to Professor ANDRÉ JULG who first brought the question of hydrides up. Probably, the individuality of the hydrogen atoms in compounds is much less pronounced than is the case for other elements. Thus, four hydride anions in CH_4 would show considerable overlap. The ionic radius of $H(-I)$ is $1.3\,\text{Å}$ in LiH, but larger than of $F(-I)$ in the solid hydrides of the heavier alkali metals where the ionic radius of $H(-I)$ is at least $1.5\,\text{Å}$. The interprotonic distance in CH_4 is only $1.094\sqrt{3} = 1.90\,\text{Å}$. It might also be argued that in carbon compounds, one should consider a valence state configuration $(2s2p^3)^{q/4}$ as first pointed out by MULLIKEN. One might define an effective differential ionization energy.

$$
I_{eff} = \frac{1}{4} I_{C2s}(z) + \frac{3}{4} I_{C2p}(z) = 56 + 87z + 12z^2
\tag{4.26}
$$

but this would not remove the paradox (4.22) though it would contribute to make most carbon compounds less ionic.

In this connection, it may be mentioned that from a spectroscopic point of view, hydrogen has a much higher electronegativity than inferred from the chemical behaviour. Thus, the ionization energy of the hydrogen atom is higher than of chlorine and xenon atoms, and just slightly below that of oxygen and krypton. Though the ionization energies corrected for spin-pairing energy would be higher for Cl and O, these facts are rather surprising. However, the impression of considerable positive charge on hydrogen in compounds is somewhat confused by the chemical reactions occurring in aqueous solution. Thus, HI is a very strong acid, forming hydrated H_3O^+ and I^- quantitatively, and H_2Te a fairly strong acid, though the charge separations in gaseous HI and H_2Te must be very small.

The main reason why the Madelung energy is overestimated in hydrides is probably the unusually small internuclear distances. The linear molecule CO_2 presents another case having $r_{MX} = 1.16$ Å. By the same token as eq. (4.24). the two 2p electrons get completely transferred to the oxygen ligands. The 2s shell (4.25) produces a new problem:

$$
\begin{aligned}
&I_{C2s}(2\xi) + 103 + 196\xi + 24\xi^2 \\
&-I_O(-\xi) - \ 69 + 110\xi - 14\xi^2 \\
&-3.5\xi\mu \qquad\quad -347\xi \\
&\overline{\qquad\qquad\qquad\qquad\qquad\qquad} \\
&\qquad + \ 34 - \ 41\xi + 10\xi^2 .
\end{aligned}
\tag{4.27}
$$

The root $\xi = 1.2$ of this polynomial is not appropriate because it represents a relative maximum and not a minimum of the total energy, since the polynomial is decreasing in the vinicity of the root. Hence, the minimum energy for (4.27) is represented by one of the two limits of its domain of definition, $z = 2$ or $z = 4$. It is easy to calculate the full Madelung expressions $+287 - 22 - 347 = -82$ and $+1193 + 106 - 1388 = -89$ kK in the two cases, inserting the energies

$$
\begin{array}{llll}
O(p^4 \text{ baricenter}) + \ 8 & \quad C(p^2) & + \ 5 \text{ kK} \\
O^-(p^5) \qquad\qquad -11 & \quad C^{+2}(s^2) + 287 & \\
O^{--}(p^6) \qquad\quad +53 & \quad C^{+4} \qquad +1193 &
\end{array}
\tag{4.28}
$$

However, the total energy -89 kK predicted for the fully ionic form $C^{+4}O_2^{-2}$ is not at all in agreement with the experimental value -135 kK for the heat of formation of CO_2 from gaseous atoms, and $E_{cov} = -46$ kK must be related to the empty 2p orbitals of carbon (IV) forming strong covalent bonds. Again, eq. (4.26) would be of no help and furnish a completely ionic form.

The conclusion of all these calculations is that the total energy of a heteronuclear molecule is remarkably unsensitive to the charge distribution within wide limits, the Madelung contributions rather systematically compensating the steeply increasing valence state energy as a function of increasing positive charge of the central atom. Further on, it can be concluded that the model of differential ionization energies seems somewhat more satisfactory in transition group halides than in the case of organic compounds. The stereospecific covalent bonding is much stronger in carbon compounds than in most of the heavier molecules and favoring in most cases less charge separation than predicted from considerations of the Madelung potential alone. Actually, much better agreement with experience could be obtained if one had a good excuse, a kind of local dielectric effect, for multiplying the Madelung potential by 0.7 or 0.8 in the model of differential ionization energies. This model is otherwise the least chemical of all models in the sense that it involves only data from atomic spectroscopy and internuclear distances.

We have not yet discussed the influence of empty shells having much larger average radii than the partly filled shell of transition group complexes. This problem is rather different from that of the 2s and 2p shells having comparable radii but differing $I(z)$ functions. We saw above that it is no satisfactory solution to consider the p shell as the unique source of transferred electrons until it is used up, at what moment the s shell starts to transfer. This is also one reason why SO_4^{--} was not included in Table 4.2; whereas ClO_4^- still depletes its 3p shell in the model of differential ionization energies taking the Madelung energy into account, sulphate would need the 3s shell too. Ros and SCHUIT (1966) suggested the rather extreme proposal not to take 4s and 4p electrons into account at all when evaluating the fractional atomic charge of a 3d group central atom. This proposal is much more plausible than it looks at first, because the parameters of interelectronic repulsion to be discussed in chapter 5 are only insignificantly smaller in neutral atoms $3d^q4s^2$ than in the corresponding $M^{++} 3d^q$. The physical reason is that the 4s orbital has so large an average radius that it is superposed the ligands in a typical complex. Actually, there is very little evidence available at present that the 4p orbitals plays any rôle in the covalent bonding of 3d group complexes. It is obvious that these statements must include some border-line cases; thus, both copper (I) and copper (II) form rather covalent complexes but Cu(I) has only the empty 4s orbital available, whereas a large part of the covalent bonding of Cu(II) is connected with the last, half-empty 3d orbital. In the case of lanthanide complexes, the empty 5d orbitals must have an average radius not much larger than in three d groups, and definitely contribute to the fractional charge, whereas the question is much more open for the empty 6s and 6p orbitals. We return to these questions in chapter 8, because the metallic character of many alloys is connected with the collective delocalization of such empty orbitals.

Since we saw p. 41 that the ionization energy corresponding to

$$\left(\frac{1}{2}H_2\right)_{gas} + (H_2O)_{aq} \rightarrow (H_3O^+)_{aq} + e^-_{vacuo} \qquad (4.29)$$

is close to 36 kK (and hence much lower than, say, the ionization energy 109.7 kK of a gaseous hydrogen atom) it is possible to estimate ionization energies of cations and other species in aqueous solution from *standard oxidation potentials.* We neglect here the entropy constituting the source of difference between heat (enthalpy) and free energy of a given species. (cf. ROSSEINSKY, 1965, and PHILLIPS and WILLIAMS, 1965, p. 260). This produces only a small error relative to the recognized uncertainty attached to the value 36 kK accepted for the absolute ionization energy of the hydrogen electrode. Unfortunately, there are not extremely many cases known where two aqua ions $M(H_2O)_x^{+z-1}$ and $M(H_2O)_y^{+z}$ exist in equilibrium in aqueous solution, and where, at the same time, the spectroscopic data for the ionization of gaseous M^{+z-1} to M^{+z} are available. Table 4.3 collects nearly all well established cases, and a few interesting, but somewhat uncompletely documented cases. JØRGENSEN (1956b) discussed the standard oxidation potentials of 3d group $M(H_2O)_6^{+2}$ to $M(H_2O)_6^{+3}$. The apparent oscillating behaviour disappears when correction is made for the *ligand-field stabilization* written on the conventional form (PENNEY, 1940; BJERRUM and JØRGENSEN, 1956; JØRGENSEN, 1956a; GEORGE and McCLURE, 1959; DUNITZ and ORGEL, 1960) respecting the baricenter rule (cf. also SCHUIT, 1962 and 1964). Hence, for octahedral complexes $(t_{2g})^a(e_g)^b$ (in MULLIKEN's notation; in BETHE's notation $\gamma_{5g}^a\gamma_{3g}^b$) having the energy difference Δ between the two sub-shells e_g and t_{2g} the stabilization is

$$(-0.4a + 0.6b)\, \Delta \tag{4.30}$$

which is identically zero for high-spin $d^5 (a = 3,\ b = 2)$ and for the filled d^{10}-shell $(a = 6,\ b = 4)$. It is seen that the standard oxidation potentials corrected for ligand field stabilization of the M(II) and M(III) hexa-aqua ions jump to the extent of some 5V or 40 kK at the half-filled shell d^5. The main reason is the spin-pairing energy (3.47) though a supplementary effect occurs in d^q systems $(q = 2,3,7,8)$ having both F $(L = 3)$ and P $(L = 1)$ terms of maximum S. In these cases, the F and P terms are separated $15\,B$, where B is one of Racah's parameters of interelectronic repulsion, and the F-terms are situated $-4.5\,B$ below the baricenter of states having the maximum value of S [cf. eq. (4.9)]. Thus, the ground terms of the gaseous ions are stabilized relative to the baricenter of the d^q configuration:

$$\begin{aligned}
d^2, d^8&: -\tfrac{2}{3}D - \tfrac{9}{2}B\\
d^3, d^7&: -2\,D - \tfrac{9}{2}B\\
d^4, d^6&: -4\,D\\
d^5&: \quad -\tfrac{20}{3}D\,.
\end{aligned} \tag{4.31}$$

In Table 4.3, only the ligand field stabilization $-2.4\,\Delta$ and not the spin-pairing energy $+6\,D$ going from $S = 2$ to $S = 0$ has been taken into account. The numerical value of $6D$ would be some 46 kK for gaseous Co^{+3} but is expected to be decreased by the nephelauxetic effect to some 28 kK for $Co(H_2O)_6^{+3}$. Hence, the ionization energy corrected for spin-pairing energy would be 59 kK. It is seen that the difference between the values 236 to 270 kK for the first six cases in Table 4.3 of ground state energy differences of gaseous M^{+3} and M^{+2} and the corresponding six corrected values 33.8 to 59 kK for hexa-aqua ions rather invariantly is 210 kK. Unfortunately, no reliable values can be established for $M(III) \rightarrow M(IV)$ ionization energies in the 3d group, since M(IV) is only known in oxo, hydroxo, fluoro,... complexes. The closest guess is Ti(IV) which does not occur as $Ti(H_2O)_6^{+4}$ in strong mineral acid; but the uncertain hydrolysis products should not shift the standard oxidation potential immensely. The *hydration difference* between the ionization energy in atomic spectroscopy and in chemistry seems to be 320 kK in this case. The well-established standard oxidation potential for Cu(I) to Cu(II) aqua ions is also atypical, partly because the stereochemistry of the two aqua ions is highly different, and partly because of the more serious effect that we have good arguments for believing that most complexes of PEARSON's (1963 and 1966) *soft Lewis acids* such as Cu(I), Ag(I), Au(I), Hg(II) and Tl(III) are particularly stabilized. Anyhow, the difference between the atomic spectroscopic and the chemical value is 120 kK.

The three differences, 120, 210 and 320 kK, are eminently reasonable from one point of view. If the electrostatic self-energy of a cation in aqueous solution is proportional to z^2 (and neglecting the expected inverse dependence on the ionic radius) it should change $3\varkappa$ between M(I) and M(II), $5\varkappa$ between M(II) and M(III), and $7\varkappa$ between M(III) and M(IV), giving the value 43 kK for the empirical parameter $\varkappa$ in the 3d group. Obviously, the fractional charge of the central atom in $Fe(H_2O)_6^{+3}$ *is not* as high as $+3$; however, the discussion above of differential ionization energies suggests that the partly covalent bonding undoubtedly decreasing the fractional charges at the same time produces a moderate, if not insignificant, stabilization relative to the fully ionic form. Hence, it is not so surprising that the very rough approximation of $\varkappa z^2$ gives a reasonable picture of the hydration energies of ions.

The silver (I) and silver (II) aqua ions also are somewhat peculiar; Table 4.3 suggests a hydration difference of 110 kK. That this value is slightly smaller than 120 kK for copper can be readily excused by the larger ionic radii expected to decrease $\varkappa$. By the same token, the hydration difference 313 kK for the *double* ionization from Tl(I) to Tl(III) is slightly smaller than $120 + 210 = 330$ kK would be in the 3d group.

Table 4.3. *Ionization energies of aqua ions derived from standard oxidation potentials in aqueous solution; and the corresponding ionization energies between groundstates and between configuration baricenters of gaseous ions*

	Standard oxidation potential relative to hydrogen electrode	Ionization energy (hydrogen 36 kK)	Corrected for ligand field stabilization	Energy difference between ground states of gaseous ions		Energy difference between configuration baricenters	
V(II) → V(III)	-0.25 V	34.0 kK	33.8 kK	$V^{+2}(^4F)$ → $V^{+3}(^3F)$	236 kK	$V^{+2}(d^3)$ → $V^{+3}(d^2)$	232 kK
Cr(II) → Cr(III)	-0.41	32.7	45	$Cr^{+2}(^5D)$ → $Cr^{+3}(^4F)$	250	$Cr^{+2}(d^4)$ → $Cr^{+3}(d^3)$	246
Mn(II) → Mn(III)	$+1.5$	48.1	60	$Mn^{+2}(^6S)$ → $Mn^{+3}(^5D)$	272	$Mn^{+2}(d^5)$ → $Mn^{+3}(d^4)$	259
Fe(II) → Fe(III)	$+0.77$	42.2	38.0	$Fe^{+2}(^5D)$ → $Fe^{+3}(^6S)$	247	$Fe^{+2}(d^6)$ → $Fe^{+3}(d^5)$	273
Co(II) → Co(III)	$+1.84$	50.8	[87]	$Co^{+2}(^4F)$ → $Co^{+3}(^5D)$	270	$Co^{+2}(d^7)$ → $Co^{+3}(d^6)$	283
Ti(III) → Ti(IV)	0	36	29	$Ti^{+3}(^2D)$ → $Ti^{+4}(^1S)$	350	$Ti^{+3}(d^1)$ → $Ti^{+4}(d^0)$	350
Cu(I) → Cu(II)	$+0.17$	37.4	45	$Cu^+(^1S)$ → $Cu^{+2}(^2D)$	165	$Cu^+(d^{10})$ → $Cu^{+2}(d^9)$	165
Ag(I) → Ag(II)	$+1.99$	52.1	64	$Ag^+(^1S)$ → $Ag^{+2}(^2D)$	175	$Ag^+(d^{10})$ → $Ag^{+2}(d^9)$	175
Tl(I) → Tl(III)	$+1.25$	2(46.1)	92.2	$Tl^+(^1S)$ → $Tl^{+3}(^1S)$	405	$Tl^+(s^2)$ → $Tl^{+3}(d^{10})$	405
Ru(II) → Ru(III)	$+0.22$	37.8	[45]	$Ru^{+2}(^5D)$ → $Ru^{+3}(^6S)$	229	$Ru^{+2}(d^6)$ → $Ru^{+3}(d^5)$	245
Ce(II) → Ce(III)	—	5	5	$Ce^{+2}(^3H_4)$ → $Ce^{+3}(^2F_{5/2})$	162	$Ce^{+2}(f^2)$ → $Ce^{+3}(f)$	156
Pr(II) → Pr(III)	—	13	13	$Pr^{+2}(^4I_{9/2})$ → $Pr^{+3}(^3H_4)$	175	$Pr^{+2}(f^3)$ → $Pr^{+3}(f^3)$	166
Eu(II) → Eu(III)	-0.43	32.5	32.5	—		—	
Yb(II) → Yb(III)	-1.15	27	27	$Yb^{+2}(^1S_0)$ → $Yb^{+3}(^2F_{7/2})$	205	$Yb^{+2}(f^{14})$ → $Yb^{+3}(f^{13})$	209
Ce(III) → Ce(IV)	$+1.9$	51	51	$Ce^{+3}(^2F_{5/2})$ → $Ce^{+4}(^1S_0)$	296	$Ce^{+3}(f^1)$ → $Ce^{+4}(f^0)$	295
Th(III) → Th(IV)	—	20	—	$Th^{+3}(^2F_{5/2})$ → $Th^{+4}(^1S_0)$	232	$Th^{+3}(f^1)$ → $Th^{+4}(f^0)$	230

Finally, the oxidation potential of $Ru(H_2O)_6^{+2}$ to $Ru(H_2O)_6^{+3}$ has recently been established by MERCER and BUCKLEY (1965). The value in Table 4.3 was only corrected for the ligand field stabilization (4.30). The spin-pairing energy $+6D$ is smaller than for $Co(H_2O)_6^{+3}$, some 21 kK, and hence, the hydration difference 205 kK is just a tiny bit below that prevailing in the 3d group.

The 4f group presents an extremely interesting set of hydration differences. Unfortunately, the Ce(IV) aqua ion is much more hydrolyzed than the Th(IV) aqua ion (the only M(IV) aqua ions known to occur in 1 molar $HClO_4$ are formed by the 5f group) but the standard oxidation potential of Ce(III) cannot be wrong by many tenths of a volt. Hence, the hydration difference is slightly above 240 kK and much smaller than the value 320 kK characterizing the 3d group. The M(II)/M(III) oxidation potentials are not known for the same 4f elements as the M^{+2}/M^{+3} ionization energies. Fortunately, the general theory for spin-pairing energy and other effects in f^q systems has been shown to be in excellent agreement with electron transfer spectra of M(III) complexes (JØRGENSEN, 1962c; RYAN and JØRGENSEN, 1966) as well as the opposite problem of $4f^q \rightarrow 4f^{q-1}5d$ transitions (JØRGENSEN, 1962c) which have also been studied for M(II) (McCLURE and KISS, 1963) and M(III) (LOH, 1966) in dilute solid solution in CaF_2. Actually, the stabilization relative to the $4f^q$-baricenter of the lowest J-level of the lowest term (having the highest L compatible with the maximum value of S) is

$$
\begin{aligned}
f^1, f^{13} &- 2\zeta \\
f^2, f^{12} &- \tfrac{8}{13}D - 9E^3 - 3\zeta \\
f^3, f^{11} &- \tfrac{24}{13}D - 21E^3 - \tfrac{7}{2}\zeta \\
f^4, f^{10} &- \tfrac{48}{13}D - 21E^3 - \tfrac{7}{2}\zeta \\
f^5, f^9 \ \ &- \tfrac{80}{13}D - 9E^3 - 3\zeta \\
f^6, f^8 \ \ &- \tfrac{120}{13}D - 2\zeta \\
f^7 \qquad &- \tfrac{168}{13}D
\end{aligned}
\tag{4.32}
$$

with the *exception* that the coefficient to ζ (an abbreviation for the Landé parameter ζ_{nf}) is *half a unit less negative* for $q > 7$. The spin-pairing energy parameter D corresponds to Racah's $(\tfrac{9}{8}E^1)$. Since $E^1 \sim 10E^3$, eq. (4.32) corresponds to a characteristic variation of the stabilization, such that the difference between f^k and f^{k+1} (appropriate for the electron transfer $4f^k \rightarrow$ (ligand M. O.)$^{-1}$ $4f^{k+1}$ or for the transition $4f^{k+1} \rightarrow 4f^k 5d$) varies

$$
f^0 > f^1 > f^2 \sim f^3 \sim f^4 > f^5 > f^6 \ll f^7
\tag{4.33}
$$

with a plateau. The jump at the half-filled shell f^7 equal to $8D$ (it can be shown generally to be $(2l+2)D$, cf. JØRGENSEN, 1962d) is enormous.

In most cases, the linear variation of the one-electron energy differences makes f^{7+k} rather similar to f^k in eq. (4.33).

The virtue of (4.33) is that the three parameters of eq. (4.32) are well known from internal transitions in the partly filled 4f shell. Thus, for the M(III) aqua ions, D is roughly 6.5 kK; E^3 increases linearly from 0.46 kK for Pr(III) to 0.67 kK for Tm(III); and ζ increases parabolically from 0.64 kK for Ce(III) to 2.95 kK for Yb(III).

The ionization energies in aqueous solution relative to the hydrogen electrode 36 kK given for Ce(II) and Pr(II) in Table 4.3 have been obtained by subtracting the calculated value for the position of the first electron transfer band of the M(III) bromide complexes in ethanol from 63 kK (JØRGENSEN, 1962c). The standard oxidation potentials of Sm(II) and Eu(II) is in agreement with this evaluation. The ionization energy of Ce^{++} is given by SUGAR (1965). The ionization energy of Pr^{++} was at first given as 187.3 kK (SUGAR, 1963) but according to kind information from Dr. JACK SUGAR, 174.5 kK is a better value. The ionization energy 205 kK for Yb^{++} given by BRYANT (1965) is admittedly somewhat uncertain. On the other hand, the hydration difference slowly increases from 157 kK for Ce(II) → Ce(III) to 178 kK for Yb(II) → Yb(III) probably related to the decreasing ionic radii which, however, are larger than for the largest 3d group members such as Sc(III).

The ionization energies of gaseous 5f group ions are very little known, at least in all cases having any connection with chemistry in aqueous solution. The ionization energy 20 kK given for Th(III) aqua ions in Table 4.3 is a very uncertain value, derived from a comparison between the optical properties of materials such as NdI_2, SmI_2 and ThI_3. The hydration difference is then only 210 kK, corresponding to the much larger ionic radii of thorium than of titanium. However, from the fact that the standard oxidation potentials of the aqua ions are

$$
\begin{aligned}
&U(III) \;\rightarrow U(IV) \;\; -0.63V \\
&Np(III) \rightarrow Np(IV) + 0.14 \\
&Pu(III) \rightarrow Pu(IV) + 0.98
\end{aligned}
\tag{4.34}
$$

it can be extrapolated that the ionization energy of the gaseous ions would be 243 kK for U^{+3}, 249 kK for Np^{+3} and 256 kK for Pu^{+3}. If the hydration difference has increased to 220 kK for plutonium, the value would be 264 kK for Pu^{+3}. Actually, eq. (4.32) suggests that the ionization energy 20 kK given for Th(III) in Table 4.3 may be slightly too small. It is also possible to interpolate the ionization energy $32 + 165 = 197$ kK for gaseous Eu^{+2}.

CONNICK (1949) was the first to emphasize the relations between ionization energies of gaseous ions and the relative stability of different oxidation states in the 4f and 5f groups. MILES (1965) compared optical transition energies with variations of the standard oxidation potentials. It is obvious that the two variations only run parallel with the correct slope ($1V = 8$ kK) if all other factors (variation of internuclear distances etc.) are comparable. However, in the 4f group, the absence of important effects of ligand field stabilization, and the regular variation of all other properties as a function of the atomic number, corresponding to a contraction of all radii from La(III) to Lu(III), makes it true that the standard oxidation potentials of the M(II) aqua ions follow the same series (with opposite sign) as established by BARNES (1964; cf. BARNES and PINCOTT, 1966).

$$Eu(III) < Yb(III) < Sm(III) < Tm(III) < \dots \qquad (4.35)$$

for electron transfer spectra of M(III) compounds. It has been realized for a long time that going from La(III) to Lu(III), the hydration difference can overcome the oxidation to Ce(IV) in aqueous solution, and that the Madelung energy can explain the ready oxidation to Pr(IV) in various mixed oxides, and the preparation of Nd(IV) in the double fluoride Cs_3NdF_7 (ASPREY and CUNNINGHAM, 1960). It was also realized that at the end of the 4f group, the electrons are bound so strongly as to make Yb(II) fairly stable in barium-like compounds such as $YbSO_4$ and $YbCO_3$ and Tm(II) accessible in TmI_2. It was much less easier to understand that the stability of half-filled shells explaining the frequent occurrence of Eu(II) and Tb(IV) isoelectronic with Gd(III) also extends to both sides to f^6-systems such as Sm(II) and f^8-systems such as Dy(IV). The quantitative form (4.32) of the spin-pairing energy explains these facts satisfactorily.

In the d-transition groups, the hydration difference overcompensates the a_1-component of the differential ionization energy (if $2x > a_1$) and the main protection against a large number of oxidation states being possible with almost ionic bonding consists in the a_2-coefficient. The situation must be different in the 4f group which behaves like aluminium in the sense that the straight oxidation up to Al(III) is more than compensated for by hydration or Madelung energy, whereas oxidation beyond to Al(IV) would be impossible, because the discontinuous differential ionization energy $I(z)$ jumps to much higher values for $z > 3$ when the neon closed-shell is attacked. However, the physical origin of the invariant oxidation state in the 4f group is another, i.e. the large difference between the huge ionization energy and small electron affinity due to effects of interelectronic repulsion as expressed by the a_1-coefficient in eq. (4.7) and A_* in eq. (4.2). Unfortunately, insufficient data are available as yet for the atomic spectra of 4f group gaseous ions (DIEKE and CROSSWHITE, 1963) to determine all three coefficients of $I(z)$.

Though the two first ionization energies of cerium are not directly involved in our argumentation, it is illustrating to consider the total energies. According to READER and SUGAR (1966) the first ionization energy is 45.6 kK and according to SUGAR and READER (1965) the second ionization energy is 87.5 kK. Hence, the energy of the groundstate of Ce^{++} relative to the groundstate of Ce^0 is 133 kK. Since the ionization energy of Ce^{++} is 161.9 kK according to SUGAR (1965), Ce^{+3} has the energy 295 kK, and Ce^{+4} 591 kK. However, for the evaluation of the differential ionization energy, the baricenters of the configurations $4f^q$ are needed. Arguments can be given (JØRGENSEN, 1969) that $4f^3$ is a highly excited configuration of Ce^+ having the baricenter approximately at 92 kK on our scale. It is easy to find from SUGAR's complete measurements (1965) that the baricenter of $4f^2$ is situated 7 kK above the ground state of Ce^{++} and hence at 140 kK. The minor relativistic effect shifts the baricenter of $4f^1$ of Ce^{+3} to 296 kK. Hence, the differential ionization energy for the 4f shell of cerium reads

$$I(z) = -85 + 48z + 15z^2 . \qquad (4.36)$$

In a way, it is rather surprising that a_1 is so small. Actually, with the exception of the much smaller a_0 coefficient, eq. (4.36) is similar to $I(z)$ for beryllium, suggesting a strong tendency toward high ionicities in the compounds. However, the 4f electrons must have a relatively large average radius in Ce^{++}, and eq. (4.36) is probably not typical for the lanthanides. It is quite conceivable that for thulium, $I(z)$ runs as high as

$$-150 + 100z + 15z^2 . \qquad (4.37)$$

However, far more data are needed for the atomic spectra. In the case of thorium, using Th^{+4} as zero-point, it is known that the baricenter $(5f^1)$ of Th^{+3} occurs at -228 kK; and arguments have been given (JØRGENSEN, 1969) for the ionization energy of Th^{++} being 150 kK, in which case the baricenter $(5f^2)$ of Th^{++} is situated at -354 kK. In analogy to the decrease of a_2 for 3p and 4d groups relative to the 2p and 3d groups, one may estimate either

$$-25 + 30z + 12z^2$$
$$\text{or } -51 + 48z + 9z^2 \qquad (4.38)$$

as reasonable polynomials going through these three points. Chemically, Th(III) is much more readily oxidized to Th(IV) than Ti(III) to Ti(IV).

In the 5f group, the influence of spin-pairing energy (4.32) is weaker than in the 4f group because all the parameters of interelectronic repulsion under equal circumstances are about 60% as large in the 5f as in the 4f group. On the other hand, the relativistic effects and consequent typical intermediate coupling are very strong, and the coefficients to ζ in (4.32) are no longer a satisfactory description (CARNALL and WYBOURNE, 1964). Chemically, it is not *much* more difficult to oxidize Cm(III) to Cm(IV) than Bk(III) to Bk(IV), though the half-filled shell $5f^7$ is represented by Cm(III) and Bk(IV). The 5f group is not as indifferent with respect to variable oxidation state as the 4d and 5d groups, but at least in the beginning, U, Np, Pu, the variability is larger than in the 3d group. Qualitatively, this can be understood by the less steep increase of I(z) for 5f compared with the corresponding 4f element. Though some evidence is available that the function (4.37) should vary at least as steeply as for the 3d group, the much more invariant oxidation state of the 4f group must be connected also with the smaller hydration difference in the 4f group relative to the 3d group (cf. Table 4.3). The classification of the electron transfer spectra makes it possible to predict that Pr(IV) by no means can be oxidized to Pr(V) in chemical compounds as once suggested, and that most lanthanides cannot be oxidized to M(IV), not even in fluorides (JØRGENSEN and RITTERSHAUS, 1967). On the other hand, there is no obvious reason why Np(VII) and Cm(VI) compounds should not be possible (HAISSINSKY and JØRGENSEN, 1966). Actually, KROT and GELMAN (1967) recently oxidized alkaline neptunates (VI) and plutonium (VI) with O_3 to species having a new spectrum and apparently containing Np(VII) and Pu(VII).

Even in cases where the differential ionization energies have been much better established than in the f groups, the variation of the hydration difference to the extent of 10 or 20 kK wipes out completely the information of chemical interest, viz. how reducing a given aqua ion is relative to the hydrogen electrode. Other electrochemical evidence is available about reversible standard oxidation potentials of other complex species, such as $+0.96$ V for $IrCl_6^{-3}/IrCl_6^{-2}$; $+0.95$ V for $IrBr_6^{-3}/IrBr_6^{-2}$; $+0.59$ V for RuO_4^{-2}/RuO_4^- and $+1.00$ V for RuO_4^-/RuO_4. However, in these cases, the ionization energies of the corresponding gaseous ions are not known from experiment. One can calculate from Table 4.1 and eq. (4.8) that the eighth ionization energy of ruthenium is 1023 kK. Hence, the chemical difference 979 kK is considerably larger than one would have expected for $15\varkappa$; but obviously, the treatment has no longer the same basis as for the 3d group hexa-aqua ions. Whereas the fractional charge of the manganese atom in $Mn(H_2O)_6^{++}$ is above 1.5, the ruthenium atom probably carries a charge well below the value 3.14 suggested in Table 4.2 for RuO_4.

One can compare higher limits for the standard oxidation potentials of unknown oxidation states with the ionization energy of the corresponding gaseous ions. Thus, it can be concluded that

	standard oxidation potential	ionization energy of gaseous ion	hydration difference
Be(I)/Be(II)	$< -1.7\,V$	146.9 kK	>124
Mg(I)/Mg(II)	$< -2.4\,V$	121.3	>108
Al(II)/Al(III)	$< -1.7\,V$	229.5	>207
Zn(I)/Zn(II)	$< -0.5\,V$	145	>110
Hf(III)/Hf(IV)	$< -\quad 2\,V$	267	>247
Hg(I)/Hg(II)	$<\quad 0.5\,V$	151.3	>110

$$(4.39)$$

and the lower limits of hydration differences are quite compatible with our results above. It is not so easy to discuss the absence of higher oxidation states because of their formation of complexes, if by nothing else, by oxide and hydroxide in aqueous solution. MnO_4^- and RuO_4 are not so readily correlated with atomic spectroscopy as the low oxidation states in (4.39). JØRGENSEN (1956b) concluded from Table 4.3 that $Sc(H_2O)_6^{++}$ disproportionates to scandium metal and Sc(III) in aqueous solution; that $Ti(H_2O)_6^{++}$, if it exists, is much more reducing then frequently believed [this was confirmed by OLVER and ROSS (1963) in the case of Ti(II) in acetonitrile solution]; and that Ni(III) and Cu(III) is slightly less and Zn(III) somewhat more oxidizing than F_2. However, the latter argument has not much connection with the fact that Ni(IV) readily form heteromolybdates and crystalline periodates such as $KNiIO_6$ and that Cu(III) and Ag(III) are known in tellurate and periodate complexes (cf. JØRGENSEN, 1963).

PHILLIPS and WILLIAMS (1965 and 1966) gave a very interesting discussion of the different attempts possible for predicting the hydration differences. Already SCROCCO and SALVETTI (1954) pointed out that Zn(II), Ag(I), Hg(II) and various other d^{10} aqua ions are slightly more stable than expected from an electrostatic model, and considered this as a proof of partly covalent bonding. That $Ni(H_2O)_6^{++}$ is even more stable than expected from an extrapolation between $Mn(H_2O)_6^{++}$ and $Zn(H_2O)_6^{++}$ is due to the ligand field stabilization (4.30) which already is built in as a smaller ionic radius of Ni(II); but it is worth remembering that additional strong stabilization exists for closed-shell systems. There has been some worry why one could not interpolate along a line of comparable slope between Sc(III) and Fe(III) and between Fe(III) and Ga(III). One reason may be that the two latter species occur as hexa-aqua ions, whereas the former seems to be $Sc(H_2O)_9^{+3}$ or at least $Sc(H_2O)_8^{+3}$ (GEIER, 1965). COTTON and VERDIER (1956) attempted to evaluate the bond energies of complexes such as 3d group $M(NH_3)_6^{+2}$ and $Co(NH_3)_6^{+3}$ and obtained fair agreement with the experimental values. PEARSON and MAWBY (1967) discussed this problem further with respect to halide complexes MX_N^{+Z-N}.

It is an interesting question whether there exists a first-order expression for the purely electrostatic part of the hydration energy. LATIMER was the first to suggest to insert the conventional ionic radii for anions in eq. (4.3) and to add 0.82 Å to the ionic radii of cations. This is obviously an empirical approach, and it works remarkably well in the sense that the huge hydration energies of M^{+Z} are not many percent wrong. However, the difficulty is that minor deviations of 10 to 20 kK are of great chemical interest. In this connection, it may be noted that formation energies from ionic species always are much larger than from gaseous atoms, and in one sense, the expression including the high energies of gaseous ions are somewhat artificial. Thus, the dissociation energy of gaseous HF to neutral atoms is 47.3 kK, whereas the dissociation to H^+ and F^- needs 129.2 kK and to H^- and F^+ 171.5 kK. The reason why the fractional charge of fluorine in gaseous HF is rather negative is essentially the Madelung potential, though the concept of non-overlapping ions is not applicable to hydrides. Actually (JØRGENSEN, 1962b, p. 236) one can calculate hydration energies for cations from the neutral, gaseous atoms. Once a value close to 36 kK is accepted for the hydrogen electrode, the hydration energies from atoms turn out to be highly negative (i.e. ΔH highly positive) with the one exception of Li (-3.8 kK). Other representative values of ΔH are in kK (the value of Ag(I) has been corrected according to kind information from Dr. S. AHRLAND):

$$
\begin{array}{llllll}
K(I) & +\ 7.4 & Cu(I) & +11.9 & Ag(I) & +17.8 & Tl(I) & +\ 20.9 \\
Mg(II) & +20.6 & Fe(II) & +30.7 & Zn(II) & +48.2 & Hg(II) & +\ 82 \\
Al(III) & +37.8 & Fe(III) & +70.0 & La(III) & +15.3 & Tl(III) & +101.9 .
\end{array} \quad (4.40)
$$

These values tend to be more positive when the bonding in the aqua ion is relatively more covalent, showing that the external electrons are particularly firmly bound in the gaseous atoms of those elements forming rather covalent aqua ions. Hence, one cannot talk about bond-energies in aqua ions in the same sense as defined in neutral molecules since the values in (4.40) systematically have the wrong sign. This is only an apparent paradox; electrons are not available at zero energy in an aqueous solution; the loosest bound electrons have the energy relative to vacuo -36 kK if the solution is in equilibrium with a hydrogen electrode, and -44 kK if it is 1 volt more oxidizing.

Another important aspect of standard oxidation potentials is that they refer to adiabatic processes, where the internuclear distances are allowed to re-arrange, if the oxidized and reduced form are most stable at different internuclear distances. This is a striking difference from optical excitations which are so rapid that FRANCK and CONDON's principle does not permit the internuclear distances to re-arrange significantly. Quite generally, the information obtained from visible and ultraviolet spectroscopy is strongly influenced by this instantaneous picture obtained (cf. JØRGENSEN, 1967a). We cannot at all expect absorption bands to correspond to photo-ionization of aqua ions in solution at the low wavenumbers, above 30 kK, obtained from the standard oxidation potentials.

Discussing PEARSON's concept of soft and hard Lewis acids and bases, KLOPMAN (1968) applied LATIMER's proposal for hydration energies in a treatment based on KLOPMAN's semi-empirical theory (1964) for molecular bonding energies and differential ionization energies of the constituent atoms. In chapter 7, we return to this question of general interest for chemistry.

Recent progress in the understanding of absorption spectra of complexes containing a partly filled d or f shell has shown the intimate relations with the theory of atomic spectra. In this chapter, we have mainly concentrated on the ionization energies whereas the structure of the ground configuration is discussed in chapter 5 and the energy differences to higher configurations in chapter 6. The unfortunate situation for chemistry that hydration energies or formation energies of oxide and halide complexes cannot be accurately predicted prevents use of many arguments which would otherwise be available. The ionization energies of monatomic entities is attenuated tremendously in aqueous solution allowing many more oxidation states to occur for aqua ions than expected by an atomic spectroscopist. The conditions for thermodynamic stability in acidic solution include that the ionization energy is above 36 kK (otherwise, H_2 is evolved) and the electron affinity below 46 kK (the limit for evolving O_2). Obviously, these two conditions are only necessary and not sufficient; the aqua ion may loose protons to form hydroxo or oxo complexes, it may disproportionate [as does Cr(IV) and Tl(II)] or rearrange in other ways (such as Hg(I) dimerizing to Hg_2^{++}). However, it is quite clear that the chemist is familiar with a much smaller range of ionization energies than the atomic spectroscopist, since the kinetic barriers so frequently encountered against the achievement of thermodynamic equilibrium rarely show up as exorbitant over-voltages for hydrogen or oxygen evolution.

The characteristic transition group behaviour is connected with a series of adjacent oxidation states being brought within the permissible range of chemical ionization energies. At this point, the five transition groups show a certain individuality. The 3d group is most stable as M(II) and M(III) though higher oxidation states are more stable at the beginning, of titanium and vanadium, and lower show up in the end such as copper (I). The 4d and 5d groups exhibit a much greater variety and higher average oxidation state; they also are known to form more complexes with other ligands than water in aqueous solution. By judicious choice of the ligand, either very low or very high oxidation states can be obtained. The 5f group tends, like the four other transition groups, toward high oxidation states in the beginning and lower at the end. However, as pointed out by HAISSINSKY, there is a certain change in character before and after americium. The elements Ac, Th, Pa are nearly as invariant as Lu, Hf, Ta; and U, Np, Pu have high and variable oxidation states roughly to the same extent as the 4d group. There is a subsequent strong, but not exclusive, tendency to form M(III), but not enough to defend completely the objectionable name "actinides". Finally, the 4f group is in the extreme position that most of the elements only are known as M(III) in aqueous solution, and that the aqua ions have ionization energies far above 46 kK and electron affinities far below 36 kK (cf. Table 4.3). The tremendous difference between ionization energy and electron affinity of the partly filled 4f shell is connected with its unusually small average radius. Actually, as we shall see in chapter 8, there is no clear-cut theoretical reason why the 4f group aqua ions fix their oxidation state at M(III) rather than at M(II) or M(IV); on the other hand, the large value of A_* in eq. (4.1) is sufficient to insure that a *constant* oxidation state is obtained. The reason why Ce(III) is far more stable than Ti(III) though the consecutive ionization energies of Ti are larger than of Ce is essentially the larger ionic radii of the 4f group compared with the 3d group producing more hydration energy in the latter case. Outside of the transition groups, the oxidation state tends to be determined by all the electrons being counted off either to the previous closed-shell noble gas or to a filled s^2-shell. In the case of the aqua ions, this is caused by the hydration energy overcompensating the spectroscopic ionization energies until a strong discontinuity such as a filled shell is encountered. However, the tendency to form complexes other than aqua ions can be extraordinarily strong outside the transition groups; and there is a close analogy between the existence of SO_4^{--} and ClO_4^- and of MnO_4^- and OsO_4. In particular at the beginning of each transition group (excepting 4f) there is such a trend toward achieving the maximum oxidation state corresponding to the side-group number.

In the transition groups 3d and 4f having partly filled shells with small average radii, the spin-pairing energy is particularly important. This is also an effect of differing interelectronic repulsion such as A_* mentioned above. One of the main conclusions of this chapter is that the spin-pairing stabilization is not an exclusive quality of the half-filled shells d^5 and f^7 but occurs, to a somewhat smaller extent, for the adjacent numbers of electrons in the partly filled shell. The spin-pairing energy parameter D contained in the contribution $- DS(S+1)$ in eq. (3.47) is mainly evaluated from internal transitions to be discussed in the next chapter; it is an order of magnitude smaller than $A*$ but is much more available for direct experimental determination in chemical compounds.

Bibliography

AHRLAND, S.: Helv. Chim. Acta **50**, 306 (1967).
ASPREY, L. B., and B. B. CUNNINGHAM: Progr. Inorg. Chem. **2**, 267 (1960).
BARNES, J. C.: J. Chem. Soc. 3880 (1964).
—, and H. PINCOTT: J. Chem. Soc. (A) 842 (1966).
BJERRUM, J., and C. K. JØRGENSEN: Rec. Trav. Chim. **75**, 658 (1956).
BRYANT, B. W.: J. Opt. Soc. Am. **55**, 771 (1965).
CARNALL, W. T., and B. G. WYBOURNE: J. Chem. Phys. **40**, 3428 (1964).
CONNICK, R. E.: J. Chem. Soc. **S**, 235 (1949).
COTTON, F. A.: Acta Chem. Scand. **10**, 1520 (1956).
DIEKE, G. H., and H. M. CROSSWHITE: Appl. Opt. **2**, 675 (1963).
DUNITZ, J. D., and L. E. ORGEL: Advan. Inorg. Radiochem. **2**, 1 (1960).
FAJANS, K.: Struct. Bonding **3**, 88 (1967).
FERREIRA, R.: J. Phys. Chem. **68**, 2240 (1964).
GEIER, G.: Ber. Bunsenges. Physik. Chem. **69**, 617 (1965).
GEORGE, P., and D. S. McCLURE: Progr. Inorg. Chem. **1**, 382 (1959).
HAISSINSKY, M., et C. K. JØRGENSEN: J. Chim. phys. **63**, 1135 (1966).
JØRGENSEN, C. K.: J. Inorg. Nucl. Chem. **1**, 301 (1955).
— Acta Chem. Scand. **10**, 887 (1956a).
— Acta Chem. Scand. **10**, 1505 (1956b).
— Energy Levels of Complexes and Gaseous Ions. Copenhagen: Gjellerup 1957.
— Solid State Phys. **13**, 375 (1962a).
— Absorption Spectra and Chemical Bonding in Complexes. Oxford: Pergamon 1962b.
— Mol. Phys. **5**, 271 (1962c).
— Orbitals in Atoms and Molecules. London: Academic Press 1962d.
— Inorganic Complexes. London: Academic Press 1963.
— Chimica Teorica, VIII. Corso Estivo di Chimica Milano 1963, Fondazione Donegani, Accademia Nazionale dei Lincei, Rome, p. 63 (1965).
— Struct. Bonding **1**, 234 (1966).
— Advances in Chemistry Series no. 62 p. 161. Washington: American Chemical Society, 1967a.
— Struct. Bonding **3**, 106 (1967b).
—, and E. RITTERSHAUS: Mat. Fys. Medd. Dan. Vid. Selskab **35**, no. 15 (1967).
—, S. M. HORNER, W. E. HATFIELD, and S. Y. TYREE: Int. J. Quantum Chem **1**, 191 (1967).
— Lanthanides and 5f Elements. London: Academic Press 1969.
KLOPMAN, G.: J. Am. Chem. Soc. **86**, 1463 and 4550 (1964).
— J. Am. Chem. Soc. **90**, 223 (1968).
KROT, N. N., and A. D. GELMAN: Dokl. Akad. Nauk. SSSR (Chem. Sect.) **177**, 124 (1967).
LADD, M. F. C., and W. H. LEE: Progr. Solid State Chem. **1**, 37 (1964).
— — ibid. **2**, 378 (1965).
LATIMER, W. M.: J. Chem. Phys. **23**, 90 (1955).

LOH, E.: Phys. Rev. **147**, 332 (1966).

LOTZ, W.: J. Opt. Soc. Am. **57**, 873 (1967).

McCLURE, D. S., and Z. KISS: J. Chem. Phys. **39**, 3251 (1963).

MERCER, E. E., and R. R. BUCKLEY: Inorg. Chem. **4**, 1692 (1965).

MILES, J. H.: J. Inorg. Nucl. Chem. **27**, 1595 (1965).

MORRIS, D. F. C.: Struct. Bonding **4**, 63 (1968).

MULLIKEN, R. S.: J. Chem. Phys. **43S**, 2 (1965).

OLVER, J. W., and J. W. ROSS: J. Am. Chem. Soc. **85**, 2565 (1963).

ORGEL, L. E.: J. Chem. Soc. 4186 (1958).

PAULING, L.: J. Chem. Soc. 1461 (1948).

PEARSON, R. G.: J. Am. Chem. Soc. **85**, 3533 (1963).

—, and H. B. GRAY: Inorg. Chem. **2**, 358 (1963).

— Science **151**, 172 (1966).

—, and R. J. MAWBY: Halogen Chemistry (Ed. V. GUTMANN) **3**, 55. (1967). London: Academic Press.

PENNEY, W. G.: Trans. Faraday Soc. **36**, 627 (1940).

PHILLIPS, C. S. G., and R. J. P. WILLIAMS: Inorganic Chemistry, Vol. I and II. Oxford: Clarendon Press 1965 and 1966.

RABINOWITCH, E., and E. THILO: Periodisches System, Geschichte und Theorie. Stuttgart: Ferdinand Enke 1930.

READER, J., and J. SUGAR: J. Opt. Soc. Am. **56**, 1189 (1966).

ROS, P., and G. C. A. SCHUIT: Theoret. Chim. Acta **4**, 1 (1966).

ROSSEINSKY, D. R.: Chem. Rev. **65**, 467 (1965).

RYAN, J. L., and C. K. JØRGENSEN: J. Phys. Chem. **70**, 2845 (1966).

SANDERSON, R. T.: J. Chem. Educ. **43**, 584 (1966).

SCHUIT, G. C. A.: Rec. Trav. Chim. **81**, 19 and 481 (1962).

— ibid. *83*, 5 (1964).

SCHWARZENBACH, G., and W. SCHNEIDER: Helv. Chim. Acta **38**, 1931 (1955).

SCROCCO, E., e O. SALVETTI: Ric. Sci. **24**, 1258 and 1478 (1954).

SHARPE, A. G.: Halogen Chemistry (Ed.: V. GUTMANN) **1**, 1 (1967). London: Academic Press.

SLATER, J. C.: J. Chem. Phys. **41**, 3199 (1964).

SUGAR, J.: J. Opt. Soc. Am. **53**, 831 (1963).

— J. Opt. Soc. Am. **55**, 33 (1965).

—, and J. READER: J. Opt. Soc. Am. **55**, 1286 (1965).

5. Internal Transitions in Partly Filled Shells

When decomposing white sunlight with a prism to the spectrum of rain-bow colours, NEWTON did not use a slit so narrow as to allow the dark absorption lines to be observed. FRAUNHOFER using refined optical equipment discovered the narrow lines. Though the emission lines in the red produced by flames containing lithium, and in the yellow of sodium, were known at FRAUNHOFER's time, it was not before 1860 BUNSEN and KIRCHHOFF studied systematically line spectra (cf. SCHEIBE, 1960). Besides assisting in the discovery of the elements rubidium, caesium, gallium, indium, thallium and all of the noble gases, the spectral analysis made an enormous impact on astronomy. Since visible light permeates space for millions of light-years, it became possible to analyze stellar at-mospheres, and the same elements were identified as in the earth's crust, though the astrophysics growing up showed that their relative abundance is very different, hydrogen and helium being far more common and the heavier elements not enriched as here. Thus, nearly all of the Fraunhofer lines of the solar spectrum are caused by elements lighter than zinc. The non-metallic elements (with exception of hydrogen) such as C, N, O, F, Ne, P, S, Cl, Ar,... show no strong absorption lines in the visible (except in the highly ionized forms occurring in stars having a much higher surface temperature than that about 6000° C of the sun) and one had to wait for satellite-borne spectrographs in order to observe the lines in the far ultraviolet (UNDERHILL and MORTON, 1967).

The reflection and transmission spectra of most coloured substances are not particularly interesting when regarded in a spectroscope (i.e. that the intensity is not measured by a photo-electric device permitting the recording of broad absorption bands). Hence, it was a great surprise for LECOQ DE BOISBAUDRAN and H. BECQUEREL to detect narrow bands in solutions, glasses and crystals containing rare earths (actually, GLAD-STONE had already observed such bands in 1857 of the aqueous solution of didymium which is now known to be a mixture of Nd and Pr). In a few cases, these bands are nearly as narrow as absorption lines of gases containing isolated atoms. They assisted in the discovery and in observing the separation of Pr, Nd, Sm, Ho, Er, and Tm. Some rare earths are genuinely colourless, such as Y, La and Lu. The narrow lines of Ce and Yb occur in the infra-red and were only detected much later. On the other hand, URBAIN observed the narrow bands of Gd in the ultraviolet;

this element has no excited levels at lower energy in nearly all Gd(III) compounds. CROOKES invented the cathode-ray tube and detected numerous emission lines in the cathodo-luminescence of colourless materials containing small amounts of lanthanides. URBAIN demonstrated 1909 that several of CROOKES' conclusions regarding new trace elements were wrong, and that in particular the red cathodo-luminescence (quite recently used for colour television in YVO_4 lattice) of Eu(III) was responsible for some of the emission spectra observed. Whereas absorption spectra in most cases follow BEER and LAMBERT's law that the optical density $\log_{10}(I_0/I)$ is proportional to the concentration of the coloured substance, phosphorescense and many other forms of luminescence may be very sensitive to trace elements, and efficient energy transfer is possible to luminescent chromophores or to quenching impurities.

In the beginning of this century, other compounds were found to have narrow absorption bands. Thus, uranium (IV) has such bands which are now taken as evidence for the configuration $5f^2$. It was also found that the ruby consisting of octahedral chromophores $Cr(III)O_6$ dispersed in colourless Al_2O_3 has narrow bands in the red and in the blue-green. This observation tended to obscure the fundamental difference between the f and d groups. It was argued by some authors that some aspects of the absorption spectrum of gaseous Cr^{+3} might be recognized in Cr(III) compounds. Thus, the (frequently fluorescent) transition in the red might be between the groundterm ^4F and the excited term ^2G both belonging to the configuration $3d^3$. We return below to the question why one cannot directly compare the d^q-levels of gaseous ions and of compounds.

On the other hand, the problem was the opposite in the 4f group. It is completely true that the narrow absorption bands correspond to transitions to levels having the same distribution as the J-levels of $4f^q$; but until quite recently, no atomic spectra were interpreted of 4f-group M^{+3}. The influence of the ligands surrounding a given central atom M(III) consists of two main effects: each J-level is split into sub-levels each containing one or more (in the case of q odd, always an even number according to KRAMERS) of the $(2J + 1)$ states; and the differences between the baricenters of the J-levels are slightly decreased in relatively covalent compounds. The former effect corresponding to energy separations usually about 0.1 kK, and normally below 0.5 kK, of the sub-levels of a given J-level, became one of the main subjects for ligand field theory. It is not possible to discuss this problem here; several books may be consulted (JUDD, 1963; WYBOURNE, 1965; JØRGENSEN, 1969). The electrostatic model of the ligand field was introduced by BETHE (1929) and the applications to the 4f group were systematized by HELLWEGE (1948) and PRATHER (1961). However, the success of this phenomenological treatment is based only upon the consistent subsistence of seven definite, slightly different, one-electron energies of the partly filled f shell (JØR-GENSEN, 1962e) and there is little doubt that a major part of the energy differences are caused by formation of molecular orbitals (M.O.) having the 4f central atom orbital as the main constituent but having a radial node because of the need for orthogonalization on the filled orbitals of the ligands. It is possible to describe such weak anti-bonding effects by the *angular overlap model* (JØRGENSEN et al., 1963) and the kinetic oper-ator (3.17) acting in the bond region explains the increased energy (JØRGENSEN, 1967b). This model has been applied with satisfactory results on $ErPO_4$ and $ErVO_4$ (KUSE and JØRGENSEN, 1967).

For the chemist, the second effect of decreased J-level separation is perhaps the most interesting. The first theoretical remarks about this effect seems to be made by HOFMANN and KIRMREUTHER (1910) observing that all the narrow line groups of Er_2O_3 systematically occur at slightly lower wavenumbers than of other Er(III) compounds. These authors suggested that the valence electrons proposed by STARK moved in larger orbits in the relatively more covalent Er_2O_3. It must be remembered that this statement was made three years before N. BOHR's hydrogen atom. EPHRAIM and BLOCH (1926 and 1928) made an extensive study of these band shifts in Pr(III), Nd(III) and Sm(III) compounds. The shift toward lower wavenumbers corresponds to the chemist's general ideas of increas-ing tendency of covalent bonding:

$$PrF_3 > Pr(H_2O)_9^{+3} > PrCl_3, 6\,H_2O > PrCl_3, 8\,NH_3 >$$
$$PrCl_3 > PrBr_3 > PrI_3 > Pr_2O_3 > Pr_2S_3 \tag{5.1}$$

with the exception of the unexpected covalent character of the oxide.

At EPHRAIM's time, it was not yet realized that the main part of the energy differences producing narrow absorption bands of $4f^q$ systems $(2 \leq q \leq 12)$ are caused by differing interelectronic repulsion, i.e. the average value $\langle r_{12}^{-1} \rangle$ being larger for the excited levels than for the groundstate, whereas the electronic density in our three-dimensional space is roughly the same. Hence, EPHRAIM suggested that the excited levels correspond to electronic orbits having a larger radius than the groundstate. He then argued that strong electrovalent bonding contract-ed these orbits, whereas covalent bonding expands them. We can only subscribe to the second half of this explanation, and only with the modi-fication of the average $\langle r^{-1} \rangle$ being proportional to the parameters of interelectronic repulsion. Hence, gaseous Pr^{+3} must occupy a position even before PrF_3 in eq. (5.1).

Because of the technical difficulties of interpreting the many thou-sands of spectral lines of 4f group M, M^+, M^{+2} and M^{+3}, the progress of atomic spectroscopy was extremely slow at this point, and the last, volume 4, of CHARLOTTE MOORE-SITTERLY's "Atomic Energy Levels" is far from completion. The late Professor DIEKE at Johns Hopkins University and his collaborators made a considerable contribution, and their results are expected soon to be published. The first 4f M^{+3} to be investigated was Ce^{+3} by LANG (1936) showing the excited configuration 5d at 49.74 and 52.23 kK. The lower energy of this configuration in Ce(III) complexes is discussed in chapter 6. The ground level of 4f is $^2F_{5/2}$; the only other J-level belonging to this configuration, $^2F_{7/2}$ occurs at 2.25 kK. It was very valuable that LANG determined the large value of the Landé parameter $\zeta_{4f} = (2.25 \text{ kK}/3.5) = 0.64$ kK. It might have been argued from the Sommerfeld equation for hydrogenic one-electron systems

$$\zeta_{nl} = 5.82 \text{ K} \cdot \frac{z_0^4}{n^3 l \, (l+\frac{1}{2}) \, (l+1)} \tag{5.2}$$

(where 1 ry/$137^2 = 0.00582$ kK and 137 is the velocity of light in atomic units) that ζ_{4f} would be very small unless the effective charge Z_0 was close to the atomic number Z. However, the central field U(r) in lanthanides is so different from a Coulomb potential $(-Z_0/r)$ that eq. (5.2) does not apply. Soon after LANG's measurements, GOBRECHT (1938) evaluated ζ_{4f} in all the trivalent lanthanides [with exception of Pm(III)] from the first-order width $(L+\frac{1}{2})\,\zeta_{4f}$ of the ground term [cf. eq. (4.32)] and found it to be smoothly increasing until 2.95 kK for the $4f^{13}$ system Yb(III). BRYANT (1965) argued from statistical considerations that the energy difference between the ground level $^2F_{7/2}$ and the excited level $^2F_{5/2}$ is 10.09 kK for Yb^{+3} which is unexpected because many Yb(III) compounds suggest values close to 10.25 kK. Actually, the former value seems too low because MEGGERS (1947) studied Yb^+ having the ground-state $^2S_{1/2}$ constituting the configuration $4f^{14}6s$, whereas $4f^{13}6s^2$ is represented with $^2F_{7/2}$ at 21.42 kK and $^2F_{5/2}$ at 31.57 kK giving $\zeta_{4f} = 2.90$ kK. Further on, BRYANT (1965) indicates values for ζ_{4f} between 2.91 and 2.95 kK for the configurations $4f^{13}6s$, $4f^{13}5d$, $4f^{13}6p$ and $4f^{13}7s$ of Yb^{++}. The minute discrepancies $\sim 1\%$ would have no interest for chemists if it was not for demonstrating the well-defined values of the Landé parameter in what is called Yb[III] in chapter 8. There is some evidence available that this parameter is changed less by chemical combination than the parameters of interelectronic repulsion in the partly filled 4f shell.

SUGAR (1965) found twelve of the thirteen possible J-levels of $4f^2$ for gaseous Pr^{+3} and finally confirmed the general expectation that the parameters of interelectronic repulsion are larger than even for PrF_3 and $Pr(H_2O)_9^{+3}$ in eq. (5.1). However, it was a surprise to many physicists that the difference is as large as 4%. Table 5.1 gives the J-level baricenters of Pr^{+3}, Pr(III) in LaF_3 (CASPERS et al., 1965), $Pr(H_2O)_9(C_2H_5SO_4)_3$ [this and the following compounds were discussed by CROSSWHITE, DIEKE and CARTER (1965) who also believe that they have found the last level $^1S_0$ of $4f^2$ for gaseous Pr^{+3}], Pr(III) in $LaCl_3$ and in $LaBr_3$ containing the chromophores $Pr(III)X_9$. MAKOVSKY et al. (1962) identified $^1S_0$ of Pr(III) in $LaCl_3$ by X-ray induced luminescence. Pr(III) in $LaBr_3$ was also studied by WONG and RICHMAN (1962) and KIESS and DIEKE (1966).

Table 5.2 gives the J-level baricenters of $Nd(H_2O)_9(C_2H_5SO_4)_3$ (measured by GRUBER and SATTEN, 1963) and of Nd(III) in $LaCl_3$ (CARLSON and DIEKE, 1961; WONG, 1961), in $LaBr_3$ (RICHMAN and WONG, 1962) and in Y_2O_3 (CHANG, 1966). Since the J-levels of Nd^{+3} have not yet been published, it is important to compare the Nd(III) J-level baricenters with the isoelectronic gaseous Pr^{++} (SUGAR, 1963; TREES, 1964). Actually, the energy differences in the Nd(III) aqua ion are remarkably close to 23% higher than those of Pr^{++}. The comparison between Pr^{+3} and the Pr(III) aqua ion suggests that Nd^{+3} has J-energy differences 26% higher than Pr^{++}.

Table 5.3 gives the J-level baricenters of $Er(H_2O)_9(C_2H_5SO_4)_3$ (ERATH, 1961), [Data above 27 kK are for the aqueous solution of Er(III) showing unusually sharp absorption bands; the assignments are obtained in analogy with those for Er_2O_3 (JØRGENSEN, 1957b)] Er(III) in $LaCl_3$ and YCl_3 (RAKESTRAW and DIEKE, 1965), in $LaBr_3$ (KIESS and DIEKE, 1966) and in Y_2O_3 (KISLIUK, KRUPKE and GRUBER, 1964) and finally

Table 5.1. *Energy levels of gaseous Pr^{+3} and J-level baricenters of Pr(III) in various solids given in kK (= 1000 cm^{-1}). The lowest sub-level is also given in the latter cases*

J-level	Pr^{+3}	Pr(III) in LaF_3	$Pr(H_2O)_9^{+3}$	Pr(III) in $LaCl_3$	Pr(III) in $LaBr_3$
$^3H_4$	0	0	0	0	0
$^3H_5$	2.152	2.16	—	2.12	2.087
$^3H_6$	4.389	4.26	—	4.30	4.174
$^3F_2$	4.997	5.01	—	4.85	4.817
$^3F_3$	6.415	6.35	6.22	6.24	6.203
$^3F_4$	6.855	6.83	6.70	6.68	6.660
$^1G_4$	9.921	9.80	9.68	9.70	9.669
$^1D_2$	17.334	16.65	16.73	16.64	16.580
$^3P_0$	21.390	20.725	20.551	20.385	20.287
$^3P_1$	22.007	21.3	21.15	20.987	20.889
$^1I_6$	22.212	—	21.40	21.32	21.269
$^3P_2$	23.161	22.55	22.305	22.142	22.045
$^1S_0$	50.090	—	—	48.71	—
lowest $^3H_4$ sub-level	—	−0.200	−0.135	−0.090	−0.072

undiluted Er_2O_3 (GRUBER et al., 1966). Dr. W. J. CARTER was so kind as to inform the writer about the identified levels of Er^{+3}. Those given with four figures are slightly less certain. Surprisingly enough, the nephelauxetic effect is slightly *negative* for several levels compared with the aqua ion.

The numbers of excited levels increase strongly on going from the two extremes f^1 and f^{13} toward the middle f^7 of the lanthanides. At the same time, a gap develops between the lowest term having the maximum value for S (cf. eq. 4.32) and the first excited terms (i.e. the groups of levels characterized by definite quantum numbers L and S) and the spectra tend to move toward the ultraviolet. This is particularly true for Gd(III) having the ground level $^8S_{7/2}$ 32.1 kK below the first excited

Table 5.2. *J-level baricenters of Nd(III) in various solids (notation as in Table 5.1) and of the isoelectronic ($4f^3$) gaseous ion Pr^{++}*

J-level	$Nd(H_2O)_9^{+3}$	Nd(III) in LaCl_3	Nd(III) in LaBr_3	Nd(III) in Y_2O_3	Pr^{++}
$^4I_{9/2}$	0	0	0	0	0
$^4I_{11/2}$	(1.93)	1.881	—	1.879	1.398
$^4I_{13/2}$	(3.98)	3.864	—	3.856	2.893
$^4I_{15/2}$	(6.10)	—	—	5.898	4.454
$^4F_{3/2}$	11.368	11.292	11.238	11.028	9.371
$^4F_{5/2}$	12.404	12.329	12.274	12.021	10.138
$^2H_{9/2}$	12.525	12.46	12.446	12.283	10.033
$^4F_{7/2}$	13.367	13.293	13.235	13.075	10.859
$^4S_{3/2}$	13.454	13.383	13.334	13.126	10.950
$^4F_{9/2}$	14.640	14.576	14.527	14.350	11.762
$^2H_{11/2}$	15.842	15.797	15.763	15.560	12.495
$^4G_{5/2}$	17.118	16.981	16.882	16.698	14.187
$^2G_{7/2}$	17.239	17.12	17.05	16.747	13.887
$^2K_{13/2}$	(18.78)	18.47	—	—	16.089
$^4G_{7/2}$	18.996	18.885	18.816	18.364	15.443
$^2G_{9/2}$	19.408	19.294	19.220	18.939	15.705
$^2K_{15/2}$	(20.78)	19.439	19.388	—	17.642
$^4G_{9/2}$	20.982	20.896	20.840	—	16.764
$^2D_{3/2}$	21.116	21.028	20.972	—	17.096
$^4G_{11/2}$	21.438	21.289	21.184	—	17.410
$^2P_{1/2}$	23.180	23.068	22.996	22.634	18.694
$^2D_{5/2}$	23.730	23.641	23.571	23.175	19.046
$^2P_{3/2}$	26.102	26.004	25.948	25.526	20.857
$^4D_{3/2}$	28.049	27.828	27.680	—	23.092
$^2I_{11/2}$	(28.46)	—	—	—	24.358
$^4D_{5/2}$	(28.48)	—	—	—	23.246
$^4D_{1/2}$	(28.82)	—	—	—	23.465
$^2L_{15/2}$	(29.23)	—	—	—	25.245
$^2I_{13/2}$	(29.77)	—	—	—	25.392
lowest sub-level	−0.178	−0.148	−0.143	−0.276	—

level $^6P_{7/2}$. For these reasons, the determination of J-levels from a study
of sub-levels of M(III) in transmission spectra of crystals at low temperature becomes much more difficult than in the case of 4f², 4f³ and 4f¹¹
given in Tables 5.1—5.3 and is not at all finished at present. Some tabula-
tions are compiled by JØRGENSEN (1969). Because of the deviations
(intermediate coupling) from the limiting case of Russell-Saunders
coupling (i. e. well-defined L and S) the term symbols are sometimes
arbitrary for some excited 4fq-levels, and only the value of J (and the
odd or even parity of q) remain good quantum numbers. Thus, it can
be discussed whether the symbols $^2G_{7/2}$ and $^4G_{7/2}$ should not be exchanged for Pr⁺⁺ and Nd(III), or $^4I_{9/2}$ and $^4F_{9/2}$ for Er(III).

Table 5.3. *Energy levels of gaseous Er⁺³ (kind communication from Dr. W. J. CARTER)
and J-level baricenters of Er(III) in various solids (notation as in Table 5.1)*

J-level	Er⁺³	Er(H₂O)$_9^{+3}$	Er(III) in LaCl₃	Er(III) in YCl₃	Er(III) in LaBr₃	Er(III) in Y₂O₃	Er₂O₃
$^4I_{15/2}$	0	0	0	0	0	0	0
$^4I_{13/2}$	6.486	6.5	6.480	6.467	6.475	6.458	6.458
$^4I_{11/2}$	10.124	10.118	10.111	10.074	10.101	10.073	10.082
$^4I_{9/2}$	12.35	12.366	12.340	12.330	12 339	12.288	12.274
$^4F_{9/2}$	15.183	15.207	15.176	15.096	15.150	15.071	15.062
$^4S_{3/2}$	18.30	18.327	18.291	18.159	18.261	18.071	18.062
$^2H_{11/2}$	19.011	19.087	19.03	18.949	19.00	18.931	18.925
$^4F_{7/2}$	20.49	20.457	20.4	20.3	—	20.266	20.255
$^4F_{5/2}$	22.18	22.122	22.068	21.959	22.022	21.894	21.890
$^4F_{3/2}$	22.45	22.461	22.409	22.332	22.367	22.207	22.284
$^2H_{9/2}$	24.475	24.515	24.43	24.360	24.4	24.303	24.294
$^4G_{11/2}$	26.377	26.348	26.27	26.105	26.180	26.073	26.062
$^4G_{9/2}$	27.319	27.45	27.36	27.144	27.16	—	26.966
$^2K_{15/2}$	27.585	—	27.68	27.5	—	—	27.203
$^2G_{7/2}$	27.825	28.0	27.99	27.7	—	—	27.713
$^2P_{3/2}$	31.41	31.55	31.385	—	31.285	31.186	31.178
$^2K_{13/2}$	32.972	—	—	—	—	—	32.504
$^4G_{5/2}$	33.09	—	—	—	—	—	32.869
$^2P_{1/2}$	33.24	—	—	—	—	—	33.151
$^4G_{7/2}$	33.849	34.0	—	—	—	33.697	33.688
$^2D_{5/2}$	34.73	34.82	—	—	—	—	34.406
$^2G_{9/2}$	36.48	36.4	—	—	—	—	36.159
$^4D_{5/2}$	38.41	38.5	—	—	—	—	37.989
$^4D_{7/2}$	39.268	39.18	—	—	—	—	38.647
$^2I_{11/2}$	—	41.1?	—	—	—	—	40.450
$^2L_{17/2}$	—	41.65?	—	—	—	—	40.780
$^4D_{3/2}$	41.85	42.35?	—	—	—	—	41.782
$^2D_{3/2}$	42.43	43.46?	—	—	—	—	42.332
lowest sub-level	—	−0.147	−0.108	−0.126	−0.089	−0.203	−0.203

The J-level baricenters given in Tables 5.1—5.3 are obtained from refined spectroscopic techniques; and it is necessary to identify all the sub-levels reliably of a given J-level before its position is known. However, very valuable evidence can be obtained by a simpler technique: to measure the baricenter of intensity of each absorption band (or group of closely adjacent bands) and compare with the aqua ions known from $M(H_2O)_9(C_2H_5SO_4)_3$ studied at liquid helium temperature. BOULANGER (1952) studied reflection spectra of powdered samples, and JØRGENSEN (1956b) spectra of aqueous solutions of M(III) complexes of chloride and organic anions and comparing with EPHRAIM's results (5.1). JØRGEN-SEN, PAPPALARDO and RITTERSHAUS (1964 and 1965) and JØRGENSEN and RITTERSHAUS (1967) studied the strong *nephelauxetic effect* (the origin of this word for the decrease of parameters of interelectronic repulsion will be discussed below) in mixed oxides, either stoichiometric such as the pyrochlore $Er_2Ti_2O_7$ or the perovskite $LaErO_3$ (the coordination numbers are respectively Er(III) 8, Ti(IV) 6; La(III) 12, and Er(III) 6) or non-stoichiometric admixtures in cubic ZrO_2, ThO_2 etc. We proposed a relation

$$\sigma - \sigma_{aqua} = d\sigma - (d\beta)\,\sigma_{aqua} \qquad (5.3)$$

between the baricenters σ of the J-levels of the compound studied (usually obtained from baricenters of intensity of absorption bands) and of the aqua ion σ_{aqua}. Whereas some authors have advocated to write $-(d\beta)\sigma_{aqua}$ alone, the first term $d\sigma$ of eq. (5.3) accounts for the difference between the average sub-level population of the ground level of the compound studied at a definite temperature and the (negative) energy of the lowest sub-level of the aqua ion relative to the baricenter of the ground J-level. In 1964 when eq. (5.3) was proposed, the exact baricenters of compounds other than aqua ions where only known in very few cases. When the material of Tables 5.1—5.3 became available, it turned out that both $d\sigma$ and the nephelauxetic parameter $d\beta$ had been somewhat exaggerated by eq. (5.3). There are a variety of reasons (JØRGENSEN, 1969), the main being that ζ_{4f} changes much less in compounds than do the parameters of interelectronic repulsion. Table 5.4 gives some of the many sets of $d\sigma$ (in K, i.e. cm^{-1}) and $d\beta$ (in percent) which have been determined. The data of Tables 5.2 and 5.3 have been treated by plotting the best straight line going through the baricenter of the ^4I term (cf. eq. 4.32) situated at $\frac{7}{2}\zeta_{4f} = 3.1$ kK in Nd(III) and at $3\zeta_{4f} = 7.1$ kK in Er(III) and having the slope $(d\beta)_*$ for $(\sigma - \sigma_{aqua})$ as a function of σ_{aqua}. The scattering of the numerous points around this straight line is perceptible, but does not impair the reproducibility of $d\beta$-values within 0.2%.

JØRGENSEN, PAPPALARDO and FLAHAUT (1965) applied eq. (5.3) to reflection spectra of 4f-group sulphides measured at liquid air temperature, and found somewhat, but not extremely, higher values of $d\beta$ for sulphides than for oxides. A comparison was made with the cyclo-pentadienides $M(C_5H_5)_3$ which have also been studied by FISCHER and FISCHER (1967). RYAN and JØRGENSEN (1966) prepared the 4f-group hexachlorides MCl_6^{-3} and hexabromides MBr_6^{-3} having very characteristic spectra. Each J-level correspond to six narrow vibrational components surrounding the electronic transition which may be so weak itself that it is not observed. In most cases, the total area of the absorption band, and hence the oscillator strength, is much smaller than for the aqua ions, with the exception of the hypersensitive pseudoquadrupolar transitions to be discussed p. 229. It must be emphasized that so low a coordination number N as six is highly unusual in the 4f group, and that octahedral chromophores $M(III)X_6$ only constitute a minority of these rare examples. Actually, MCl_6^{-3} and MBr_6^{-3} can only be studied in solvents such as acetonitrile and succinonitrile, whereas even minute amounts of water, alcohols or other hydroxyl-function-containing solvents decompose the hexahalide complexes giving solvates with N higher than 6.

The nephelauxetic effect of Er(III) in chlorides varies considerably, as seen in Table 5.4. In LaCl$_3$, not only is $N = 9$, but the Er-Cl distances must be unusually large, the ionic radius of Er(III) being 0.1 Å smaller

Table 5.4. *Nephelauxetic parameter $d\beta$ and $d\sigma$ (in cm^{-1}) from equation (5.3) and $(d\beta)_*$ from J-level baricenters assuming invariant Landé parameter ζ_{4f}*

	$d\sigma$ (K)	$d\beta$ (%)	$(d\beta)_*$ (%)
Nd(III) in LaCl$_3$	$-$ 60	0.6	0.75
NdCl$_6^{-3}$	$+$ 50	2.2	$-$
Nd(III) in LaBr$_3$	$-$	$-$	1.2
NdBr$_6^{-3}$	$+$ 50	2.3	$-$
Nd$_2$O$_3$ (type A)	$+200$	3.6	$-$
Nd(III) in Y$_2$O$_3$ (type C)	$-$	$-$	3.1
Nd(III) in Yb$_2$O$_3$ (type C)	$+150$	3.2	$-$
Nd(III) in ThO$_2$	$+150$	3.3	$-$
BaNd$_2$S$_4$	$+250$	4.2	$-$
Nd(C$_5$H$_5$)$_3$	$+400$	3.8	$-$
Er(III) in LaCl$_3$	$-$ 30	0.3	0.4
Er(III) in YCl$_3$	$+$ 50	1.1	1.3
ErCl$_6^{-3}$	$+100$	1.2	$-$
Er(III) in LaBr$_3$	$-$	$-$	0.7
Er$_2$O$_3$ (type C)	$+200$	1.6	1.6
Er(III) in Y$_2$O$_3$ (type C)	$+200$	1.6	1.5
Er$_2$Ti$_2$O$_7$ (pyrochlore)	$+250$	1.3	$-$
Er$_2$Zr$_2$O$_7$ (fluorite)	$+300$	1.7	$-$
Er(III) in ThO$_2$	$+250$	1.6	$-$
Er(III) in CdGa$_2$S$_4$	$+300$	3.6	$-$
Er(C$_5$H$_5$)$_3$	$+350$	2.8	$-$

than that of La(III). In YCl_3 studied by RAKESTRAW and DIEKE (1965), $N = 6$ (but the coordination is not regular octahedral) and in $ErCl_6^{-3}$ the nephelauxetic effect is much more pronounced. Thus, it can be argued that the main reason for the extremely strong nephelauxetic effect in $M(III)O_7$ in A-type oxides such as Nd_2O_3 and in the two sites, both having the constitution $M(III)O_6$, in C-type oxides is very short $M-O$ distances, and not just a strongly reducing character of the oxide ligands. Actually, JØRGENSEN and RITTERSHAUS (1967) showed that pyrochlores such as $Er_2Ti_2O_7$ containing $Er(III)O_8$ with longer and $Ti(IV)O_6$ with shorter $M-O$ distances have a more moderate nephelauxetic effect than C-type Er_2O_3 as is also true for $Er(III)O_8$ in $ErPO_4$ and $ErVO_4$ (KUSE and JØRGENSEN, 1967). Mixed oxides with statistically disordered fluorite lattice such as $M_{0.2}Zr_{0.8}O_{1.9}$ were also shown to have a smaller $d\beta$ than M_2O_3 though the fluorite unit cell edge is shorter. Thus, the average $M-O$ distance is smaller in the mixed zirconium oxide; but there may occur local distortions such that $M(III)O_8$ and $M(III)O_7$ (adjacent to vacancies in the anion lattice) have longer $M-O$ distances than in C-type M_2O_3.

The case for well defined oxidation states in 4f group compounds being neither black semi-conductors nor metals (we return to these cases in chapter 8) is overwhelming. A multitude of excited J-levels can be determined in most elements between Pr(III) and Tm(III), and the extent of the nephelauxetic effect shows that the partly filled 4f shell not only has a sharply defined preponderant configuration with two to twelve electrons, but also that its radial extension cannot be grossly different in all transparent compounds of M(III). Further on, the atomic spectra has shown the same to be true for Pr^{+3} and Er^{+3}, and it is undoubtedly true for all 4f group M^{+3}. Actually, no other transition group has so solid a fundament for defining spectroscopic oxidation states M(II), M(III) and M(IV).

The situation is just a tiny bit less clear-cut in the 5f group. In a way, the narrow absorption bands leave no doubt about the preponderant configuration being $5f^q$ with only one partly filled shell. In chapter 6, the reasons will be given for the only other conceivable ground configuration, $5f^{q-1}6d$, having higher energy than the lowest level of $5f^q$ in Pa(IV), U(III) and in all complexes of heavier elements. Some doubt may reside in the case of Th(III) which is only known in the dark coloured, but non-metallic ThI_3. However, the sub-level splittings of each J-level can be considerably larger in the 5f group, and the identification of individual J-levels has progressed less rapidly. It does not seem that the strong deviations from spherical symmetry are so much a case of higher oxidation state [compare M(IV) aqua ions studied by CONWAY (1964) with M(III) aqua ions studied by CARNALL and WYBOURNE (1964)] as the first elements with one or two 5f electrons having much stronger chemical-effects than $q > 2$. Octahedral coordination is nearly as rare in the 5f group as in the 4f group. However, the $5f^1$-systems $PaCl_6^-$, UF_6^-, UCl_6^- and NpF_6 are well established now, as are the $5f^2$-systems UCl_6^-, UBr_6^-, UI_6^-, NpF_6^- and PuF_6. Uranium(IV) hexahalides are only stable in non-hydroxylic solvents; and RYAN and JØRGENSEN (1963) and PAPPALARDO and JØRGENSEN (1964) noted a red-shift of some 3% between UCl_6^- and UBr_6^- in much the same way as a red-shift of some 4% between $PuCl_6^-$ and $PuBr_6^-$. The individual sub-levels have been identified and their symmetry type Γ_J determined by JOHNSTON et al. (1966) in many cases for the two former species and are given in Table 5.5. Though most of the J-level baricenters are not completely known, a systematic shift toward lower wavenumbers in UBr_6^- is quite perceptible. The "ligand field" effects are comparable in UX_6^- is with the energy differences between adjacent J-levels, and hence, the quantum number J may be not particularly good for certain sub-levels. Thus, the strong absorption peaks noted in nitromethane solution at room temperature (JØRGENSEN, 1963a) for UCl_6^- at 22.12 and 24.69 kK and for UBr_6^- at 21.60 and 23.81 kK (cf. also DAY and VENANZI, 1966) are both identified as transitions to Γ_3 sub-levels by JOHNSTON, SATTEN, SCHREIBER and WONG, probably involving a mixture of $^1I_6$ and $^3P_2$ characters. Unfortunately, it is not possible to compare with the uranium (IV) aqua ion of which neither N nor the symmetry is known. In view of the relatively large difference between $d\beta$ for UCl_6^- and $d\beta$ for UBr_6^-,

Table 5.5. *Sub-levels of octahedral uranium (IV) hexahalide complexes for which the symmetry type Γ_J has been identified. 1 kK $= 1000$ cm^{-1}*

	UCl_6^-	UBr_6^-
$^3H_4$	Γ_1 0; Γ_4 0.924; Γ_3 —; Γ_5 —	Γ_1 0; Γ_4 0.70; Γ_3 —; Γ_5 —
$^3F_2$	Γ_5 4.899; Γ_3 5.060	Γ_5 4.717; Γ_3 4.940
$^3H_5$	Γ_4 6.343; Γ_5 7.011; Γ_4 7.267; Γ_3 8.197	Γ_4 6.261; Γ_5 6.908; Γ_4 7.221; Γ_3 8.003
$^3F_4$	Γ_1 8.467; Γ_5 —; Γ_4 9.232; Γ_3 —	Γ_1 8.484; Γ_5 —; Γ_4 9.015; Γ_3 —
$^3F_3$	Γ_2 —; Γ_4 10.065; Γ_5 —	Γ_2 —; Γ_4 9.905; Γ_5 —
$^3H_6$	Γ_3 —; Γ_5 —; Γ_2 12.128; Γ_5 —; Γ_4 —	Γ_3 —; Γ_5 —; Γ_2 11.796; Γ_5 —; Γ_4 —
$^3P_0$	Γ_1 14.789	Γ_1 14.699
$^1G_4$	Γ_3 15.213; Γ_4 15.754; Γ_5 —; Γ_1 16.797	Γ_3 14.973; Γ_4 15.378; Γ_5 —; Γ_1 16.143
$^1D_2$	Γ_3 —; Γ_5 —	Γ_3 —; Γ_5 —
$^3P_1$	Γ_4 18.824	Γ_4 18.251
$^1I_6$	Γ_1 —; Γ_4 —; Γ_5 —; Γ_2 —; Γ_5 21.814; Γ_3 22.183	Γ_1 —; Γ_4 —; Γ_5 —; Γ_2 —; Γ_5 21.086; Γ_3 21.629
$^3P_2$	Γ_5 23.329; Γ_3 24.700	Γ_5 22.797; Γ_3 23.892

one would expect somewhat larger parameters of interelectronic repulsion for the aqua ion and much larger parameters for U^{+4} than for UCl_6^{--}, but we have to wait for these results.

Even in cases where it is not possible to assign definite J-levels to the absorption bands, a striking regularity is frequently noted that isoelectronic systems with the oxidation state z have wavenumbers of closely analogous bands roughly proportional to $(z+2)$ as we saw in Table 5.2 for Pr^{++} and Nd(III), or as was previously known for Sm(II) and Eu(III). Thus, Np(III) can be compared with Pu(IV), or Am(III) with Cm(IV) (JØRGENSEN, 1959). This strong dependence on the oxidation state is in disagreement with the weak variation predicted by eq. (5.2) assuming the large screening constants compatible with weak increase of parameters of interelectronic repulsion in sequences of *constant* oxidation state such as Ce^{++} to Er^{++} or Pr(III) to Tm(III). Actually, GRUEN (1952) was the first to compare the atomic spectrum of $5f^2$ Th^{++} with the isoelectronic U(IV), NpO_2^+ and PuO_2^{++}, suggesting parameters of interelectronic repulsion proportional to $(Z-58)$. However, we know today that already U(IV) has the J-levels of $5f^2$ spread over an energy range at least 30% larger than for Th^{++}. GRUEN's note also brought up the other problem to what extent the J-levels are recognizable in the 5f group dioxo complexes, i.e. whether the energy differences between sub-levels produced by chemical bonding effects may be larger, perhaps much larger, than the distances between adjacent J-levels. We are not going here to discuss this problem which has not been completely clarified. Also in the case of UF_6^- and NpF_6, it is possible that the chemical effects on the $5f^1$-levels are rather large, as discussed by REISFELD and CROSBY (1965). Further references treating this problem are reviewed by JØRGENSEN (1967 c).

In the three d-groups, a few transitions correspond to narrow absorption bands showing a systematic shift toward lower wavenumbers when the partly covalent bonding is expected to become more pronounced, constituting an analogy to the series (5.1). When analyzed according to ligand field theory, these transitions occur between states belonging, at least approximately, to the same M.O. configuration. Such cases are most common in the high symmetries, such as O_h for regular octahedral MX_6 or T_d for regular tetrahedral MX_4. However, they can occur in lower symmetries, and there is nothing contradictory in their occurrence even in the lowest symmetry C_1 since a given distribution of at least two and at most eight electrons on the five d-orbitals can produce levels having different total spin quantum number S. On the other hand, such transitions seem always to be spin-forbidden in the limit of RUSSELL-SAUNDERS coupling, going from S to $S-1$. This is not because several levels having the maximum value of S cannot belong to the same M.O.

configuration (this is frequently the case for T_1 and T_2 terms in octahedral or tetrahedral symmetry) but because another M.O. configuration having the highest S-value present only once then has lower energy.

In octahedral chromophores, the lower d-sub-shell is formed by the three orbitals $d\pi c$, $d\pi s$ and $d\delta s$ of eq. (3.5) having angular functions proportional to (xz/r^2), (yz/r^2) and (xy/r^2). The higher, σ-anti-bonding, sub-shell occurs at Δ higher energy [some authors use the symbols 10Dq or (E_1-E_2)] and consists of the orbitals $d\sigma$ and $d\delta c$ having angular functions proportional to

$$\frac{\sqrt{3}}{r^2}\left(z^2-\frac{1}{3}r^2\right) \quad \text{and} \quad (x^2-y^2)/r^2,$$

both directed against the ligands placed on the Cartesian axes. The variation of the value of Δ as a function of central atoms and of the ligands is called the *spectrochemical series*. To the first approximation (JØRGENSEN, 1962b),

$$\Delta = f \text{ (ligands)} \cdot g \text{ (central ion)} \tag{5.4}$$

where the function f has been fixed to 1.00 for the hexa-aqua ions and is

Br$^-$	0.72	H$_2$O	1.00
Cl$^-$	0.78	NCS$^-$	1.02
dsep$^-$ = (C$_2$H$_5$O)$_2$PSe$_2^-$	0.8	p-CH$_3$C$_6$H$_4$NH$_2$	1.15
N$_3^-$	0.83	CH$_3$NH$_2$	1.17
dtp$^-$ = (C$_2$H$_5$O)$_2$PS$_2^-$	0.83	gly$^-$ = NH$_2$CH$_2$CO$_2^-$	1.18
F$^-$	0.9	CH$_3$CN	1.22
dtc$^-$ = (C$_2$H$_5$)$_2$NCS$_2^-$	0.90	NH$_3$	1.25
dmso = (CH$_3$)$_2$SO	0.91	en = NH$_2$CH$_2$CH$_2$NH$_2$	1.28
urea = (NH$_2$)$_2$CO	0.92	den = NH(CH$_2$CH$_2$NH$_2$)$_2$	1.29
CH$_3$COOH	0.94	NH$_2$OH	1.30

(5.5)

C$_2$H$_5$OH   0.97   dip =   1.33

dmf = (CH$_3$)$_2$NCHO   0.98   phen =   1.34

ox^{--} = C$_2$O$_4^{--}$ 0.99   CN$^-$ ~ 1.7

In the case of polyatomic ligands, the ligating atoms have been printed with fat type. It may be added that ambidentate ligands can have rather different f-values, thus 0.75 for SCN$^-$ and 1.25 for as well NCSH as NCSHg$^+$. The value for pyridine is still uncertain, but seems to be close to 1.23. CN^- is known from precipitated double cyanides and has $f \sim 1.15$. The g-function is measured in kK and is

$3d^5$	Mn(II)	8.0		$4d^6$	Ru(II)	20
$3d^8$	Ni(II)	8.7		$3d^3$	Mn(IV)	23
$3d^7$	Co(II)	9		$4d^3$	Mo(III)	24.6
$3d^3$	V(II)	12.0		$4d^6$	Rh(III)	27.0
$3d^5$	Fe(III)	14.0		$4d^3$	Tc(IV)	30
$3d^3$	Cr(III)	17.4		$5d^6$	Ir(III)	32
$3d^6$	Co(III)	18.2		$5d^6$	Pt(IV)	36

$$(5.6)$$

showing the general ordering M(II) <M(III) <M(IV) and 3d <4d <5d.

We are not going here to present a general outline of ligand field theory (introductory books are ORGEL, 1960; JØRGENSEN, 1962b and 1962e) but there is little doubt that the main reason for variation of Δ is that this sub-shell energy difference is the *difference between σ-anti-bonding and π-anti-bonding effects* on the partly filled d shell. Thus, nitrogen- or sulphur-containing anions having more than one lone-pair have lower Δ values than ligands containing only one lone-pair such as NH_3 and SO_3^- because the ligands with two or more lone-pairs *also* are π-anti-bonding. One reason why cyanide does not have a definite value for f in eq. (5.5) may be π-bonding to empty orbitals of the ligand; thus, $Mn(CN)_6^{-5}$ is colourless and must have Δ above 27 kK, whereas Δ is 32.2 kK for the isoelectronic $Fe(CN)_6^{-4}$, 33.5 kK for $Co(CN)_6^{-3}$ but only 26.7 kK for $Cr(CN)_6^{-3}$ (the latter value may be compared with $\Delta = 20.8$ kK reported by DUNKEN and MARX (1966) for $Cr(CH_3)_6^{-3}$ presumably only σ-anti-bonding). BECK and FELDL (1965) indicate the values $\Delta = 27.0$ kK for the fulminate complex $Fe(CNO)_6^{-4}$ and $\Delta = 26.1$ kK for the isoelectronic $Co(CNO)_6^{-3}$.

However, besides the effects of π-anti-bonding and much less frequent effects of π-bonding, another contribution to Δ can come from anomalous internuclear distances. Though the anti-bonding effects probably are an exponential function of the distance R, it has frequently been argued that Δ is proportional to R^{-5}. Obviously, this expression has only been verified as a differential quotient, 1% decreased distance corresponding to Δ increasing 5%. Thus, experiments at high pressure above 150000 atm. have indicated Δ some 20 to 30% higher than under normal pressure (DRICKAMER and ZAHNER, 1962; DRICKAMER, 1965).

By substitution of a colourless cation in a mineral or synthetic syncrystallized mixture, it is often possible to vary the R-value of a transition group central ion to a considerable extent. The effects on the reflection spectra of solid fluorides and chlorides are perceptible but not dramatic (CLARK, 1964; JØRGENSEN, 1967c). On the other hand, mixed

oxides may show a wide variation of Δ values as first pointed out by
NEUHAUS (1960) based on a note by ORGEL (1957) on the difference
between green Cr_2O_3 and the red ruby containing $Cr(III)O_6$ in Al_2O_3
having $\Delta = 16.6$ and 18.1 kK, respectively. POOLE (1964) reviewed the
ligand field treatment of $Cr(III)O_6$ in various compounds. SCHMITZ-
DUMONT and KASPER (1964) pointed out that Δ is as low as 6.0 kK for
$Ni(II)O_6$ in the relatively dilated lattice of $CdTiO_3$, actually the lowest
value known at that time for any octahedral nickel (II), whereas $\Delta = 7.3$
kK in $MgTiO_3$ and 8.6 kK in MgO, the visible spetrum of $Ni_xMg_{1-x}O$ for
small values of x being very similar to that of $Ni(H_2O)_6^{++}$. Undiluted
NiO has a slightly larger $\Delta = 8.8$ kK somewhat in contrast to Cr_2O_3.
REINEN (1966) studied many other mixed oxides of $Ni(II)$, such as the
perovskite Ba_2NiWO_6 $(\Delta = 7.4$ kK) and $BaNi_{0.05}Ca_{0.45}W_{0.5}O_3$ $(\Delta = 6.3$ kK) where Ni, W and Ca are octahedrally coordinated and Ba has
cubic 12–coordination. REINEN pointed out that though $Sr_3NiNb_2O_9$
and $Sr_3NiSb_2O_9$ have the same unit cell parameter, Δ is 8.6 and 6.9 kK,
respectively. REINEN argues that the d^{10} constituents such as Sb(V) and
Te(VI) allows the oxide anions to be more polarized and produce smaller
Δ values for Ni(II) than would oxide anions bound to $d°$ constituents
such as Ta(V) and W(VI). The extremely low value $\Delta = 4.8$ kK was
obtained for $BaNi_{0.05}Ca_{0.45}Te_{0.5}O_3$. Other perovskites present normal
values of Δ, such as 8.95 kK for $LaNi_{0.2}Mg_{0.3}Ti_{0.5}O_3$. The rutile modification of $NiNb_2O_6$ has $\Delta = 8.2$ kK whereas the columbite modification
has $\Delta = 7.2$ kK, and the two trirutiles $NiTa_2O_6$ and $NiSb_2O_6$ have $\Delta = 7.8$
and 8.9 kK, respectively. Unfortunately, most sulphides are black semi-
conductors, if not metallic, and show strong absorption bands due to
non-stoichiometry or due to collective effects such as the undiluted halides
studied by CLARK (1964). It may be noted that $Mn(II)S_6$ in the green
modification (NaCl type) of MnS shows $\Delta = 7.1$ kK which may be compared with $\Delta = 9.4$ kK for $Mn(II)O_6$ in MnO and 8.5 kK for $Mn(H_2O)_6^{++}$.

If only the lower sub-shell (xy, xz, yz) contains electrons, narrow
and weak bands corresponding to intra-sub-shell transitions are observed.
STEVENS (1953) noted the similarity between p^{6-q} in spherical symmetry
and $(t_{2g})^q$ in octahedral symmetry. If the relativistic effects can be neglected, as they can in the 3d group for this purpose, the effect is to produce
one excited level ($^1A_{1g}$ for d^2 and d^4, $^2T_{2g}$ for d^3) 5 K above the groundstate ($^3T_{1g}$ for d^2 and d^4, $^4A_{2g}$ for d^3) where $K = (3B + C)$ (cf. JØRGENSEN,
1962e) is closely related to the spin-pairing energy parameter $D = \frac{7}{6}$
$(\frac{5}{2}B + C)$. In actual atoms, C is approximately $4B$. The lowest sub-shell
configuration $(t_{2g})^2$ or $(t_{2g})^4$ has excited levels ^1E_g and $^1T_{2g}$ at 2K above
the groundstate and $(t_{2g})^3$ has ^2E_g and $^2T_{1g}$ at 3 K above the groundstate.
The degeneracy between the two excited levels can either be removed by
mixing with other sub-shell configurations or by relaxing the constraint

of the separability of an angular function having $l=2$. However, according to spectroscopic observations, the degeneracies remain remarkably close.

$V(III)O_6$ in Al_2O_3 has the narrow bands at 8.8, 9.7 and 21.02 kK (PRYCE and RUNCIMAN, 1958) and low-spin $Mn(CN)_6^{-3}$ the first group around 9.5 kK (JONES and RUNCIMAN, 1960). The corresponding values of $K \sim 7B \sim 4.5$ kK are distinctly smaller than for gaseous V^{+3}(6.0 kK) and Mn^{+3} (6.8 kK). Similar conclusions can be drawn from RuF_6 according to WEINSTOCK and GOODMAN (1965) having the narrow bands at 5.9, 7.8 and 11.9 kK. The corresponding $K \sim 2.1$ kK (corrected for relativistic effects) is far smaller than the value 6 kK extrapolated for gaseous Ru^{+6}. The situation is entirely analogous for the $5d^2$-system OsF_6 and the $5d^4$-systems $OsCl_6^{--}$, $OsBr_6^{--}$ and PtF_6 (MOFFITT et al., 1959; JØRGENSEN, 1962c) but the strong relativistic effects and the co-excited vibrations make the evaluation of K slightly complicated.

The octahedral d^3-chromophores are far better known. Table 5.6 gives the intra-sub-shell transitions, the derived value for K and the *nephelauxetic ratio* β_{55} between the value of K for the complex and $7B$

Table 5.6. *Spin-forbidden intra-sub-shell transitions in octahedral d^3-chromophores. The abbreviations for the ligands as in eq. (5.5); mal^{--} = malonate; aca$^-$ = acetyl-acetonato. The parameter of interelectronic repulsion K (in kK) and the nephelauxetic ratio β_{55} are discussed in the text*

	$^2E_g, ^2T_{1g}$	$^2T_{2g}$	$K = (3B+C)$	β_{55}
V(II) in MgO	11.50	17.24	4.3	0.83
CrF_6^{-3}	15.67, (16.4)	(22.0)	6.0	0.93
$CrF_5(H_2O)^{--}$	15.13	—	5.7	0.89
$CrF_3(H_2O)_3$	15.0	—	5.6	0.87
$CrF(H_2O)_5^{++}$	14.9	—	5.5	0.86
$Cr(H_2O)_6^{+3}$	15.0	(21.0)	5.5	0.86
$Cr(NH_3)_6^{+3}$	15.3	—	5.37	0.84
Cr en$_3^{+3}$	14.9, 15.5	—	5.23	0.82
Cr en F$_4^-$	15.0	—	5.45	0.85
Cr(III) in $Be_3Al_2Si_6O_{18}$	14.7	—	5.4	0.84
Cr urea$_6^{+3}$	14.4, (15.1)	(20.95)	5.33	0.83
Cr dmso$_6^{+3}$	14.3	—	5.2	0.81
Cr dmf$_6^{+3}$	14.45	—	5.2	0.81
Cr mal$_3^{-3}$	14.45	—	5.15	0.80
Cr ox$_3^{-3}$	14.35	—	5.1	0.79
Cr(III) in Al_2O_3	14.45	21.0	5.2	0.81
Cr_2O_3	13.7, 14.2	19.6	4.7	0.73
Cr(III) in Mg_2TiO_4	14.15	—	5.0	0.78

Table 5.6. (continued)

	^2E_g, $^2T_{1g}$	$^2T_{2g}$	$K = (3B + C)$	β_{55}
Cr dip$_3^{+3}$	13.75	—	4.75	0.74
Cr phen$_3^{+3}$	13.7	—	4.7	0.73
Cr(CH$_3$)$_6^{-3}$	13.4?	—	4.6?	0.71
Cr(N$_3$)$_6^{-3}$	(13.2), (13.9)	—	4.65	0.72
Cr(NCS)$_6^{-3}$	13.0, 13.3	—	4.6	0.71
Cr dtp$_3$	13.15, (13.6)	18.6, 19.05	4.6	0.71
Cr aca$_3$	12.95	—	4.55	0.71
Cr dsep$_3$	(12.6), (13.1)	(18.7)	4.35	0.68
Cr(CN)$_6^{-3}$	12.45	(18.5), (18.8)	4.2	0.65
MnF$_6^{--}$	16.3	—	5.7	0.77
Mn(IV) in Mg$_2$TiO$_4$	15.3	—	5.2	0.70
Mn(IV) in MgO	15.05	—	5.15	0.69
Mn(IV) in Al$_2$O$_3$	14.8	—	5.1	0.68
Mo urea$_6^{+3}$	10.0	15.85	3.5	0.82
Mo urea$_3$Cl$_3$	9.85	15.1	3.4	0.80
MoCl$_6^{-3}$	9.65	14.8	3.35	0.79
MoBr$_6^{-3}$	9.55	14.45	3.3	0.77
Mo(NCS)$_6^{-3}$	—	12.1	2.7	0.63
TcF$_6^{--}$	11.0, (11.5)	17.6, (18.1)	3.9	0.79
TcCl$_6^{--}$	9.5	14.15	3.2	0.65
TcBr$_6^{--}$	8.7, 9.2	13.4	3.05	0.62
TcI$_6^{--}$	8.5	—	2.9	0.59
RuF$_6^{-}$	10.4	15.4	3.4	0.61
ReF$_6^{--}$	9.0, 10.9	17.7, 18.9	3.8	0.83
ReCl$_6^{--}$	7.6, 9.4	14.1, 15.4	3.1	0.68
ReBr$_6^{--}$	7.4, 9.2	13.3, 15.0	2.9	0.64
ReI$_6^{--}$	7.3, 8.5	—	2.85	0.62
OsF$_6^{-}$	10.4, 11.0	18.0	3.7	0.73
IrF$_6$	8.5, 9.1	12.6, 15.6	2.65	0.43

for the corresponding gaseous ion. Most of the bands have previously been compiled (JØRGENSEN, 1962b and 1963b). The lowest excited level frequently is luminescent, at least at low temperature in vitreous or solid matrices. Thus, STURGE (1963) studied the line emission of V(II) in MgO, and SCHLÄFER et al. (1967a, 1967b) many Cr(III) compounds, and (1966) a few Mo(III) complexes. DEARMAND and FORSTER (1963) made a systematic investigation of substituted Cr(III) acetylacetonates, and FORSTER (in press) recently wrote a review on Cr(III) spectra. PORTER and SCHLÄFER (1964) reviewed d-group luminescence in general.

The absorption bands reported after 1963 and given in Table 5.6 are from JØRGENSEN and SCHWOCHAU (1965) in the case of technetium (IV) and rhenium (IV) hexahalides, BECK et al. (1967) for $Cr(N_3)_6^{-3}$, Mo(III) urea complexes according to KOMORITA, MIKI and YAMADA (1965) and private communication from these authors, RuF_6^- and OsF_6^- were reported by BROWN, RUSSELL and SHARP (1966) and IrF_6 by MOFFITT et al. (1959). $Mo(NCS)_6^{-3}$ has a rather peculiar spectrum with an asymmetric peak at 12.1 kK surrounded by weaker bands at 10.1 kK and a shoulder at 14 kK. This complex is perhaps trigonally distorted (LEWIS et al., 1961; CHRIST, 1965; SCHMIDTKE, 1967; and private communication from Mr. D. GARTHOFF).

The most interesting feature of Table 5.6 is perhaps that all the values of β_{55} observed fall in the interval 0.9 to 0.6 with the exception of CrF_6^{-3} (above) and TcI_6^{--} and IrF_6 (below the limits). When the oxidation state increases, the nephelauxetic ratio decreases, and in a way, this dependence is more important than the other variation, viz. the less pronounced nephelauxetic effect for fluorides than for iodides. However, rather than comparing with the corresponding gaseous ion, one may also note a certain invariance in the numerical value of K in a given isoelectronic series. As discussed below, this should not be taken as evidence for PAULING's electroneutrality principle; the fractional charge most favoured by most central atoms is rather in the interval $+1$ to $+2$. There is also another cancellation operating; across a transition group, a sequence of ions having the same ionic charge increase their parameters of interelectronic repulsion corresponding to decreasing average radius of the partly filled shell. In the d groups (but not in the f groups), this increase is rather accurately compensated by the increasing tendency toward covalent bonding.

We saw above that the nephelauxetic effect has been known in 4f group compounds at least since HOFMANN and KIRMREUTHER (1910) and was systematically studied by EPHRAÏM after 1926. The suggestion that parameters of interelectronic repulsion may be much more decreased in the d groups was definitely made by STEVENS (1953). However, isolated cases, in particular hexa-aqua ions, had also been discussed by ABRAGAM and PRYCE (1951), TANABE and SUGANO (1954) and OWEN (1955). C. E. SCHÄFFER pointed out to the writer that the variation of parameters of interelectronic repulsion as a function of the reducing character of the ligands is very striking when a series of chromium(III) or cobalt(III) complexes is compared. SCHÄFFER and JØRGENSEN (1958) and JØRGENSEN (1962g) reviewed the variation as a function of the ligands and of the central atom. Professor K. BARR in Copenhagen kindly proposed the Greek word *nephelauxetic* for the *cloud-expanding effect* detected as a decrease of the parameters of interelectronic repulsion relative to the corresponding gaseous ion. If one takes the assumption of *linear combinations of atomic orbitals* (L.C.A.O.) seriously, one writes the M.O. constituting the partly filled shell of an octahedral d-group complex MX_6

$$e_g: \quad a_3\,\psi_M - b_3 \sum \psi_x$$
$$t_{2g}: \quad a_5\,\psi_M - b_5 \sum \psi_x \tag{5.7}$$

where the sub-scripts 3 and 5 are reminders of the BETHE nomenclature γ_3 and γ_5 for MULLIKEN's e and t_2 orbitals. The central atom orbital ψ_M contributing to the upper e_g-sub-shell has the angular function either proportional to (x^2-y^2) or to $(3z^2-r^2)$ and the linear combination $\sum \psi_x$ of ligand orbitals has the corresponding symmetry type and each ψ_x has cylindrical σ-symmetry around the axis $M-X$. On the other hand, the lower t_{2g}-sub-shell has ψ_M proportional to (xy), (xz) or (yz), and $\sum \psi_x$ consists of ligand orbitals ψ_x having π-symmetry with respect to $M-X$. If the ligands have only one available lone-pair such as H^-, CH_3^-, NH_3, SO_3^{--} etc., no π-anti-bonding effects are expected, and $a_5=1$ and $b_5=0$. The normalization conditions for the orbitals (5.7) including the overlap integrals S_3 and S_5 are

$$a_3^2 + b_3^2 - 2\,a_3b_3S_3 = 1 \qquad a_5^2 + b_5^2 - 2\,a_5b_5S_5 = 1 \tag{5.8}$$

It may be noted that the corresponding bonding, filled M.O. of high ionization energy do not have exactly $c_n=b_n$ and $d_n=a_n$ in the expressions

$$c_n\,\psi_M + d_n \sum \psi_x \tag{5.9}$$

because of the conditions of mutual orthogonalization

$$a_nc_n - b_nd_n + (a_nd_n - b_nc_n)\,S_n = 0 \tag{5.10}$$

and normalization

$$c_n^2 + d_n^2 + 2\,c_nd_nS_n = 1 . \tag{5.11}$$

As already pointed out by STEVENS (1953), the parameters of inter-electronic repulsion depend on squares or products of electronic densities, and hence on *fourth powers* of the *delocalization coefficients* a_n and b_n. However, we have to remember that the *radial* function belonging to the constituent ψ_M may have been modified, and in particular expanded because we consider a smaller fractional charge than the oxidation state (JØRGENSEN, 1958 and 1966). The latter feature may be represented by an effective charge Z_* being smaller for the complex than for the corresponding gaseous ion. In this case, we have *three different nephelauxetic* *ratios*, viz. β_{55} for the interaction between two electrons both in the lower sub-shell, β_{35} for the interaction between an electron in the lower and an electron in the upper sub-shell, and β_{33} for the interaction between two electrons in the upper sub-shell. To the first approximation,

$$\beta_{55} = a_5^4 \quad (Z_{*\,com}/Z_{*\,gas})$$
$$\beta_{35} = a_5^2 a_3^2 \,(Z_{*\,com}/Z_{*\,gas}) \qquad\qquad (5.12)$$
$$\beta_{33} = a_3^4 \quad (Z_{*\,com}/Z_{*\,gas})\ .$$

KOIDE and PRYCE (1958) were the first to propose that the three nephelauxetic ratios might be different, and since a_3 always is expected to be smaller than a_5, the variation $\beta_{55} > \beta_{35} > \beta_{33}$ is predicted, as confirmed by experiment. The approximation (5.12) further on gives $(\beta_{35})^2 = \beta_{55}\beta_{33}$.

Actually, a closer analysis of this problem has been provided (JØRGENSEN, 1966). The Tanabe-Sugano diagrams are valid for modified radial functions, but on the conditions that all the three ratios are identical, e. g. because of $a_5 = a_3$ in eq. (5.12). The introduction of three independent nephelauxetic ratios is only possible to the extent that the diagonal elements of energy E_{sf} (the so-called "strong field" diagonal elements in conventional ligand field theory) for well-defined sub-shell M.O. configurations are reasonably good approximations to the eigenvalues E observed. In that case, it is possible to use second-order perturbation theory

$$E = E_{sf} - k\, B^2/\Delta \qquad\qquad (5.13)$$

putting $C = 4B$ and evaluating k as the sum of the squares of all non-diagonal elements connecting the groundstate with excited levels having the same S and symmetry type Γ_n in Tanabe and Sugano's determinants. Usually, as a compromise, the Racah parameter B is then derived from β_{35} data. Thus, the three excited levels of the ground sub-shell configuration $(t_{2g})^3$ of octahedral d^3 complexes have the energy relative to the ground level $^4A_{2g}$:

$$
\begin{aligned}
&^2E_g \quad && 21\,B -\ && 90\,B^2/\Delta \\
&^2T_{1g} \quad && 21\,B -\ && 24\,B^2/\Delta \qquad\qquad (5.14)\\
&^2T_{2g} \quad && 35\,B -\ && 176\,B^2/\Delta\ .
\end{aligned}
$$

However, as pointed out by Dr. W. SCHNEIDER ETH, Zürich (cf. also SCHNEIDER, 1963), eq. (5.14) for ^2E_g is not a typical case for second-order perturbation theory because nearly all the sub-shell configuration intermixing is caused by states having much larger distance than Δ, and hence, it is a better approximation to cut the coefficient 90 to half the value. The values for $K(=7B)$ in Table 5.6 were obtained along these lines.

Octahedral and tetrahedral d^5-systems have the same Tanabe-Sugano diagram. The groundstate of high-spin complexes is $^6A_{1g}$ belonging to the M.O. configuration $(t_{2g})^3(e_g)^2$ having two half-filled sub-shells. This is not the *lowest* sub-shell configuration; but the interelectronic repulsion included in the diagonal element E_{sf} is so relatively small because of the spin-pairing energy eq. (3.47) that $^6A_{1g}$ remains the groundstate until rather high values (~ 25) of the independent variable (Δ/B) in the Tanabe-Sugano diagrams[1]. In low-spin octahedral d^5-complexes, the ground level $^2T_{2g}$ belongs to the sub-shell configuration $(t_{2g})^5$ (having k of eq. (5.13) equal to 140) and low-spin tetrahedral d^5-complexes, if such entities exist, are expected to be $^2T_2$ of $(e)^4(t_2)^1$ (the symbol of parity g or u is removed in the symmetry T_d not possessing a centre of inversion). The two first excited levels $^4A_{1g}$ and ^4E_g of $(t_{2g})^3(e_g)^2$ [or $(e)^2(t_2)^3$ in tetrahedral chromophores] share the unusual property of not having *any* non-diagonal elements (i.e. $k=0$) with other d^5-sub-shell configurations with the high-spin groundstate $^6A_{1g}$. Consequently, for $\Delta=0$, the energy difference $10B+5C$ ($=30B$ in our approximation) corresponds to a definite term distance, between ^6S and ^4G, in spherical symmetry. This is the closest analogy to the 4f group behaviour shown by 3d group spectroscopy; not only are the absorption bands corresponding to these two (degenerate) transitions narrow, but their states have $L=0$ and 4. If icosahedral chromophores MX_{12} were known of d group compounds, the group-theoretical necessity of five degenerate d-like orbitals would produce a comparable situation for all configurations d^q. On the other hand, the narrow bands produced by the excited levels in Table 5.6 correspond to transitions to mixtures of terms such as ^2G, ^2H, $^2D, \ldots$ and the diagonal element E_{sf} contains the large contributions $21B$ and $35B$ in eq. (5.14). The levels constituting the term ^2G occurring in V^{++} at 11.97 and 12.19 kK and in Cr^{+3} at 15.06 and 15.41 kK have the lower energy $4B+3C$ ($\sim 16B$) relative to the ground term ^4F of $3d^3$ in spherical symmetry. Because of the nephelauxetic effect, one understands why it had been suggested that the fluorescent lines of the ruby are due

---

[1] Dr. CLAUS SCHÄFFER proposed (1967) the symbol $\sum$ for the *ligand field strength* Δ/B. This happens to be the first letter of the Greek word σφοδρότης meaning force applied against a certain resistance, here represented by the parameters of interelectronic repulsion and, in particular, by the spin-pairing energy. On the other hand, Δ can be thought of as the first letter of δύναμις meaning force as a potential ability. The nephelauxetic ratio β might have got its symbol from "frog" βάτραχος referring to the common indo-european legend about the frog wanting to blow himself up to the size of an ox. However, this mnemotechnic derivation is a posteriori

to the term ^2G only slightly perturbed by the neighbour atoms in analogy to the rare earths.

Table 5.7 compares the transitions $^6S - {}^4G$ in monatomic entities with the narrow absorption band just described in high-spin Mn(II) and Fe(III) complexes. ORGEL (1955) emphasized the close similarity between the weak (spin-forbidden) absorption bands produced by the octahedral chromophores $Mn(II)F_6$ in crystalline MnF_2 (each fluoride is coordinated to three manganese atoms) and $Mn(II)O_6$ in $Mn(H_2O)_6^{++}$ in aqueous solution and solid salts. In the ORGEL diagram (having Δ as variable, assuming fixed parameters of interelectronic repulsion) or in the TANABE-SUGANO diagram (having Δ/B as variable), this close similarity is simply explained by $\sum = (\Delta/B)$ alone determining the complicated spectrum (all ten excited quartet levels have been identified for aqueous manganese (II) perchlorate, cf. JØRGENSEN, 1957a; HEIDT, KOSTER and JOHNSON 1958). However, ORGEL (1955) made another, very important, remark about the analogous decrease of the term distance $^6S - {}^4G$ when adding two 4s electrons to Mn^{++} and when forming $Mn(II)F_6$ or $Mn(II)O_6$ from gaseous Mn^{++}. Actually, ORGEL did not at all suggest that the Mn(II) central atom is approximately neutral, but rather that accumulated electron density at a distance comparable with the lone-pairs of the adjacent ligand atoms has a moderate effect on the radial extension of the partly filled 3d shell.

Most entries in Table 5.7 have been discussed previously (JØRGENSEN, 1958; 1962b; 1963b). Recently, LOHR (1966) discussed general aspects of M.O. calculations for manganese (II) complexes, and reported data for crystalline $MnCO_3$. LAWSON (1966) measured the transmission spectrum of $MnCl_2, 2H_2O$ known from the crystal structure to contain the chromophore trans-$Mn(II)O_2Cl_4$, each chloride coordinating two manganese atoms. LAWSON (1967) also studied solids containing tetrahedral $MnCl_4^-$ and the perovskite $N(CH_3)_4MnCl_3$ containing octahedral Mn(II) Cl_6. Previously, FORSTER and GOODGAME (1964) compared the cyanate complex $Mn(NCO)_4^{--}$ with $MnBr_4^{--}$. The data given for salts containing $Mn(NCS)_6^{-4}$ and $Mn(NCS)_4^{--}$ are unpublished measurements by Mr. D. GARTHOFF, and salts and solutions containing Mn den$_2^{++}$ by Mr. G. CASARO.

Most iron (III) complexes in Table 5.7 were discussed by JØRGENSEN (1958), many of the measurements are unpublished results by Dr. C. E. SCHÄFFER. Tetrahedral $Fe(III)O_4$ was studied by BROWN (1963) as the dodecatungstoferrate, probably $FeW_{12}O_{40}^{-5}$. Some years ago, it was argued that the narrow band of $FeCl_4^-$ at 18.8 kK and the corresponding band at 16.4 kK of $FeBr_4^-$ first observed by GILL (1961) is the $^6S - {}^4G$ transition. However, GINSBERG and ROBIN (1963) argued that already the bands at 15.6 and 13.3 kK, respectively, represent this transition. The problem is not easy to resolve; among the experimental difficulties are the fact that $CoCl_4^{--}$ has bands in the red, four orders of magnitude stronger than those of $FeCl_4^-$. However, even when the trace impurities of cobalt are avoided, a much more serious problem is that all inter-sub-shell transitions of tetrahedral halide complexes tend to be split as if the effective symmetry is lower than T_d (JØRGENSEN, 1967c). Hence, it cannot be argued that the third excited level of MnX_4^- and FeX_4^- is necessarily the $^4A_1$ and ^4E components. This problem is connected with the existence of octahedral $FeCl_6^{-3}$ in a few solid salts. SCHLÄFER (1955) reports a weakly pronounced shoulder at 19.0 kK of the latter species, and HATFIELD et al. (1963) reported a band of $FeCl_6^{-3}$ at 18.73 and a broader band at 22.08 kK, ascribing ^4G to the latter level. The preceeding excited level is at much lower energy, and actually, BALT and VERWEY (1967) suggest ^4G at 13.2 kK, and at 14.2 kK for $FeCl_5(H_2O)^{--}$. However, the general tendency for tetrahedral chromophores to have slightly smaller wavenumbers than the corresponding octahedral chromophores (cf. Table 5.7) makes is rather plausible that ^4G actually occurs at 19 kK in the hexa-chloro and at 18.8 kK in the tetra-chloro complex.

SCHLÄFER (1956) studied carefully the small shifts of the absorption bands of an aqueous solution of $MnCl_2$, $4H_2O$ as a function of the concentration, presumably forming complexes such as $MnCl(H_2O)_5^+$. They are all octahedral and not tetrahedral. The crystalline substance is known to contain the chromophore cis-$Mn(II)O_4Cl_2$.

The nephelauxetic effect does not vary more for the manganese (II) complexes given in Table 5.7 than within the interval defined by the two isoelectronic gaseous ions Cr^+ and Mn^{++}. However, when making such a comparison, it must not be neglected that the atomic number of Mn(II) remains 25. If one asks the question what the term distance $^6S - {^4G}$ would be for d^5 in Mn^+ if the twenty-fourth electron were somewhere else, but allowing the 3d shell to behave as it would for $3d^6$ in Mn^+, it can be shown [see eq. (5.26)] that B would be 0.058 kK higher for Mn^+ than for Cr^+, and consequently, $^6S - {^4G}$ would be 22.25 kK, between the values characterizing manganese (II) iodides and sulphides.

It may be noted that $\beta_{35} = 0.79$ for the *least* nephelauxetic iron (III) complex, FeF_6^{-3}, is nearly the same as the value, 0.785, for the *most* nephelauxetic manganese (II) chromophore, in ZnSe. Actually, the numerical values of $30B$, the $^6S - {^4G}$ distance, are roughly the same for the least nephelauxetic cases, such as $Mn(II)F_6$ and $Fe(III)F_6$, or also $Mn(II)O_6$ compared with the corresponding $Fe(III)O_6$. On the other hand, when the covalent bonding becomes more pronounced, the actual parameters of interelectronic repulsion are *smaller* for the iron (III) than for the isoelectronic manganese (II) complexes. Thus, even the higher limit 18.8 kK for $^6S - {^4G}$ in $FeCl_4^-$ is 21% below the value for $MnCl_4^{--}$. In this connection, it may be mentioned that the hypothetical value

Table 5.7. *Energy difference ^6S—^4G observed for d^5-systems in spherical symmetry and the corresponding narrow absorption bands (in kK) for high-spin manganese(II) and iron(III) complexes*

Cr^+	$3d^5$	20.516
Mn^{++}	$3d^5$	26.846
Fe^{+3}	$3d^5$, extrapolated	32.0
Mn^0	$3d^5\,4s^2$	25.279
Mo^+	$4d^5$	15.375
Re^0	$5d^5\,6s^2$	15.84
$Mn(II)F_6$	in $KMnF_3$	25.3
	in MnF_2	25.3
$Mn(II)O_6$	in $Mn(H_2O)_6^{++}$	25.0
	in $Mn\,dmf_6^{++}$	24.6
	in $MnCO_3$	24.6
	in MnO	23.8
cis-$Mn(II)O_2Cl_4$	in $MnCl_2$, 4 H_2O	24.6
$trans$-$Mn(II)O_2Cl_4$	in $MnCl_2$, 2 H_2O	24.32
$Mn(II)Cl_6$	in $N(CH_3)_4MnCl_3$	23.75
	in $MnCl_2$	23.7
$Mn(II)Cl_4$	in $MnCl_4^{--}$	23.1
$trans$-$Mn(II)N_2Cl_4$	in $Mn\,py_2Cl_2$	23.7
$Mn(II)N_6$	in $Mn\,en_3^{++}$	23.65
	in $Mn\,den_2^{++}$	23.5
	in $Mn(NCS)_6^{-4}$	24.0
$Mn(II)N_4$	in $Mn(NCO)_4^{--}$	23.3
	in $Mn(NCS)_4^{--}$	22.7
$Mn(II)Br_6$	in $MnBr_2$	23.3
$Mn(II)Br_4$	in $MnBr_4^{--}$	23.0
$Mn(II)I_4$	in MnI_4^{--}	22.5
$Mn(II)S_6$	in MnS	22.2
$Mn(II)S_4$	in ZnS	21.65
$Mn(II)Se_4$	in $ZnSe$	21.1
$Fe(III)F_6$	in FeF_6^{-3}	25.35
$Fe(III)O_6$	in $Fe(H_2O)_6^{+3}$	24.45
	in $Fe\,urea_6^{+3}$	23.25
	in $Fe\,mal_3^{-3}$	22.8
	in $Fe(HCOO)_6^{-3}$	22.75
	in $Fe\,ox_3^{-3}$	22.2
$Fe(III)O_4$	in dodecatungstoferrate	24.4
$Fe(III)Cl_6$	in $FeCl_6^{-3}$	19.0 or 13.2
$Fe(III)Cl_4$	in $FeCl_4^-$	18.8 or 15.6
$Fe(III)Br_4$	in $FeBr_4^-$	16.4 or 13.3

for Fe^+ defined above as $30B$ would be 24.0 kK. It is worth noting in Table 5.7 that usually, the nephelauxetic ratio is only some 2 to 3% lower in tetrahedral MX_4 than in octahedral MX_6. Thus, there is no intrinsic tendency toward *much* stronger covalent bonding in tetrahedral complexes; the main reason for the small difference may simply be the slightly shorter internuclear distances known from crystallography for tetrahedral compounds.

Though the d^3-system in Table 5.6 and the d^5-systems in Table 5.7 represent the two most numerous categories of narrow absorption bands of d-group compounds, there exist many other cases. Thus, octahedral nickel (II) complexes have a very conspicuous transition to the first excited level ^1E_g of the sub-shell configuration $(t_{2g})^6(e_g)^2$ to which also the groundstate $^3A_{2g}$ belongs (JØRGENSEN, 1955 and 1956a; BOSTRUP and JØRGENSEN, 1957; REINEN, 1965; REEDIJK, VAN LEEUWEN and GROENEVELD, 1968). When Δ/B increases, first a^3T_{1g} and later $^3T_{2g}$ cross the energy of ^1E_g remaining roughly constant. The result at the crossing-points is that two narrow bands (corresponding to transitions to the two Γ_3 components having mixed character of $S=0$ and 1 due to the effects of intermediate coupling) at the distance of about 1.2 kK are superposed the broad band corresponding to the three other Γ_J components having nearly exclusive triplet character and having a more anti-bonding character than the ground sub-shell configuration consequently showing the large band-width. However, when the diagonal energy of ^1E_g is extracted by correction for the intermixing with the triplet Γ_3 component, the nephelauxetic ratio β_{33} is found to be roughly the same as the value β_{35} derived from the baricenters of the spin-allowed transitions.

The other level $^1A_{1g}$ of the ground sub-shell configuration of octahedral nickel (II) complexes has nearly twice as high an excitation energy as ^1E_g. The corresponding absorption band is extraordinarily weak, but has been detected by DINGLE and PALMER (1966) at 21.36 kK in a crystal containing $Ni\ en_3^{++}$ measured at low temperature. In this connection, it may be mentioned that the weak band of $Ni(H_2O)_6^{++}$ at 18.4 kK, to which no electronic transition can be assigned, was shown by PIPER and KOERTGE (1960) to be due to co-excited $O-H$ vibrations, and had a lower wavenumber in the deuterium-substituted $Ni(D_2O)_6^{++}$. Co-excited solvent vibrations are exceedingly rare; JØRGENSEN (1962c) noted some cases for $OsCl_6^{--}$. Another case may be the narrow band of $Mn(H_2O)_6^{++}$ at 26.54 kK observed by HEIDT et al. (1958), this band has the lowest molar extinction coefficient $\varepsilon = 0.002$ known (with the possible exception of the highly forbidden $4f^6$-transition $^7F_0 - ^5D_0$ of europium (III) aqua ions). However, the band may also be due to a intra-sub-shell doubly spin-forbidden transition to a doublet level.

The lowest sub-shell configuration $(e)^4(t_2)^3$ of tetrahedral cobalt (II) complexes contains the ground term $^4A_2$ and the excited terms ^2E, $^2T_1$ and $^2T_2$ in close analogy to the octahedral d^3-compounds studied in Table 5.6. The spin-allowed transition to b^4T_1 coincides with ^2E and $^2T_1$ in nearly all cases and gives raise to effects of intermediate coupling comparable to those discussed for octahedral nickel (II) complexes. However, this problem is difficult to treat quantitatively in the case of cobalt (II) because already the spin-allowed transition to a^4T_1 usually corresponds to three bands in the near infra-red separated by a larger distance, some $2\,kK$, than can be explained by spin-orbit coupling alone. This fact seems to indicate that the instantaneous picture obtained by spectroscopic studies of CoX_4^{--} does not correspond exactly to the high symmetry T_d. Similar effects are observed for the spin-allowed transitions of NiX_4^{--} and VX_4^- (JØRGENSEN, 1967c).

The blue tetrahedral $3d^2$-system MnO_4^{-3} was studied by ORGEL (1964) and KINGSLEY, PRENER and SEGALL (1965). The sub-shell configuration $(e)^2$ comprises the groundstate $^3A_2$, the excited level ^1E at $8.7\,kK$ from which luminescent line emission can be detected, and the level ^1A at $13.6\,kK$. The next sub-shell configuration $(e)(t_2)$ has $^3T_2$ and $^3T_1$ corresponding to two broad bands having maxima at 11.0 and $14.8\,kK$. From the spin-forbidden transitions, $(4B+C)=4.4\,kK$ and $\beta_{33}=0.48$ can be derived, whereas the spin-allowed bands indicate $B=370\,K$ and $\beta_{35}=0.32$.

Other narrow bands occur in the visible of high-spin $3d^6$-systems such as $Fe(H_2O)_6^{++}$ and of high-spin $3d^4$-complexes (frequently of highly distorted stereochemistry) of chromium (II) and manganese (III). Thus, the anhydrous halides were studied by CLARK (1964) and OELKRUG (1966) and the narrow peak at $19.8\,kK$ superposed the broad spin-allowed band of the Mn(III) ethylenediaminetetraacetate Mn enta$^-$ was reported by JØRGENSEN (1957a) and YOSHINO et al. (1961). In all of these cases, the exited levels have essentially the same sub-shell configuration as the groundstate, but the theoretical treatment is rendered difficult by the uncertainty regarding the symmetry. In this connection, it may be notee that Mn(H_2O) enta^{--} has a narrow band at $24.3\,kK$ and might have been included in Table 5.7 if it had not been for the case that the complex is known to be seven-coordinated in solids and probably also in solution. However, high-spin manganese (II) complexes is the closest one can come to a valid "weak-field diagonal element" description involving the excited term ^4G. It is a pity that electron transfer bands prevent the detection of the narrow, weak bands in most iron (III) complexes.

Accepting the "vertical" transitions (not changing the internuclear distances according to FRANCK and CONDON's principle) corresponding to maxima of broad absorption bands, one obtains a very coherent set of β_{35} values for a large number of octahedral complexes. The wavenumbers

^{6}S — ^{4}G in Table 5.7 represent $30\,B$, and one can obtain β_{35} directly by dividing the observed wavenumbers by 26.85 kK in the case of Mn(II) or by 32.0 kK in the case of Fe(III). It is not argued here that one cannot obtain reliable values of B and β_{35} for Jahn-Teller-unstable chromophores, and in particular, many $3d^2$ vanadium (III) and $3d^7$ cobalt (II) complexes yield excellent determinations. However, for the clarity of the discussion, Table 5.8 has been restricted to octahedral d^3, low-spin d^6 and high-spin d^8 chromophores.

The diagonal sum rule applied to Tanabe and Sugano's determinants for octahedral d^3 and d^8 systems gives

$$\sigma_2 + \sigma_3 - 3\,\sigma_1 = 15\,B \qquad (5.15)$$

where the baricenter of intensity of the first spin-allowed band situated at the wavenumber σ_1 indicates Δ, and where $15B$ also is the term distance $^3F - {}^3P$ for d^2 and d^8 and $^4F - {}^4P$ for d^3 and d^7 systems in spherical symmetry. Whereas one has to find the term baricenters corrected for relativistic effects before finding B (in $K = cm^{-1}$):

$$
\begin{array}{llll}
V^{+3}\ 861 & V^{++}\ 755 & Cr^{+3}\ 918 & Mn^{+4}\ 1064 \\
Ni^{++}\ 1041 & Nb^{+3}\ 602 & Rh^+\ 541 & Pd^{++}\ 683
\end{array} \qquad (5.16)
$$

for the gaseous ions, the octahedral complexes in Table 5.8 fortunately have no relativistic splitting of their groundstate. Such first-order effects occur for octahedral V(III) and Co(II). In most Cr(III) complexes, only σ_1 and σ_2 can be observed, and one either uses the implicit equation inherent in TANABE and SUGANO's results

$$\sigma_2 - \sigma_1 = 12\,B - \chi \quad \text{where} \quad \chi = 36\,B^2/(\Delta - 9\,B + \chi) \qquad (5.17)$$

or the equivalent explicit equation first proposed by POOLE (1964)

$$B = (2\,\sigma_1 - \sigma_2)\,(\sigma_2 - \sigma_1)/(27\,\sigma_1 - 15\,\sigma_2) \qquad (5.18)$$

or, as a third alternative, one can tabulate $(\sigma_2 - \sigma_1) = \varkappa B$ as a function of the variable (σ_2/σ_1) which is the most practical form if many values should be evaluated (JØRGENSEN, PAPPALARDO and RITTERSHAUS, 1965). Actually, when a third band of Cr(III) is observed it does not exactly obey eq. (5.15), and in Table 5.8, all values for Cr(III) complexes have been obtained from σ_1 and σ_2. In the case of V(II), a mean value of the two estimations has been given.

According to Dr. CLAUS SCHÄFFER, eq. (5.13) is in excellent agreement with the positions of the three first singlet levels of octahedral low-spin

Table 5.8. Octahedral d^3, d^6 and d^8 chromophores. Racah's parameter B of interelectronic repulsion in K (cm^{-1}), the nephelauxetic ratio β_{35} the fractional charge Z_{min} assuming no symmetry-restricted covalency, the compromise Z_{root} corresponding to equal importance of central-field and symmetry-restricted covalency, and the sub-shell energy difference Δ observed and derived from eq. (5.5), in kK

		B	β_{35}	Z_{min}	Z_{root}	Δ obs.	Δ eq. (5.5)
$3d^8$							
V(II)O$_6$	V(H$_2$O)$_6^{++}$ solution	640	0.85	1.25	1.5	11.8	12.0
	(NH$_4$)$_2$Zn(H$_2$O)$_6$(SO$_4$)$_2$	650	0.86	1.3	1.55	12.35	12.0
V(II)Cl$_6$	CsCdCl$_3$	600	0.80	1.0	1.4	7.6	9.4
	VCl$_2$	590	0.78	0.95	1.35	9.0	9.4
Cr(III)F$_6$	CrF$_6^{-3}$	820	0.89	2.1 (2.4)	2.5	15.2	15.7
Cr(III)O$_6$	Cr(H$_2$O)$_6^{+3}$	725	0.79	1.4 (1.85)	2.1	17.4	17.4
	Cr urea$_6^{+3}$	660	0.72	1.05 (1.7)	1.8	16.0	16.0
	Cr dmso$_6^{+3}$	670	0.73	1.1 (1.6)	1.85	15.9	15.8
	Cr dmf$_6^{+3}$	680	0.74	1.15 (1.6)	1.85	16.9	17.0
	Cr ox$_3^{-3}$	620	0.68	0.85 (1.45)	1.6	17.5	17.2
	Al$_2$O$_3$	650	0.71	1.0 (1.6)	1.7	18.0	—
	Cr$_2$O$_3$	480	0.52	0.3 (1.15)	1.0	16.6	—
	YCrO$_3$	540	0.59	0.55	1.3	16.45	—
	LaCrO$_3$	510	0.56	0.45	1.15	16.35	—
	MgAl$_2$O$_4$	730	0.80	1.45	2.15	18.2	—
	MgCr$_2$O$_4$	610	0.67	0.8	1.55	17.2	—
	Al$_2$BeO$_4$	600	0.66	0.75	1.5	17.35	—
	Be$_3$Al$_2$Si$_6$O$_{18}$	730	0.80	1.45 (1.85)	2.15	16.5	—
Cr(III)N$_6$	Cr(NH$_3$)$_6^{+3}$	650	0.71	1.0 (1.75)	1.7	21.6	21.8
	Cr en$_3^{+3}$	620	0.68	0.9 (1.65)	1.6	21.9	22.3
	Cr(NCS)$_6^{-3}$	570	0.62	0.65 (1.65)	1.4	17.8	17.7
	Cr(N$_3$)$_6^{-3}$	470	0.51	0.3 (1.1)	1.0	14.9	14.4

7*

Table 5.8 (continued)

		B	β_{35}	Z_{min}	Z_{root}	Δ obs.	Δ eq. (5.5)
Cr(III)Cl$_6$	CrCl$_6^{-3}$	560	0.61	0.6 (1.2)	1.4	13.2	13.6
	CrCl$_3$	510	0.56	0.45	1.15	13.8	13.6
Cr(III)Br$_6$	CrBr$_3$	370	0.40	0.1	0.65	13.4	12.5
Cr(III)C$_6$	Cr(CH$_3$)$_6^{-3}$	520	0.57	0.5 (1.0?)	1.2	20.8	—
	Cr(CN)$_6^{-3}$	530	0.58	0.5 (0.75)	1.25	26.7	29.6
Cr(III)S$_6$	Cr dtp$_3$	410	0.45	0.2 (1.0)	0.85	14.4	14.4
	Cr dtc$_3$	430	0.47	0.2	0.9	15.6	15.7
Cr(III)Se$_6$	Cr dsep$_3$	340	0.37	0.05 (0.85)	0.6	13.7	13.9
	Cr dsec$_3$	390	0.42	0.15	0.7	14.8	—
Mn(IV)F$_6$	MnF$_6^{--}$	600	0.56	0.55 (1.6)	1.5	21.8	20.7
3d^6							
Fe(II)C$_6$	Fe(CNO)$_6^{-4}$	410	0.44	−0.05	0.45	27.0	—
	Fe(CN)$_6^{-4}$	400	0.43	−0.1	0.4	32.2	—
Co(III)O$_6$	Co(H$_2$O)$_6^{+3}$	670	0.61	0.4	1.2	18.2	18.2
	Co(CO$_3$)$_3^{-3}$	540	0.49	0.1	0.7	17.3	—
	Co ox$_3^{-3}$	540	0.49	0.1	0.7	18.0	18.0
Co(III)N$_6$	Co(NH$_3$)$_6^{+3}$	620	0.56	0.3	1.0	22.9	22.8
	Co en$_3^{+3}$	590	0.53	0.2	0.95	23.2	23.3
Co(III)C$_6$	Co(CNO)$_6^{-3}$	450	0.41	−0.1	0.5	26.1	—
	Co(CN)$_6^{-3}$	460	0.42	−0.1	0.55	33.5	31
Co(III)S$_6$	Co dtp$_3$	400	0.36	−0.15	0.35	14.2	15.1
	Co dtc$_3$	380	0.34	−0.2	0.3	16.3	16.4
Ni(IV)F$_6$	NiF$_6^{--}$	450	0.36	−0.1	0.5	19.9	—

Table 5.8 (continued)

		B	β_{35}	Z_{min}	Z_{root}	Δ obs.	Δ eq. (5.5)
$3d^8$							
$Ni(II)F_6$	$KNiF_3$	960	0.92	1.4	1.7	7.5	7.8
	NiF_2	950	0.91	1.35	1.65	7.7	7.8
$Ni(II)O_6$	$Ni(H_2O)_6^{++}$	940	0.90	1.3	1.6	8.5	8.7
	$Ni(C_2H_5OH)_6^{++}$	910	0.87	1.15	1.5	8.35	8.4
	$Ni(CH_3COOH)_6^{++}$	920	0.88	1.2	1.55	8.15	8.2
	$Ni\ dmf_6^{++}$	890	0.86	1.0	1.45	8.55	8.5
	$Ni\ dmso_6^{++}$	910	0.87	1.15	1.5	7.85	7.9
	$Ni\ urea_6^{++}$	910	0.87	1.15	1.5	8.05	8.0
	MgO	865	0.83	0.85	1.35	8.65	—
	NiO	800	0.77	0.65	1.2	8.8	—
	$MgTiO_3$	840	0.81	0.8	1.3	7.3	—
	$NiTiO_3$	825	0.79	0.75	1.25	7.4	—
	$CdTiO_3$	830	0.80	0.75	1.25	6.0	—
	$Sr_3NiNb_2O_9$	850	0.82	0.8	1.3	8.6	—
	$Sr_3NiSb_2O_9$	860	0.83	0.85	1.35	6.9	—
	Ba_2CaTeO_6	850	0.82	0.8	1.3	4.8	—
$Ni(II)N_6$	$Ni(CH_3CN)_6^{++}$	890	0.86	1.0	1.45	10.6	10.6
	$Ni(NH_3)_6^{++}$	890	0.86	1.0	1.45	10.8	10.9
	$Ni\ en_3^{++}$	850	0.82	0.8	1.3	11.5	11.1
$Ni(II)Cl_6$	$RbNiCl_3$	780	0.75	0.55	1.1	7.2	6.8
	$NiCl_2$	750	0.72	0.5	1.0	7.6	6.8
$Ni(II)Br_6$	$RbNiBr_3$	770	0.74	0.55	1.05	6.9	6.3
	$NiBr_2$	730	0.70	0.45	0.9	7.3	6.3

Table 5.8 (continued)

		B	β_{35}	Z_{min}	Z_{root}	Δ obs.	Δ eq. (5.5)
4d³	Mo(III)O₆ — Mo urea$_6^{+3}$	450	0.74	0.95 (1.3)	1.55	25.9	22.6
	Mo(III)Cl₆ — MoCl$_6^{-3}$	440	0.72	0.9 (1.15)	1.4	19.2	19.2
	Mo(III)Br₆ — MoBr$_6^{-3}$	390	0.64	0.7 (1.1)	1.25	17.7	17.7
	Tc(IV)F₆ — TcF$_6^{-}$	530	0.76	1.3 (1.6)	2.2	28.4	27
4d⁶	Ru(II)O₆ — Ru(H₂O)$_6^{++}$	490	0.79	0.85	1.25	19.7	20.0
	Ru(II)N₆ — Ru en$_3^{++}$	440	0.71	0.6	1.0	28.0	25.6
	Rh(III)O₆ — Rh(H₂O)$_6^{+3}$	510	0.71	0.8	1.5	27.0	27.0
	Al₂O₃	400	0.56	0.35	0.9	26.4	–
	Rh(III)N₆ — Rh(NH₃)$_6^{+3}$	430	0.60	0.5	1.1	34.1	33.8
	Rh en$_3^{+3}$	420	0.58	0.45	1.0	34.6	34.6
	Rh(N₃)$_6^{-3}$	270	0.37	-0.05	0.4	21.3	22.4
	Rh(III)Cl₆ — RhCl$_6^{-3}$	350	0.49	0.2	0.7	20.3	21.1
	Rh(III)Br₆ — RhBr$_6^{-3}$	280	0.39	0.0	0.5	19.0	19.4
	Rh(III)S₆ — Rh dtp₃	210	0.29	-0.1	0.3	22.0	22.4
	Pd(IV)F₆ — PdF$_6^{-}$	340	0.42	0.1	0.7	25.9	–
5d⁶	Ir(III)N₆ — Ir(N₃)$_6^{-3}$	170	0.26	–	–	25.6	26.6
	Ir(III)Cl₆ — IrCl$_6^{-3}$	300	0.45	–	–	25.0	25.0
	Ir(III)Br₆ — IrBr$_6^{-3}$	250	0.38	–	–	23.1	23.0
	Ir(III)S₆ — Ir dtp₃	160	0.24	–	–	26.6	26.6
	Ir(III)Se₆ — Ir dsep₃	140	0.21	–	–	25.0	25.6
	Pt(IV)F₆ — PtF$_6^{-}$	380	0.51	–	–	32.6	32.4

d^6-systems when the eigenvalues of TANABE and SUGANO's determinants are found in all of the appropriate range of $\sum = (\Delta/B)$ values for $C = 4B$. Since

$$
\begin{array}{ll}
{}^1A_{1g} & -120\, B^2/\Delta \\
{}^1T_{1g} & \Delta - C - 34\, B^2/\Delta \\
{}^1T_{2g} & \Delta + 16\, B - C - 118\, B^2/\Delta
\end{array}
\qquad (5.19)
$$

(where only the second-order perturbation result (5.13) has been retained for the groundstate) the two first spin-allowed transitions are situated at

$$
\begin{array}{ll}
\sigma_1 & \Delta - 4\, B + 86\, B^2/\Delta \\
\sigma_2 & \Delta + 12\, B + 2\, B^2/\Delta \,.
\end{array}
\qquad (5.20)
$$

When both absorption bands can be observed (i.e. that they are not covered by electron transfer bands or split by deviations from octahedral symmetry) it is easy to evaluate Δ and B from eq. (5.20) either from a table of $(\sigma_2 - \sigma_1) = \varkappa B$ as a function of (σ_2/σ_1) or by an iteration which converges very rapidly in actual practice.

Besides the question whether spin-forbidden, superposed transitions or deviations from octahedral symmetry interfere with the ready determination of σ_1 and σ_2, another, more general, question is whether the absorption maxima really are representative for these two wavenumbers. Two dissenting arguments can be given: The quantity (ε/σ) has its maximum a little earlier than ε, and it can be argued that the "vertical" transition corresponds to the former maximum (cf. JØRGENSEN, 1962b). Further on, the necessary co-excitation of odd parity vibrations shifts all Laporte-forbidden bands of systems lacking a centre of inversion to slightly higher wavenumbers. However, these two effects are small and tend to cancel. Since the band-widths of σ_1 and σ_2 are roughly the same, because both transitions are essentially one-electron jumps from the lower (t_{2g}) to the higher, σ-anti-bonding (e_g) sub-shell, the difference $(\sigma_2 - \sigma_1)$ should remain a good measure for a multiple of B corrected by eq. (5.13) or eq. (5.17) for the effects of sub-shell configuration intermixing. This question is connected with the interesting problem why most broad absorption bands are Gaussian error curves with slightly larger half-widths $\delta(+)$ toward higher than $\delta(-)$ toward lower wavenumbers. According to SCHÄFFER, this situation is closely equivalent to a Gaussian error curve in the wavelength λ. Anyhow, in the theoretical treatment of transition group complexes, there is very little use of considering the electronic origins of the broad transitions which sometimes can be detected in absorption at low temperatures. In molecular spectroscopy of gaseous molecules, the electronic origins are much more important. However, the different internuclear distances corresponding to the minima of the potential surfaces for exited states belonging to other sub-shell configurations have important consequences for the shift of broad bands toward lower wavenumbers in fluorescence.

Most of the entries in Table 5.8 have previously been compiled (JØRGENSEN, 1962b; 1962g; 1967a; 1967c). Mixed oxides of chromium (III) were discussed by SCHMITZ-DuMONT and REINEN (1959) and POOLE (1964) and recently related to crystallographic results by REINEN (1969). The B-value given for emerald $Be_3Al_2Si_6O_{18}$ by NEUHAUS (1960) may seem very high when compared with chrysoberyl Al_2BeO_4, but actually, WOOD et al. (1963) indicate an even larger value of B, and it is seen from Table 5.6 that β_{55} is also relatively high for emerald [Cr(III) coloured beryl]. It must be concluded that the green colour and corresponding low value for Δ is caused by unusual large internuclear distances $Cr - O$, whereas Cr_2O_3 though green has a much lower value for B suggesting strong π-anti-bonding effects diminuishing Δ.

In Table 5.8, the data for $Cr(N_3)_6^{-3}$ are from BECK et al. (1967), for $Cr(CH_3)_6^{-3}$ from DUNKEN and MARX (1966), for Cr dtc$_3$ private communication from SCHÄFFER, cf. JØRGENSEN (1962f), for Cr dsep$_3$ from JØRGENSEN (1962d) and for Cr dsec$_3$ from JENSEN and KRISHNAN (1967) and FURLANI et al. (1968) dsec$^-$=$(C_2H_5)_2NCSe_2^-$. SCHÄFFER (1958) and NEUHAUS (1960) found that anhydrous $CrCl_3$ has a larger Δ and a smaller B than $CrCl_6^{-3}$ studied by HATFIELD et al. (1963). It is surprising how small B is for anhydrous $CrBr_3$ according to WOOD et al. (1963). The fulminates of Fe(II) and Co(III) were investigated by BECK and FELDL (1965). The values given for NiF_6^{--} derive from WESTLAND, HOPPE and KASENO (1965) but this complex has also been studied recently by REISFELD, ASPREY and PENNEMAN (1968). Many nickel (II) fluorides were discussed by RÜDORFF, KÄNDLER and BABEL (1962). Ni(II) complexes of oxygen-containing ligands such as ethanol. acetic acid, dimethylsulphoxide and urea were studied by LEEUWEN (1967) and REEDIJK et al. (1968). The mixed oxides of Ni(II) were studied by SCHMITZ-DuMONT and KASPER (1964) and REINEN (1966). It may be seen from Table 5.8 that Δ varies relatively much more for these mixed oxides than B. The main reason seems to be varying internuclear distances. An interesting case of the effect of $M - O$ distances determined by external constituents of the complex is the heteromolybdates, probably $MMo_6O_{21}^{-3}$, where $M = Cr$ (III) has $\Delta = 18.5$ kK, larger than for the hexa-aqua ion, whereas $M = Co(III)$ has Δ 0.15 kK smaller than for $Co(H_2O)_6^{+3}$ (SCHÄFFER, 1958), Ni(II) complexes in general were discussed by BOSTRUP and JØRGENSEN (1957), and Mo(III) complexes by KOMORITA, MIKI and YAMADA (1965).

TcF_6^{--} was reported by JØRGENSEN and SCHWOCHAU (1965), Ru $(H_2O)_6^{++}$ by MERCER and BUCKLEY (1965), Ru en$_3^{++}$ by SCHMIDTKE and GARTHOFF (1966), Rh(III) in Al_2O_3 by BLASSE and BRIL (1967) showing a slightly smaller Δ than for the hexa-aqua ion in contrast to the ruby, $Rh(N_3)_6^{-3}$ and $Ir(N_3)_6^{-3}$ by SCHMIDTKE and GARTHOFF (1967), Rh dtp$_3$ and Ir dtp$_3$ by JØRGENSEN (1962f) and PdF_6^- by BROWN, RUSSELL and SHARP (1966).

When considering the 5d group, it may be argued that the exceedingly small values for β_{35} derived from eq. (5.19) for iridium (III) complexes are unrealistically small. This can be explained to some extent by the determinants of intermediate coupling for d^6 (and d^4) systems in octahedral symmetry calculated by SCHROEDER (1962) and is confirmed by the unexpectedly large difference between $^3T_{1g}$ and $^1T_{1g}$ observed for Ir(III) and Pt(IV) which should be only $8B + 36B^2/\Delta$ when relativistic effects are not taken into account. Hence, the measurements of β_{35} in the 5d group have only interest as a relative measure of the nephelauxetic effect.

Absorption spectra of solid compounds present certain problems not known for liquid solutions. At a few occasions, co-operative effects involving more than a single transition group atom can be observed. The prototype of this situation is the "basic rhodo ion" $(NH_3)_5CrOCr$ $(NH_3)_5^{+4}$ studied by SCHÄFFER (1958). Four narrow bands in the region $27-28$ kK can be ascribed to simultaneous excitation of the doublet levels of Table 5.6 in *both* Cr(III) as suggested by Dr. LESLIE ORGEL in a private conversation. In the ruby, a similar band group has a molar extinction coefficient proportional to the Cr(III) concentration (LINZ and NEWHAM, 1961; NAIMAN, private communication) caused by CrOCr groupings only present at higher chromium concentrations. REINEN (1965) pointed out that the spin-forbidden transitions in $Ni_xMg_{1-x}O$ became more intense for high x values, and he ascribed this effect to the antiferromagnetic behaviour at room temperature of undiluted NiO. However, REINEN also observes a narrow band at 26,3 kK which might be due to simultaneous excitation of ^1Eg in two Ni(II) rather than an intensified transition to $^1T_{1g}$ or a weak electron transfer band. Whether or not direct co-operative effects such as simultaneous excitations occur in the undiluted NiO and Cr_2O_3, there is not the slightest doubt that B and the nephelauxetic ratio β_{35} are unusually low. It can also be seen from Table 5.8 that B is lower for $MgCr_2O_4$ than for the highly dilute solid solutions of Cr(III) in the spinel $MgAl_2O_4$. On the other hand, the perovskites $LaAl_{1-x}Cr_xO_3$ (where each La is coordinated with 12 oxygens in a cubic tetrakaidecahedral chromophore $La(III)O_{12}$, Al and Cr are coordinated in an almost regular octahedron, and each oxygen is coordinated by octahedral *trans*-La_4M_2) have B roughly independent of x. The B values for Ni(II) in $MgTiO_3$ and $CdTiO_3$ are just insignificantly larger than for the isomorphous $NiTiO_3$. It must be concluded that the decrease of the nephelauxetic ratio β_{35} is more important in undiluted NiO and Cr_2O_3 than in most other Ni(II) and Cr(III) compounds. Many analogous cases cannot be studied because the black compounds have strong absorption bands caused by electron transfer or collective effects (which, in a way, correspond to electron transfer from one d-group atom to another).

The most nephelauxetic ligand in Table 5.8 is diethyldiselenophosphate $(C_2H_5O)_2PSe_2^-$ (JØRGENSEN, 1962 d). The alkoxo groups prevent polymerization of the chromophores $Cr(III)Se_6$ known in double selenides being black semi-conductors, as discussed in chapter 8. There is good reason to believe that black CrSb contains $Cr(III)Sb_6$; but of course, the $3d^3$ transitions cannot be detected. However, cases are known of mononuclear complexes in solution indirectly suggesting a stronger nephelauxetic effect than in $Crdsep_3$. CLARK, GREENFIELD and NYHOLM (1966) studied two triarsine complexes of Cr(III), both solution and reflection spectrum of the solid. The cubic appearance of the spectra suggests strongly chromophores fac-$Cr(III)Cl_3As_3$ having $\sigma_1 = \Delta = 15.6$ kK and $\sigma_2 = 19.9$ kK. HOWELL, VENANZI and GOODALL (1967) studied comparable chromophores fac-$Cr(III)Cl_3P_3$ having $\sigma_1 = 16.5$ kK and $\sigma_2 = 21.0$ kK. From these values, it can be argued from SCHÄFFER's principle of holohedrized symmetry and from known values for $CrCl_6^{-3}$

	$Cr(III)Cl_3P_3$	$Cr(III)Cl_3As_3$	$Cr(III)P_6$	$Cr(III)As_6$	
B	410	400	260	240	(5.21)
β_{35}	0.44	0.43	0.28	0.26	

This extrapolation shows that the tridentate phosphines and arsines have a more pronounced nephelauxetic effect than even selenium-containing ligands.

It is a good approximation to write the analogy to eq. (5.4)

$$(1 - \beta_{35}) = h \text{ (ligands)} \cdot k \text{ (central ion)} \tag{5.22}$$

for which the appropriate functions (pure numbers) are:

F$^-$	0.8	Mn(II)	0.07	
H$_2$O	1.0	V(II)	0.1	
dmf	1.2	Ni(II)	0.12	
urea	1.2	Mo(III)	0.15	
NH$_3$	1.4	Cr(III)	0.20	
en	1.5	Fe(III)	0.24	
ox^{--}	1.5	Rh(III)	0.28	(5.23)
Cl$^-$	2.0	Ir(III)	0.28	
CN$^-$	2.1	Tc(IV)	0.3	
Br$^-$	2.3	Co(III)	0.33	
N$_3^-$	2.4	Mn(IV)	0.5	
I$^-$	2.7	Pt(IV)	0.6	
dtp$^-$	2.8	Pd(IV)	0.7	
dsep$^-$	3.0	Ni(IV)	0.8	

It might be argued that $\beta_{35} = (1 - h')(1 - k')$ would be a better choice of the two functions, protecting more effectively against very small or negative β values. However, there is the truth expressed by eq. (5.22) that if a compound is expected to have an extremely large nephelauxetic effect, it is probably metallic, and the extrapolation may have no meaning anyhow.

It may be noted how sensitive the k-variable in eq. (5.23) is to the oxidation state of a given element. Thus, Mn(II) and Ni(II) show the least pronounced nephelauxetic effect, whereas Mn(IV) and Ni(IV) show some of the strongest effects observed. Quite generally, the most reducing ligands and the most oxidizing central atoms (including the general inequality $k(3d) > k(4d)$ under equal circumstances) have the most pronounced nephelauxetic effect. This is true independent of the fact that Δ is about 45% larger in the 4d group than in the corresponding 3d case, and that the Tanabe-Sugano variable Δ/B (for which SCHÄFFER proposes the name *ligand field strength* $\sum$) is far larger in the 4d group (because of the smaller B values for the gaseous ions) corresponding to nearly exclusive low-spin behaviour with exception of the high-spin Pd(II)F$_6$. By the same token as eq. (5.21), the nephelauxetic series can be extended to octahedral chromophores for which B cannot be determined. Thus, the values of β_{55} from Table 5.6 for hexafluoride complexes MF_6^{+z-6} suggest the nephelauxetic series

$$
\begin{aligned}
Cr(III) > Re(IV) > Tc(IV) > Os(IV) \sim Mn(IV) > Os(V) \\
> Ru(V) > Os(VI) > Ir(VI) > Ru(VI) > Pt(VI)
\end{aligned}
\tag{5.24}
$$

when combined with such values as $(3B + C) = 1.8$ kK and $\beta_{55} = 0.30$ for PtF$_6$ (cf. JØRGENSEN, 1962g and 1967c).

We might restrict ourselves to compare the parameters of interelectronic repulsion in compounds with those for the corresponding gaseous ions. However, we may also want to make a comparison with the values expected as a function of the *fractional charge* of the central atom. From a purely qualitative point of view, there are very few complexes, if any, having σ-anti-bonding orbitals of equal population on the central atom and on the combined set of ligands, as would be expected for fully covalent bonding. It cannot be argued that $(a_3)^2$ would be exactly 0.5 in this special case because of the normalization conditions eq. (5.8). Further on, the electronic density residing on the ligands contribute to a small extent to the parameters of interelectronic repulsion for the partly filled shell. However, there is little doubt that β_{33} would be close to 0.25 and β_{35} close to 0.5 in the extreme case of $b_5 = 0$. The fact that most nephelauxetic ratios β_{35} in Table 5.8 are above 0.5 is a striking argument for the bonding being typically heteronuclear, the ligands having higher electronegativity than the central atom. On the other hand, no well established value of β_{35} is one, and even solid manganese (II) fluorides and the aqua ion in Table 5.7 have β_{35} perceptibly decreased to 0.94 and 0.93.

The writer (1958) discussed the relative importance of the *central-field covalency*, viz. that $\beta_0 = (Z_{*com}/Z_{*gas})$ of eq. (5.12) is below 1 because of expansion of the radial function of the partly filled shell adapting to a smaller effective charge than found in the corresponding gaseous ion, and *symmetry-restricted covalency*, viz. that the Stevens delocalization coefficients a_3 and possibly also a_5 are below one. We can re-write eq. (5.12) on the more compact form:

$$\beta_{55} = a_5^4 \beta_0 \qquad \beta_{35} = a_5^2 a_3^2 \beta_0 \qquad \beta_{33} = a_3^4 \beta_0 . \tag{5.25}$$

It is obvious that the minimum value of β_0 is obtained if it is assumed that $a_5 = a_3 = 1$. In Table 5.8, the column Z_{min} corresponds to the value of β_0 thus obtained. We have to discuss the dependence of B and related parameters on the fractional charge of the central atom. If the ionic charge is written $(Z_0 - 1)$ (i.e. $Z_0 = 4$ for M^{+3}) it is a very good approximation (JØRGENSEN, 1967a) in the 3d group to write (in $K = cm^{-1}$)

$$B = 384 + 58\, q + 124\, Z_0 - \frac{540}{Z_0} \tag{5.26}$$

and in the 4d group having q electrons in the partly filled shell

$$B = 472 + 28\, q + 50\, Z_0 - \frac{500}{Z_0} \tag{5.27}$$

as may be compared with eq. (5.16). It has frequently been assumed that 4d-group ions M^{+z} have B 66% of the values for the corresponding 3d-group ion, and 5d-group ions B 60% of the 3d-value. Surprisingly enough, these values do not deviate much from eq. (5.27) for most common ionic charges. We now extend eqs. (5.26) and (5.27) to non-integral values for Z_0; and further on, we extrapolate the functions outside the interval $2 \leq q \leq 8$ for which they conceivably could have observable consequences.

The question is now whether we should choose one of two alternatives:

1. to consider isoelectronic series, keeping q constant but interpolating between differing elements.

2. to keep the element constant and study the variation of B, allowing q to decrease one unit for each unit increase of Z_0. It may be practical re-writing $B = 384 + 58(Z - 17) + 66 Z_0 - (540/Z_0)$.

The numerical consequences can be seen for an example, viz. Cr(III) with $q = 3$, using B values evaluated according to eq. (5.26):

$$
\begin{array}{llllll}
1. & \text{Sc}^0\ 142 & \text{Ti}^+\ 536 & \text{V}^{++}\ 750 & \text{Cr}^{+3}\ 919 & \text{Mn}^{+4}\ 1070 \\
2. & \text{Cr}^0\ 316 & \text{Cr}^+\ 652 & \text{Cr}^{++}\ 808 & \text{Cr}^{+3}\ 919 & \text{Cr}^{+4}\ 1012
\end{array}
\tag{5.28}
$$

[it may be interesting to compare with the empirical values eq. (5.16)]. Of these alternatives, the second is the most sensible for our purposes. Actually, the element remains constant when going from the gaseous ion to the complex, and the reasonable standard with which to compare the influence of the invading electronic density of the ligands on the partly filled shell is to compare it with the effect on the gaseous ion of adding further d-electrons. This is an important point to make; the fractional charges derived from the nephelauxetic effect are those obtained by comparison with electronic density (e.g. the bonding M.O. of different symmetry types) penetrating to the same radial distribution as the partly filled shell. Essentially external electrons such as 4s electrons in the 3d group or even the 5d electrons in the 4f group have only a moderate influence on the estimate of the fractional charge. There is no lower limit on how little influence sufficiently external electrons can have on the radial extension of the partly filled shell.

In Table 5.8, the value Z_{min} is that value of $(Z_0 - 1)$ which corresponds to the observed value of B for the complex according to the second alternative above. Thus, $\text{Cr(NH}_3)_6^{+3}$ having $B = 650$ K is strictly comparable with Cr^+ in eq. (5.28), and hence, $Z_{min} = 1.0$. By the same token, $B = 725$ K for $\text{Cr(H}_2\text{O})_6^{+3}$ is found by comparison with a graph representing eq. (5.26) to have $Z_{min} = 1.4$. Obviously, the fractional charge of chromium (III) seen from the point of view of the partly filled shell must be somewhere in the interval between Z_{min} and the oxidation number $+3$. The intra-sub-shell transitions observed in d^3-systems constitute an opportunity of narrowing down this interval to some extent. The values of $K = 7B$ and β_{55} given in Table 5.6 can be used for the evaluation of (higher) Z_{min} values given in parentheses in Table 5.8. At this point, exclusively σ-bonding ligands having only one lone-pair and no low-lying (or at least no orbitals highly effective for π-backbonding, such as CO) orbitals such as NH_3, en and CH_3^- should have $a_5 = 1$ and $\beta_{55} = \beta_0$. In sofar this picture has any physical significance, this particular class of σ-bonded d^3-complexes should have the fractional charge close to Z_{min} given in parenthesis in Table 5.8, and a_3^2 should be (β_{35}/β_{55}) according to eq. (5.25).

In principle, there is another technique available for separating the effects of central-field and symmetry-restricted covalency, viz. the relativistic nephelauxetic effect, i. e. the decrease of the Landé parameter ζ_{nl} in complexes compared with the corresponding gaseous ion. Thus, for intra-sub-shell transitions in $(t_{2g})^q$ one expects $\beta_{rel} = a_5^2\beta_0^2$ differing from $\beta_{55} = a_5^4\beta_0$. However, the evidence for the relativistic nephelauxetic effect is not particularly reliable, and in the 3d group, it comes mainly from the g-factors obtained by electron spin resonance (OWEN, 1955) and from the relative intensities of spin-forbidden and spin-allowed absorption bands (JØRGENSEN, 1956a and 1962b). Actually, the band-splittings of terms in octahedral and tetrahedral symmetry are not in satisfactory agreement with the theory for spin-orbit coupling and may have other origins as well. There is no doubt that ζ_{5d} for rhenium, osmium and iridium complexes ~ 3 kK is about half as large as the values expected for the corresponding gaseous ions; but one can only conclude qualitatively (JØRGENSEN and SCHWOCHAU, 1965) that the behaviour of the 5d group hexahalide complexes is compatible with β_0 and a_5^2 having the same order of magnitude.

In Table 5.8, we also give the value Z_{root} introduced (JØRGENSEN, 1967a) as a compromise, ascribing equal importance to central-field and symmetry-restricted covalency by putting $\beta_0 = a_5^2a_3^2$ in eq. (5.25). Whereas Z_{min} was obtained by a direct comparison of the observed value of B with eqs. (5.26) or (5.27), Z_{root} is obtained by a comparison with $B_0(\beta_{35})^{1/2}$ where B_0 is the value predicted for the gaseous ion and β_{35} is the nephelauxetic ratio relative to a slightly different B_{gas}. The deviation is already perceptible for Mn^{++} having $B_{gas} = 890$ K according to Table 5.7 and $B_0 = 866$ K according to eq. (5.26). Actually, the values of B_{gas} used in Table 5.8 are

V^{++} 755	Fe^{++} 930	Ni^{++} 1041	Ru^{++} 620	Ir^{+3} 660
Cr^{+3} 918	Co^{+3} 1100	Mo^{+3} 610	Rh^{+3} 720	Pt^{+4} 750 (5.29)
Mn^{+4} 1064	Ni^{+4} 1250	Tc^{+4} 700	Pd^{+4} 810	

showing some discrepancies in the case of B_0 according to eq. (5.27) being 631 K for Mo^{+3}, 623 K for Ru^{++}, 715 K for Rh^{+3} and 790 K for Pd^{+4}. The choice between the B_{gas} values previously applied for the 4d and 5d group ions and the new extrapolation of B_0 for the 4d group is not easy, because the highest ionic charge experimentally known in the end of the 4d group is Pd^{++} 683 K, whereas eq. (5.27) is based mainly on the iso-electronic series $4d^2$.

However, these variations of one to three percent of the B value for the gaseous ion are not important relative to the other approximations built in our treatment. It is satisfactory that the values of Z_{root} in Table 5.8 are close to the β_{55}-values of Z_{min} in parentheses for the d^3-systems of σ-bonding ligands such as $Cr(NH_3)_6^{+3}$. The 82 values of Z_{root} are distributed:

0.3—0.45	0.5—0.95	1—1.45	1.5—1.95	2.0—2.5	
6	16	34	21	5	(5.30)

with a broad maximum around 1.3. The 82 chromophores considered should be a fairly representative selection of transition group compounds with one exception, however: such complexes having the electron transfer bands at so low wavenumbers that the two spin-allowed inter-subshell transitions cannot be detected are expected to have relatively low values of β_{35} and Z_{root}. It is striking that PAULING's principle of electroneutrality is in disagreement with eq. (5.30); most complexes have Z_{root} in the interval 0.5 to 2 and by no means in the interval 0 to 1.

Of course, one can discuss to what extent eq. (5.7) is a good approximation to the one-electron functions of the partly filled shell. It is quite obvious that the deformation in the bond region between the central atom and the ligands may not be exactly represented by L.C.A.O. of only two orbitals and having definite radial functions belonging to fractional charges of the constituent atoms obtained by interpolation between the radial function appropriate for integral ionic charges. In this sense, the nephelauxetic ratio β_{35} may be more closely connected with an unspecified expansion of the partly filled shell, than is the concept of Z_{root}. However, the argumentation against the electroneutrality principle seems to have a sufficient safety margin to remain valid under all circumstances for most of the complexes in Table 5.8. Much other evidence (e.g. hyperfine structure of ligand nuclei observed in electron spin resonance curves) is available for a moderate delocalization, i.e. b_3 for all complexes and b_5 at least for halide ligands being positive in the region 0.2 to 0.4. On the other hand, it is not so easy to find clear-cut evidence for the effective charge of the central atom being much below the oxidation number. However, the SCHRÖDINGER equation hardly would permit the loosest bound electrons of a transition group complex to retain the same contracted radial function as in the corresponding gaseous ion, and Z_{root} seems finally a sensible compromise.

Another question we have to discuss is whether parameters of interelectronic repulsion such as B really are proportional to the average reciprocal radius $\langle r^{-1} \rangle$ of the partly filled shell. JØRGENSEN (1967a) discussed how WATSON and FREEMAN's 3d, 4d and 4f radial functions in a sequence of identical ionic charge have values of $\langle r^{-1} \rangle$ being linear, and in most cases even proportional, functions of the empirical parameters of interelectronic repulsion. Thus, it is approximately true that

$$d: \ D \sim 7.6\,B \sim (3.0\ \text{kK})\,\langle r^{-1} \rangle$$
$$f: \ D = \tfrac{9}{8}E^1 \sim (2.4\ \text{kK})\,\langle r^{-1} \rangle \tag{5.31}$$

where the reciprocal radius is measured in reciprocal Ångström units. It may be remembered that the product $\langle r \rangle \langle r^{-1} \rangle$ is $^7/_6$ for hydrogenic 3d and $^9/_8$ for hydrogenic 4f orbitals, whereas WATSON and FREEMAN's

Hartree-Fock radial functions have this product between 1.25 and 1.32 showing a certain similarity with hydrogenic 2p functions for which the product is 1.25.

However, the Hartree-Fock functions suggest too large parameters of interelectronic repulsion, and many regularities are known regarding these deviations, such as the *Watson effect* (JØRGENSEN, 1962a and 1962e) that in an isoelectronic series, most term distances are decreased in the actual gaseous ion by an amount independent of the ionic charge. Consequently, the linear relation between B or D and $\langle r^{-1} \rangle$ is much better for highly positive ions than for neutral atoms having the (excited) configuration $[Ar]3d^q$ (JØRGENSEN, 1967a). It may be argued that the Watson effect is built in the hyperbolic parts $(-540/Z_0)$ and $(-500/Z_0)$ of eqs. (5.26) and (5.27) producing a dramatic decrease of parameters of interelectronic repulsion when the neutral atoms with $Z_0 = 1$ are approached. This may also be connected with the fact that mononegative anions $[Ar]3d^q$ having $Z_0 = 0$ undoubtedly have zero ionization energy. If we accepted modifications of eqs. (5.26) and (5.27) involving only linear and quadratic terms of Z_0, we would obtain lower values of Z_{min} and Z_{root} in Table 5.8. However, the non-linear behaviour going from M to M^+ to M^{++} cannot be disputed away and seems to be something very characteristic for the transition groups. On the other hand, the variation across a transition group, varying q for a sequence of identical ionic charge, seems to be perfectly linear.

The comparative study of parameters of interelectronic repulsion has shown the tendency toward increasing average radii $\langle r \rangle$

$$4f < 3d < 5f < 4d < 5d \tag{5.32}$$

under comparable circumstances (ionic charge and number q of electrons in the partly filled shell). It cannot be argued that the extent of the interaction with the ligands is well represented by eq. (5.32). Actually, the ionic radii are much larger in the 4f group protecting the 4f shell against strong anti-bonding effects, and the ionic radii are unusually small in the 3d group. Though Δ increases in the direction $3d < 4d < 5d$, the nephelauxetic effect is slightly more pronounced in the 3d than in the 4d group. This is connected with the more oxidizing character of the 3d central atoms to be discussed in chapter 7.

Table 5.9 gives average radii in Å derived from eq. (5.31) and the assumption $\langle r \rangle \langle r^{-1} \rangle = 1.30$ under the (admittedly only approximate) form $\langle r \rangle = (514 \, K)/B$. It is seen that $\langle r \rangle$ for gaseous Mn^{++} and Ni^{++} are smaller than the ionic radii and increase only slowly when the ligands are strongly reducing. $\langle r \rangle$ for Cr^{+3}, Co^{+3} and Rh^{+3} are nearly the same as the ionic radii, and the variation of the nephelauxetic effect is more pronounced. In the 4f group, $\langle r \rangle$ is some 0.5 to 0.6 Å smaller than Gold-schmidt's ionic radii.

Already WERNER (1912) pointed out the hypsochromic influence of the nearest neighbour atoms on the colour of Co(III) complexes

$$I < Br < Cl < O < S < N < C \qquad (5.33)$$

later forming the basis for TSUCHIDA's *spectrochemical series* (cf. SHIMURA and TSUCHIDA, 1956) which is now understood as the variation of the sub-shell energy difference Δ as a function of the ligands. However, WERNER used NH_3 and aliphatic amines as the examples for the ligating atom N in eq. (5.33) and SO_3^- for S. If ligands having more than one lone-pair are considered, the values of Δ can be much smaller, as can be seen for N_3^-, dtp$^-$ and dtc$^-$ in Table 5.8 which have a spectrochemical position between Cl$^-$ and F$^-$ as also expressed by the function f in eq. (5.5). The spectrochemical variation of sulphur-containing ligands is particularly impressive (JØRGENSEN, 1962f).

Table 5.9. *Average radius $\langle r \rangle$ of the partly filled d shell in Å estimated from the expression (514 K)/B, and compared with the ionic radii. Ionic radii in parentheses have been derived from hexahalide complexes by* GOLDSCHMIDT's *method, and values with one decimal are extrapolated*

M^{+z}	6 F	6 NH_3	6 Cl	6 S	GOLDSCHMIDT's ionic radius
Cr(III)	0.56	0.63	0.79	0.92	1.25	0.64
Mn(IV)	0.48	0.86	—	—	—	0.52
Mn(II)	0.58	0.61	0.64	0.65	0.69	0.91
Fe(III)	0.48	0.61	—	0.81	—	0.67
Co(III)	0.47	—	0.83	—	1.28	0.47
Ni(IV)	0.41	1.14	—	—	—	(0.38)
Ni(II)	0.49	0.54	0.58	0.66	—	0.78
Mo(III)	0.84	—	—	1.17	—	0.8
Tc(IV)	0.73	0.95	—	—	—	0.7
Rh(III)	0.71	—	1.19	1.46	2.44	0.68
Pd(IV)	0.63	1.51	—	—	—	(0.49)
Ir(III)	0.78	—	—	1.71	3.2	0.7
Pt(IV)	0.68	1.35	—	—	—	(0.53)

The fact that central atoms such as Cr(III) or Co(III) form octahedral complexes of all conceivable colours, green, blue, violet, purple, red, orange and yellow, show clearly that the ligands have a profound influence on the energy levels of d^3 and d^6 quite different from the situation for f^q systems. However, with the exception of the $^6S - {}^4G$ transition in high-spin d^5 systems compiled in Table 5.7, the direct evaluation of the parameters of interelectronic repulsion and their decrease, the nephelauxetic effect, is not generally possible in d^q systems. The parameters such as B and the ratios such as β_{35} are only obtained after a certain mathematical treatment of the observed wavenumbers of the maxima of the absorption bands. We cannot subscribe to TSUCHIDA's original idea that the origin of the two first spin-allowed bands of Cr(III) and Co(III) complexes is different. However, it is very interesting that KIDA and YONEDA (1955) in their pioneer work on the absorption spectra of complexes of sulphur-containing ligands such as xanthates and dithiocarbamates pointed out that $(\sigma_2 - \sigma_1)$ is unusually small for Cr(III) and Co(III) compared with the situation for more common ligands.

In principle, there is no absolute difference between the ligand field description of complexes containing a partly filled p and a partly filled d shell. Most p group complexes are low-spin. JØRGENSEN (1967c) discussed the behaviour of linear p^4-systems MX_2 where the two lone-pairs form an equatorial belt around the central atom M. It is possible that one can identify Laporte-forbidden $5p\pi \to 5p\sigma$ transitions

$$ICl_2^- \quad 29.8\ kK \qquad IBr_2^- \quad 27.0\ kK \qquad XeF_2 \quad 43.0\ kK \qquad (5.34)$$

but the difficulty is that these species sometimes are in equilibrium with diatomic molecules XY having weak transitions (between M.O. delocalized on both atoms) occurring at comparable wavenumbers. Quadratic p^2-systems MX_4 also seem to have Laporte-forbidden transitions, this time from the lone-pair $5p\sigma$ to the two empty, σ-anti-bonding orbitals mainly consisting of $5p\pi$:

$$ICl_4^- \quad 29.4\ kK \qquad XeF_4 \quad 38.8\ and\ 43.9\ kK \qquad (5.35)$$

but this time, the difficulty is to make a distinction from electron transfer from filled M.O. mainly localized on the four ligands to the two empty $5p\pi$ orbitals of the central atom. Anyhow, eq. (5.35) gives a lower limit to the sub-shell energy difference in these quadratic low-spin complexes of iodine (III) and xenon (IV).

Under certain circumstances, the p-shell seems to be much less influenced by the neighbour atoms. Thus, the atomic transition $^1S_0(6s^2) \to {}^3P_1(6s6p)$ of mercury atoms near 39.41 kK can be recognized in many solvents and solid matrices. DELBECQ, GHOSH and YUSTER (1967) were able to produce thallium atoms by irradiation of Tl(I) in potassium halide crystals. Two weak, narrow absorption bands in the infra-red correspond to the transition between the two possible J-levels $^2P_{1/2}$

and $^2P_{3/2}$ belonging to the configuration $6s^26p$. The decreased wave-numbers (in kK)

Tl⁰ gaseous	KCl	KBr	KI	
7.793	6.80, 7.70	6.33, 6.99	5.90, 6.45	(5.36)

indicate a weak expansion and/or delocalization of the 6p orbitals.

N. J. Bjerrum et al. (1967a, 1967b) studied the species Bi^+ in molten mixtures of $AlCl_3$ and $NaCl$ and of $AlBr_3$ and $NaBr$. The measurements are difficult because of the tendency toward formation of strongly coloured cluster-complexes such as Bi_5^{+3}. Bi^+ is a high-spin $6p^2$-system but is nevertheless diamagnetic of relativistic reasons because the non-degenerate groundstate $^3P_0$ is highly separated from the other levels of the configuration $6p^2$. The narrow absorption bands occur (in kK)

	Bi⁺ gaseous	Bi(I) chloride	bromide	
$^3P_1$	13.324	11.1, 14.4	10.8, 13.9	(5.37)
$^3P_2$	17.030	17.1	16.4	
$^1D_2$	33.936	30.0, 32.5	—	

The small splittings of $^3P_1$ and, in particular, the nephelauxetic effect indicate a certain effect of the neighbour atoms. Actually, the Landé parameter ζ_{6p} and the parameter of interelectronic repulsion (cf. Table 3.5) $K_{av}(6p, 6p) = \frac{3}{25} F^2 = 3 F_2$ have the values in kK

	Bi⁺ gaseous	Bi(I) chloride	bromide	
ζ_{6p}	11.68	10.72	10.40	(5.38)
$K_{av} (6p, 6p)$	3.52	3.17	2.93	

again suggesting a moderate expansion and/or delocalization of the 6p shell.

Species such as Bi(I) fall into the same line of description as the transition group complexes containing a partly filled d or f shell. It may be asked where are the limits for the application of ligand field theory to such cases. If it were argued that the separability of the orbitals in the product of a radial function and a hydrogenic angular function was necessary, already most d-group complexes would not be satisfactory examples. However, the delocalization of the M.O. formed conserves the number "l" of angular node-planes, and in many ways, the characteristic features of the distribution of excited levels remain recognizable.

Probably, the moment where ligand field theory seriously breaks down is when the chemical bonding effects are comparable with the energy differences between different nl-shells. Thus, in CO, the chemical bonding energy is not negligible compared with the energy difference between 2s and 2p, and in M.O. having the appropriate symmetry (σ and not higher λ values) the intermixing of 2s and 2p character is known to be considerable. By the same token, the oxidation states N(III), P(III), S(IV), Cl(V), As(III), Se(IV), Br(V), Sn(II), Sb(III), Te(IV), I(V), Xe(VI), Tl(I), Pb(II) and Bi(III) have the preponderant configuration s^2 of the corresponding gaseous ions N^{+3},... but the stereochemistry suggests a frequent strong admixture of s and p character. As discussed in the next chapter (cf. also WALTON et al., 1967) one frequently observes a weaker and a stronger transition in such complexes, which to a good approximation can be described as going from $^1S_0$ (s^2) to $^3P_1$ and $^1P_1$ belonging to (sp). It is even possible (cf. JØRGENSEN, 1962a and 1962b) to evaluate parameters ζ_{np} and $K(s, p)$ indicating a more pronounced nephelauxetic effect than for Bi(I) in eq. (5.38). However, the separation in s and p orbitals is certainly much less clear-cut in the s^2-family than, say, the separation between 4f and 5d orbitals in the lanthanides. On the other hand, partly filled p-shells can be said to occur in the post-transition elements though nearly all complexes known are low-spin. Actually, an interesting analogy occurs between quadratic p^2- and d^8-chromophores having a lone-pair ($p\sigma$ or $d\sigma$) perpendicular to the plane of the four ligands.

Bibliography

ABRAGAM, A., and M. H. L. PRYCE: Proc. Roy. Soc. (London) A **206**, 173 (1951).

BALT, S., and A. M. A. VERWEY: Spectrochim. Acta **23 A**, 2069 (1967).

BECK, W., and K. FELDL: Z. Anorg. Allgem. Chem. **341**, 113 (1965).

—, W. P. FEHLHAMMER, P. PÖLLMANN, E. SCHUIERER, and K. FELDL: Chem. Ber. **100**, 2335 (1967).

BETHE, H.: Ann. Physik [5] **3**, 133 (1929).

BJERRUM, N. J., C. R. BOSTON, and G. P. SMITH: Inorg. Chem. **6**, 1162 (1967a).

—, H. L. DAVIS, and G. P. SMITH: Inorg. Chem. **6**, 1603 (1967b).

BLASSE, G., and A. BRIL: J. Electrochem. Soc. **114**, 1306 (1967).

BOSTRUP, O., and C. K. JØRGENSEN: Acta Chem. Scand. **11**, 1223 (1957).

BOULANGER, F.: Ann. Chim. (Paris) [12] **7**, 732 (1952).

BROWN, D. H.: Spectrochim. Acta **19**, 1683 (1963).

—, D. R. RUSSELL, and D. W. A. SHARP: J. Chem. Soc. (A) 18 (1966).

BRYANT, B. W.: J. Opt. Soc. Am. **55**, 771 (1965).

CARLSON, E. H., and G. H. DIEKE: J. Chem. Phys. **34**, 1602 (1961).

CARNALL, W. T., and B. G. WYBOURNE: J. Chem. Phys. **40**, 3428 (1964).

CARTER, W. J.: private communication.

CASPERS, H. H., H. E. RAST, and R. A. BUCHANAN: J. Chem. Phys. **43**, 2124 (1965).

CHANG, N. C.: J. Chem. Phys. **44**, 4044 (1966).

CHRIST, K.: Spektralphotometrische und magnetische Untersuchungen an Molybdän(III) Komplexverbindungen. Inaugural-Dissertation, Frankfurt am Main, 1965.

CLARK, R. J. H.: J. Chem. Soc. 417 (1964).

—, M. L. GREENFIELD, and R. S. NYHOLM: J. Chem. Soc. (A) 1254 (1966).

CONWAY, J. G.: J. Chem. Phys. 41, 904 (1964).

CROSSWHITE, H. M., G. H. DIEKE, and W. J. CARTER: J. Chem. Phys. 43, 2047 (1965).

DAY, J. P., and L. M. VENANZI: J. Chem. Soc. (A) 197 (1966).

DE ARMOND, K., and L. S. FORSTER: Spectrochim. Acta 19, 1393 and 1403 (1963).

DELBECQ, C. J., A. K. GHOSH, and P. H. YUSTER: Phys. Rev. 154, 797 (1967).

DINGLE, R., and R. A. PALMER: Theoret. Chim. Acta 6, 249 (1966).

DRICKAMER, H. G., and J. C. ZAHNER: Advan. Chem. Phys. 4, 161 (1962).

— Solid State Phys. 17, 1 (1965).

DUNKEN, H., u. G. MARX: Z. Chem. 6, 436 (1966).

EPHRAIM, F., u. R. BLOCH: Ber. Deut. Chem. Ges. 59, 2692 (1926) and ibid. 61, 65 and 72 (1928).

ERATH, E. H.: J. Chem. Phys. 34, 1985 (1961) [and erratum, ibid. 38, 1787 (1963)].

FISCHER, R. D., and H. FISCHER: J. Organometall. Chem. 8, 155 (1967).

FORSTER, D., and D. M. L. GOODGAME: J. Chem. Soc. 2790 (1964).

FORSTER, L. S.: Transition Metal Chem., in press.

FURLANI, C., E. CERVONE, and F. D. CAMASSEI: Inorg. Chem. 7, 265 (1968).

GILL, N. S.: J. Chem. Soc. 3512 (1961).

GINSBERG, A. P., and M. B. Robin: Inorg. Chem. 2, 817 (1963).

GOBRECHT, H.: Ann. Physik [5] 31, 181 and 755 (1938).

GRUBER, J. B., and R. A. SATTEN: J. Chem. Phys. 39, 1455 (1963).

, J. R. HENDERSON, M. MURAMOTO, K. RAJNAK, and J. G. CONWAY: J. Chem. Phys. 45, 477 (1966).

GRUEN, D. M.: J. Chem. Phys. 20, 1818 (1952).

HATFIELD, W. E., R. C. FAY, C. E. PFLUGER, and T. S. PIPER: J. Am. Chem. Soc. 85, 265 (1963).

HEIDT, L. J., G. F. KOSTER, and A. M. JOHNSON: J. Am. Chem. Soc. 80, 6471 (1958).

HELLWEGE, K. H.: Ann. Physik [6] 4, 95, 127, 136, 143, 150 and 357 (1948).

HOFMANN, K. A., u. H. KIRMREUTHER: Z. Physik. Chem. 71, 312 (1910).

HOWELL, I. V., L. M. VENANZI, and D. C. GOODALL: J. Chem. Soc. (A) 395 (1967).

JENSEN, K. A., and V. KRISHNAN: Acta Chem. Scand. 21, 2904 (1967) and private communication.

JØRGENSEN, C. K.: Acta Chem. Scand. 9, 1362 (1955).

— Acta Chem. Scand. 10, 887 (1956a).

— Mat. Fys. Medd. Dan. Vid. Selskab 30, no. 22 (1956b).

— Acta Chem. Scand. 11, 53 (1957a).

— Acta Chem. Scand. 11, 981 (1957b).

— Discussions Faraday Soc. 26, 110 (1958).

— Mol. Phys. 2, 96 (1959).

— Solid State Phys. 13, 375 (1962a).

— Absorption Spectra and Chemical Bonding in Complexes. Oxford: Pergamon Press 1962b. U. S. distributor: Addison-Wesley.

— Acta Chem. Scand. 16, 793 (1962c).

— Mol. Phys. 5, 485 (1962d).

— Orbitals in Atoms and Molecules. London: Academic Press 1962e.

— J. Inorg. Nucl. Chem. 24, 1571 (1962f).

— Progr. Inorg. Chem. 4, 73 (1962g).

— Acta Chem. Scand. 17, 251 (1963e).

— Advan. Chem. Phys. 5, 33 (1963b).

—, R. PAPPALARDO, and H. H. SCHMIDTKE: J. Chem. Phys. 39, 1422 (1963).

— —, and E. Rittershaus: Z. Naturforsch. 19 a, 424 (1964).
— — — Z. Naturforsch. 20 a, 54 (1965).
—, and K. Schwochau: Z. Naturforsch. 20 a, 65 (1965).
—, R. Pappalardo, et J. Flahaut: J. Chim. Phys. 62, 444 (1965).
— Struct. Bonding 1, 3 (1966).
— Helv. Chim. Acta Fasciculus extraordinarius Alfred Werner, 131 (1967 a).
— Chem. Phys. Letters 1, 11 (1967 b).
—, and E. Rittershaus: Mat. Fys. Medd. Dan. Vid. Selskab 35, no. 15 (1967).
— In V. Gutmann (ed.), Halogen Chemistry 1, 265. London: Academic Press 1967 c.
— Lanthanides and 5f Elements. London: Academic Press 1969.
Johnston, D. R., R. A. Satten, C. L. Schreiber, and E. Y. Wong: J. Chem. Phys. 44, 3141 (1966).
Jones, G. D., and W. A. Runciman: Proc. Phys. Soc. 76, 996 (1960).
Judd, B. R.: Operator Techniques in Atomic Spectroscopy. New York: McGraw-Hill 1963.
Kida, S., and H. Yoneda: Nippon Kagaku Zasshi 76, 1059 (1955).
Kiess, N. H., and G. H. Dieke: J. Chem. Phys. 45, 2729 (1966).
Kingsley, J. D., J. S. Prener, and B. Segall: Phys. Rev. 137, A 189 (1965).
Kisliuk, P., W. F. Krupke, and J. B. Gruber: J. Chem. Phys. 40, 3606 (1964).
Koide, S., and M. H. L. Pryce: Phil. Mag. 3, 607 (1958).
Komorita, T., S. Miki, and S. Yamada: Bull. Chem. Soc. Japan 38, 123 (1965).
Kuse, D., and C. K. Jørgensen: Chem. Phys. Letters 1, 314 (1967).
Lang, R. J.: Can. J. Res. 14 A, 127 (1936).
Lawson, K. E.: J. Chem. Phys. 44, 4159 (1966).
— J. Chem. Phys. 47, 3626 (1967).
van Leeuwen, P. W. N. M.: Inorganic Complexes of Ligands Containing the Thionyl Group. Thesis, Leiden University, 1967.
Lewis, J., R. S. Nyholm, and P. W. Smith: J. Chem. Soc. 4590 (1961).
Linz, A., and R. E. Newnham: Phys. Rev. 123, 500 (1961).
Lohr, L. L.: J. Chem. Phys. 45, 3611 (1966).
Makovsky, J., W. Low, and S. Yatsiv: Phys. Letters 2, 186 (1962).
Meggers, W. F.: J. Opt. Soc. Am. 37, 988 (1947) and J. Res. Nat. Bur. Stand. 71A, 396 (1967).
Mercer, E. E., and R. R. Buckley: Inorg. Chem. 4, 1692 (1965).
Moffitt, W., G. L. Goodman, M. Fred, and B. Weinstock: Mol. Phys. 2, 109 (1959).
Neuhaus, A.: Z. Krist. 113, 195 (1960).
Oelkrug, D.: Ber. Bunsenges. Physik. Chem. 70, 736 (1966).
Orgel, L. E.: J. Chem. Phys. 23, 1824 (1955).
— Nature 179, 1348 (1957).
— Introduction to Transition-Metal Chemistry. London: Methuen 1960. 2. Edition 1966. U. S. distributor: John Wiley.
— Mol. Phys. 7, 397 (1964).
Owen, J.: Proc. Roy. Soc. (London) A 227, 183 (1955).
Pappalardo, R., and C. K. Jørgensen: Helv. Phys. Acta 37, 79 (1964).
Piper, T. S., and N. Koertge: J. Chem. Phys. 32, 559 (1960).
Poole, C. P.: J. Phys. Chem. Solids 25, 1169 (1964).
Porter, G. B., and H. L. Schläfer: Ber. Bunsenges. Physik. Chem. 68, 316 (1964).
Prather, J. L.: Atomic Energy Levels in Crystals. Nat. Bur. Stand. Monograph no. 19, Washington, D. C. (1961).
Pryce, M. H. L., and W. A. Runciman: Discussions Faraday Soc. 26, 34 (1958).
Rakestraw, J. W., and G. H. Dieke: J. Chem. Phys. 42, 873 (1965).

REEDIJK, J., P. W. N. M. VAN LEEUWEN, and W. L. GROENEVELD: Rec. Trav. Chim. 87, 129 (1968).
REINEN, D.: Ber. Bunsenges. Physik. Chem. 69, 82 (1965).
— Theoret. Chim. Acta 5, 312 (1966).
— Struct. Bonding 6, in press (1969).
REISFELD, M. J., and G. A. CROSBY: Inorg. Chem. 4, 65 (1965).
—, L. B. ASPREY, and R. A. PENNEMAN: J. Mol. Spectr., in press (1968).
RICHMAN, I., and E. Y. WONG: J. Chem. Phys. 37, 2270 (1962).
RÜDORFF, W., J. KÄNDLER, u. D. BABEL: Z. Anorg. Allgem. Chem. 317, 261 (1962).
RYAN, J. L., and C. K. JØRGENSEN: Mol. Phys. 7, 17 (1963).
— — J. Phys. Chem. 70, 2845 (1966).
SCHÄFFER, C. E., and C. K. JØRGENSEN: J. Inorg. Nucl. Chem. 8, 143 (1958).
— J. Inorg. Nucl. Chem. 8, 149 (1958).
— Proc. Roy. Soc. (London) A 297, 96 (1967).
SCHEIBE, G.: Z. Elektrochem. 64, 784 (1960).
SCHLÄFER, H. L.: Z. Physik. Chem. 3, 222 (1955).
— Z. Physik. Chem. 6, 201 (1956).
—, H. GAUSMANN, and H. WITZKE: J. Mol. Spectr. 21, 125 (1966).
— — — J. Chem. Phys. 46, 1423 (1967).
— —, and H. U. ZANDER: Inorg. Chem. 5, 1528 (1967).
SCHMIDTKE H. H., and D. GARTHOFF: Helv. Chim. Acta 49, 2039 (1966).
— — J. Am. Chem. Soc. 89, 1317 (1967).
— Ber. Bunsenges. Physik. Chem. 71, 1138 (1967).
SCHMITZ-DU MONT, O., and D. REINEN: Z. Elektrochem. 63, 978 (1959) [and erratum, ibid. 64, 330 (1960)].
—, and H. KASPER: Monatsheft 95, 1433 (1964).
SCHNEIDER, W.: Helv. Chim. Acta 46, 1842 (1963).
SCHROEDER, K. A.: J. Chem. Phys. 37, 2553 (1962).
SHIMURA, Y., and R. TSUCHIDA: Bull. Chem. Soc. Japan 29, 311 (1956).
STEVENS, K. W. H.: Proc. Roy. Soc. (London) A 219, 542 (1953).
STURGE, M. D.: Phys. Rev. 130, 639 (1963).
SUGAR, J.: J. Opt. Soc. Am. 53, 831 (1963).
— J. Opt. Soc. Am. 55, 1058 (1965).
TANABE, Y., and S. SUGANO: J. Phys. Soc. Japan 9, 753 and 766 (1954).
TREES, R. E.: J. Opt. Soc. Am. 54, 651 (1964).
UNDERHILL, A. B., and D. C. MORTON: Science 158, 1273 (1967).
WALTON, R. A., R. W. MATTHEWS, and C. K. JØRGENSEN: Inorganica Chim. Acta 1, 355 (1967).
WEINSTOCK, B., and G. L. GOODMAN: Advan. Chem. Phys. 9, 169 (1965).
WERNER, A.: Ann. Chem. 386, 31 (1912).
WESTLAND, A. D., R. HOPPE, and S. S. I. KASENO: Z. Anorg. Allgem. Chem. 338, 319 (1965).
WONG, E. Y.: J. Chem. Phys. 34, 1989 (1961).
—, and I. RICHMAN: J. Chem. Phys. 36, 1889 (1962).
WOOD, D. L., J. FERGUSON, K. KNOX, and J. F. DILLON: J. Chem. Phys. 39, 890 (1963).
WYBOURNE, B. G.: Spectroscopic Properties of Rare Earths. New York: Interscience (John Wiley) 1965.
YOSHINO, Y., Y. TSUNODA, and A. OUCHI: Bull. Chem. Soc. Japan 34, 1194 (1961).

6. Inter-Shell Transitions

The absorption bands discussed in chapter 5 are caused by transitions essentially between levels belonging to the same electron configuration with *one* partly filled l-shell containing from 1 to $4l+1$ electrons. It is true that the width of absorption bands corresponding to change of the sub-shell occupation number at the same time indicate somewhat different spatial extension of the different sub-shells. However, the agreement of the spectra of octahedral complexes with the Tanabe-Sugano diagrams shows that the characteristics of the partly filled d-shell are retained to a large degree.

One would not at first expect $l \rightarrow l'$ inter-shell transitions to be frequently observed in condensed states of matter. In gaseous molecules, these transitions are called *Rydberg transitions* (for a recent review, see MULLIKEN, 1964) because the excited states frequently form a series having energies below the ionization limit with a roughly constant quantum defect δ of eq. (3.14). From the point of view of quantum chemistry, this can be understood because the model of the "united atom" becomes valid when the average radius of the Rydberg orbital becomes far larger than the radius of the molecule. Then, the firmly bound electrons of the molecule play the same rôle as the positive core of an alkali metal atom.

However, the 4f and 5f groups definitely show $f \rightarrow d$ excitations. FREED (1931) was the first to point out that cerium (III) in crystalline $Ce(H_2O)_9 (C_2H_5SO_4)_3$ has intense bands in the ultra-violet which can be ascribed to $4f \rightarrow 5d$ transitions. In aqueous solution, six bands occur (JØRGENSEN and BRINEN, 1963) at 33.7, 39.6, 41.7, 45.1, 47.4 and 50.0 kK. There has been some discussion as to whether the first band is due to some other aqua ion in low concentration having a smaller coordination number than $Ce(H_2O)_9^{+3}$ since its intensity is only $2-3\%$ of the following bands. However, if the angular overlap model is applied on the enneaaqua ion (JØRGENSEN et al., 1963) the most reasonable agreement is obtained by ascribing the first five bands to $5d\sigma$, the two relativistic ω-components of $5d\delta$ and the two of $5d\pi$. Actually, a definite species of arbitrary low symmetry cannot have more than five levels of a d-shell with one electron. Hence, the sixth band derives most probably from the 6s though it is surprising that it is not broader than the five first bands. It is important to note that in gaseous Ce^{+3}, the two J-levels of 5d are situated at 49.74 and 52.23 kK and 6s at 86.60 kK. The decrease of the energy difference 4f — 5d in Ce(III) compounds relative to Ce^{+3} is accentuated in complexes of organic anions such as citrate, tartrate and ethylenediaminetetraacetate (JØRGENSEN, 1956a). Actually, in most chromophores of low symmetry, all five 5d-orbitals are σ-anti-bonding to some extent. This is not the case for octahedral complexes, and $CeCl_6^{-3}$ has a strong band already at 30.3 kK and $CeBr_6^{-3}$ at 29.15 kK (RYAN and JØRGENSEN, 1966). Though it will be argued below that most inter-shell transitions have wavenumbers varying with the ligands in much the same way as the nephelauxetic series (5.23) there is no doubt that the main reason for the low wavenumbers in the hexahalide complexes is the absence of σ- (though not π-) anti-bonding effects on the lowest 5d subshell. The same is true for Ce(III) in CaF_2 having a moderately broad band at 32.5 kK developing vibrational structure and a pronounced sharp origin at 31.95 kK when measured at low temperatures (KAPLYANSKII, MEDVEDEV and FEOFILOV, 1963; STRUCK and HERZFELD, 1966). LOH (1967) reported three broader bands at higher energy for Ce(III) in

$$
\begin{array}{lll}
CaF_2 & 32.5 & 49.5, 51.1, 53.9 \\
SrF_2 & 33.6 & 49.0, 50.0, 53.4 \\
BaF_2 & 34.2 & 50, \quad 51.7, 53.5
\end{array}
\tag{6.1}
$$

and suggested that they correspond to the four 5d subshells of tetragonally distorted $Ce(III)F_8$. This is not unreasonable because the d^1-systems Sc(II), Y(II) and La(II) are known to have three broad bands in the visible when substituting Ca(II) in CaF_2 (O'CONNOR and CHEN, 1963). However, this is an effect of Jahn-Teller type, other d^1-systems such as $Ti(H_2O)_6^{+3}$ and VCl_4 are also known to have instantaneous symmetry much lower than cubic. In the writer's opinion, it is more probable that the three adjacent bands of eq. (6.1) correspond to transitions to the higher, σ-anti-bonding sub-shell (xy, xz, yz) which may be separated by a moderate distortion of the ground potential surface, not so much because of Jahn-Teller effect operating as because the ionic radii of M(II) and M(III) are not the same, making the adaption of Ce(III) in the crystal lattice less perfect. LOH's spectra are measured at so low a concentration (0.005%) of Ce(III) that the charge-compensating entities (interstitial fluoride or substitutional oxide) are not expected to occur in the immediate vicinity. The sub-shell energy difference — Δ should then be 19.0 kK in CaF_2, 17.2 kK in SrF_2 and 17.5 kK in BaF_2. The average energy of the five 5d-orbitals are at 43.5, 44.0 and 44.7 kK in the three crystals, somewhat below 51.23 kK in Ce^{+3}. LOH (1967) identified weak, broad bands ~ 70 kK of Ce(III) in CaF_2 with 4f → 6s transitions. Other weak bands of various M(III) have recently been ascribed by LOH (private communication) to 4f → 6p excitations.

Praseodymium (III) aqua ions have $4f^2 \rightarrow 4f5d$ transitions at 46.8 and 52.9 kK (JØRGENSEN and BRINEN, 1963) and terbium (III) a spin-forbidden transition from $^7F_6$ of $4f^8$ to ^9D-levels of $4f^75d$ at 39 kK and a spin-allowed transition to ^7D-levels at 45.9 kK. The two latter types of transitions can be recognized (RYAN and JØRGENSEN, 1966) in $TbCl_6^{-3}$ at 36.8 and 42.75 kK. The corresponding value of the parameter of interelectronic repulsion $K_{av}(4f, 5d)$ defined in eq. (3.35) is 0.75 kK whereas it is known to be 1.0 kK in the isoelectronic gaseous ion Gd^{++}.

The occurrence of $4f \rightarrow 5d$ transitions below 50 kK in Tb(III) is a typical half-filled shell effect quantitatively expressed in eq. (4.32). Fortunately, LOH (1966) measured the transitions in all the 4f group M(III) in CaF_2 and found a striking agreement with eq. (4.32) as seen in Table 6.1. The calculated value assumes the one-electron part of the wavenumber in kK of the first $4f^q \rightarrow 4f^{q-1}5d$ transition to be $27 + 4.3q$, and the spin-pairing energy parameter $D = 6.5$ kK and E^3 and ζ_{4f} of eq. (4.32) known from the internal transitions in $4f^q$. It may be noted that we are neglecting the quantity $K_{av}(4f, 5d)$. The largest spreading, 8 times this parameter, occurs for ^9D and ^7D of $4f^75d$. The intense transition to ^7D is situated only $K_{av}(4f,5d)$ above the result neglecting this effect, whereas ^9D is situated $7K_{av}$ below. Quite generally, in such situations, the lowest term having $S = S' + \frac{1}{2}$ is shifted $-2S'K_{av}$ and the lowest term having $S = S' - \frac{1}{2}$ is shifted $+ K_{av}$, where S' is the total spin of the partly filled shell with lowest average radius.

The fact that CaF_2 is transparent to above 80 kK is of considerable help in identifying the $4f \rightarrow 5d$ transitions. On the other hand, M(II)

Table 6.1. *Wavenumbers in kK of the first intense $4f \rightarrow 5d$ transition of M(II) and M(III) in CaF$_2$. The calculated values are derived from the parameters given in the text. If the groundstate belongs itself to $4f^{q-1}5d$ negative values are indicated*

q =		obs.	calc.		obs.	calc.
1	La(II)	<0	−16	Ce(III)	32.5	32
2	Ce(II)	<0	− 6	Pr(III)	45.6	45
3	Pr(II)	4	+ 2	Nd(III)	55.9	54
4	Nd(II)	7	+ 4	−	−	56
5	−	−	+ 5	Sm(III)	59.5	57
6	Sm(II)	16	+13	Eu(III)	68.5	66
7	Eu(II)	28	+22	Gd(III)	>78	78
8	Gd(II)	<0	− 8	Tb(III)	46.5	39
9	Tb(II)	?	+ 3	Dy(III)	58.9	54
10	Dy(II)	10.4	+11	Ho(III)	64.1	62
11	Ho(II)	11.1	+11	Er(III)	64.2	62
12	Er(II)	10.9	+11	Tm(III)	64	62
13	Tm(II)	17	+19	Yb(III)	70.7	71
14	Yb(II)	27.5	+29	Lu(III)	>80	84

have these absorption bands at much lower wavenumbers. BUTEMENT (1948) found broad, intense $4f^6 \rightarrow 4f^5 5d$ bands of the red samarium (II) aqua ion at 17.9, 21.1, 23.4 and 30.2 kK and the analogous bands of europium (II) at 31.2 and 40.4 kK and of ytterbium (II) at 28.4 and 40.6 kK. All these three species evolve H_2 with water, though Eu(II) very slowly, and there is not much hope for preparing other 4f group M(II) in aqueous solution. Hence, it is very interesting that McCLURE and KISS (1963) succeeded in making all M(II) in CaF_2. Actually, La(II), Ce(II) and Gd(II) contain one 5d electron already in the groundstate and show very broad bands, possibly due to Laporte-allowed $4f^q 5d \rightarrow 4f^{q+1}$ transitions (La^{++} has ^2F levels at 7.20 and 8.70 kK above the ^2D ground level) and Tb(II) is probably in the same situation. However, evidence from electron spin resonance indicates that the groundstate of Pr(II) in CaF_2 belongs to $4f^3$. Table 6.1 gives McCLURE and KISS' lowest value for such transitions in kK and the values calculated from the one-electron contribution $-21 + 3.8q$ and $D = 5.2$ kK. On the average, the first $4f^q \rightarrow 4f^{q-1} 5d$ transition occurs some 52 kK earlier in M(II) than in M(III). The first $4f^2 5d$ level of Pr^{++} is situated 12.85 and the first $4f^2 6s$ level 28.40 kK above the groundstate belonging to $4f^3$ (SUGAR, 1963). Quite generally, the lower 5d sub-shell seems to be some 10 kK lower in M(II) than in M^{+2} and some 15 kK lower in M(III) than in M^{+3}. DUPONT (1967) found the energy levels of the term ^7F of $4f^6$ at 0.2% higher energy in Sm^{++} than of $4f^6 6s^2$ in Sm^0. DUPONT compares with the rich experimental material available on Sm(II) in crystals. In some cases, the $^5D_0$ level of $4f^6$ (situated at 14.62 kK in SrF$_2$, 14.45 kK in LaCl$_3$ and 14.44 kK in LaBr$_3$) is slightly below the first $4f^5 5d$ level, whereas the latter has lower energy (14.37 kK) in CaF_2. Obviously, this situation has a great importance for the luminescence, though $4f^{q-1} 5d$ levels also may produce narrow line emission. Thus, Eu(II) in CaF_2 emits at 24.21 kK. According to DUPONT (1967), the first $4f^5 5d$ level of Sm^{++} occurs at 26.28 kK, KAPLYANSKII and FEOFILOV (1964) and BRON and HELLER (1964) studied the many absorption bands of Sm(II) introduced in alkali metal halide crystals. AXE and SOROKIN (1963) studied Sm(II) in SrCl$_2$ finding the luminescent level $^5D_0$ at 14.28 kK and the origin at 14.84 kK of the first broad $4f^5 5d$ band centered around 16 kK. REISFELD and GLASNER (1964) found broad $4f^6 5d$ bands of Eu(II) in KCl crystals at 29.15 and 40.13 kK and in KBr at 29.0 and 40.2 kK. They concluded in the sub-shell energy difference $\Delta = 11$ kK. WAGNER and BRON (1965) found emission bands at 23.2 kK for Eu(II) in KI and at 23.1 kK for Yb(II) in KI and absorption bands at 26.4 and 29.6 kK for Yb(II) in KBr. The origins for Eu(II) are situated at 23.67 NaCl; 24.34 KCl; 24.54 RbCl; 24.33 KBr and 23.75 KI. According to KAPLYANSKII and FEOFILOV (1962) the sub-shell energy difference $-\Delta$ does not decrease markedly across the 4f group since the two strong bands of Yb(II) are situated (in kK)

$$
\begin{array}{lll}
CaF_2 & 27.45 & 43.9 \\
SrF_2 & 27.95 & 42.7 \\
BaF_2 & 28.4 & 41.7
\end{array}
\tag{6.2}
$$

Actually, there is little doubt that the energies of the 5d and 6s orbitals of M^{++} cross or at least approach each other at the end of the 4f group. McELANEY (1967) reports the first $4f^{10}5d$ level of Ho^{++} at 19.01 kK and the first $4f^{10}6s$ level at 21.82 kK. BRYANT (1965) finds the first $4f^{13}5d$ level of Yb^{++} at 33.73 and the first $4f^{13}6s$ level at 34.66 kK. Since the configuration $4f^{13}5d$ is much wider (20.2 kK) than the $4f^{13}6s$ (10.5 kK, nearly exclusively caused by the relativistic effects in $4f^{13}$) the actual one-electron energy of 5d is slightly higher than of 6s. There is no reliable evidence for weak $4f \rightarrow 6s$ transitions in Yb(II) compounds though they may have nearly the same energy as $4f \rightarrow 5d$ transitions. PIPER, BROWN and McCLURE (1967) discussed the ligand field treatment of Yb(II) in $SrCl_2$ having prominent transitions at 27.3 and 41.1 kK. In the writer's opinion, the energy difference 13.8 kK represents mainly the sub-shell energy difference $-\Delta$. Said in other words, the strong transitions are conditioned by the approximate selection rule that the f^q core remains constant. It is true that the ten levels belonging to $(^2F_{7/2})5d$ of Yb^{++} according to BRYANT (1965) are distributed in the interval from 33.73 to 43.48 kK but the only one of these levels having $J = 1$ is situated at 39.95 kK. The $(^2F_{5/2})5d$ levels having $J = 1$ occur at 49.86 and 53.36 kK. Though these arguments are profoundly modified by the cubic symmetry of the chromophore $Yb(II)Cl_8$ it would still be true that the main part of the intensity would be concentrated in the allowed $J = 0 \rightarrow 1$ transitions. However, eighteen excited levels have the appropriate symmetry type T_{1u} of which 11 would be connected with the lower 5d sub-shell in the "strong-field" limit.

All available evidence shows that M^+ and M(I) would be far less stable as $4f^q$ than as $4f^{q-1}6s$. SPECTOR (1967) analyzed the spectrum of Tm^+ having its lowest levels belonging to $4f^{13}6s$. The first level of $4f^{12}6s^2$ occurs at 12.46 kK and the first level of $4f^{12}5d6s$ is found at 16.57 kK. Hence, there is every reason to believe that the ground configuration can only contain as many 4f electrons as characterizing M^{+2} (and in the case of gadolinium, Gd^{+3}). We return to this problem of conditional oxidation states M[II] in Chapter 8. By comparison with Table 6.1, it is easy to see that no low-lying $4f \rightarrow 5d$ transitions are expected for 4f group M(IV). Any strong absorption bands in the visible of M(IV) are caused by electron transfer, as discussed in Chapter 7.

The 5f group has 5f →6d transitions at lower energy than 4f →5d of the corresponding compound of the 4f group. There may even be one case observed for M(V) when REISFELD and CROSBY (1965) ascribe bands of crystalline $CsUF_6$ at 36, 40 and 44 kK to 5f →6d. FRIED and HINDMAN (1954) pointed out that the protactinium (IV) aqua ion has three bands very similar to the three first strong bands of Ce(III). Unfortunately, it is not known whether the aqua ion has the same symmetry D_{3h} and the same coordination number 9, nor whether it has a weak band analogous to the band of Ce(III) at 33.7 kK. BAGNALL and BROWN (1967) studied the spectrum of the yellow-green solution of Pa(IV) in 12 M HCl. Though other hexa-chlorides such as UCl_6^{--} discussed below dissociate under such circumstances, there is some reason to believe that $PaCl_6^{--}$ may occur. The unusually low wavenumber 24.3 kK of the first 5f →6d transition suggests absence of σ-anti-bonding effects on the lower sub-shell. In Table 6.2, these transitions are tabulated.

Table 6.2. *Wavenumbers in kK of intense 5f → 6d transitions*

		aqua ion	hexa-chloride	hexa-bromide
$5f^1$	Pa(IV)	36.3, 39.2, 44.8	24.3, 35.2, (36.4)	—
$5f^2$	U(IV)	48.2	31.8, 33.0, 35.7, 40.0	29.4, 30.4, 33.1
$5f^3$	Np(IV)	—	36.9, 39.1, 41.2, 42.5	33.4, 35.6, 37.6
	U(III)	25.5, 28.6, 31.2	17.4, 20.9	—
$5f^4$	Pu(IV)	—	40.9, 42.1, 43.7, 45.9	36.4, 37.6, 38.8
	Np(III)	34.5, 37.6, 43.5	26.0	—
$5f^5$	Pu(III)	∼40	32.0	—
$5f^6$	Am(III)	∼45	42.5	39.65, 42.1

Th^{+3} is iso-electronic with Pa(IV) and according to KLINKENBERG and LANG (1949) the two 5f-levels occur at 0 and 4.33 kK and the two 6d-levels at 9.19 and 14.49 kK. The energy difference 9 kK is within the reach of chemical stabilization, and it is by no means certain whether the lowest configuration of Th(III) is 5f or 6d. Whereas ThI_2 is a golden metal to be discussed in Chapter 8, SCAIFE and WYLIE (1964) report that ThI_3 is black, it shows strong violet to green dichroïsm in a polarization microscope. Though it cannot be excluded that this non-metallic material is a cluster compound like trimeric Re_3Cl_9 it is most probably a genuine $6d^1$-compound.

The $5f^2$ →5f6d transition of uranium (IV) aqua ions at 48.2 kK was reported by COHEN and CARNALL (1960). Broad bands of Np(IV) and Pu(IV) aqua ions close to 50 kK are electron transfer bands to be discussed in Chapter 7. STEWART (1952) identified the broad $5f^q$ →$5f^{q-1}6d$ bands of M(III) aqua ions. Recently, BARNARD, BULLOCK and LARK-

WORTHY (1967) reported a reflection spectrum of $U_2(SO_4)_3$, $8H_2O$ very similar to the grey solution of U(III) aqua ions in perchloric acid which forms red chloro complexes by addition of HCl (JØRGENSEN, 1956b). YOUNG (1967) studied U(III) and U(IV) in various molten fluorides. A yellow solution with bands at 23.8 and 26.3 kK seems to have a high coordination number for U(III), whereas the red to purple solutions seem to have lower coordination numbers. Dr. J. R. MORREY was so kind as to inform the writer about the analysis into Gaussian error-curves of the absorption spectrum of U(III) in molten CsCl which is thought to contain UCl_6^{-3}. Table 6.2 gives the two strongest transitions in the visible; there is a strong band at 33.2 kK. TITLE et al. (1962) indicate broad bands of U(III) in SrF_2 at 18.5, 20.8 and 23.3 kK. The octahedral UCl_6^-, UBr_6^-, UI_6^- (narrow $5f^2 \rightarrow 5f6d$ transitions at 24.5, 24.8, 25.9 and 27.9 kK), $NpCl_6^-$, $NpBr_6^-$, $PuCl_6^-$ and $PuBr_6^-$ were studied in acetonitrile solution by RYAN and JØRGENSEN (1963). Several of these complexes show electron transfer spectra as well from the reducing ligands to the partly filled 5f shell. The large number of excited levels has the same reason as in Sm(II), viz. the degeneracy of available terms is very pronounced. Since the bands of MX_6^- are so narrow, even closely adjacent transitions can be detected. The data given in Table 6.2 for $AmBr_6^{-3}$ are according to a kind, private information from Dr. JACK RYAN.

When comparing Table 6.2 with Table 6.1 it becomes obvious that large differences exist between the 4f and the 5f group. The $5f \rightarrow 6d$ transitions are rather dependent on the coordination number and the nature of the ligands. However, for the first half of the 5f group, it can be concluded that $5f \rightarrow 6d$ on the average needs 30 kK less excitation energy than $4f \rightarrow 5d$ under equal circumstances. This spectroscopic observation is connected with the chemical fact that the oxidation states are much more variable, and on the whole higher, in the 5f than in the 4f group. An interesting point is that the half-filled shell effects at f^7 dependent on the spin-pairing energy parameter D are less important than the general stabilization of the 5f shell $(E - A)$ per unit of increasing atomic number Z. Actually, diligent tracer experiments with mendelevium $(Z = 101)$ have recently shown that Md(III) is more readily reduced to Md(II) and trapped in $BaSO_4$ than Eu(III) is reduced to Eu(II) (MALÝ and CUNNINGHAM, 1967; HULET at al., 1967). Einsteinium $(Z = 99)$ and fermium $(Z = 100)$ remain as Es(III) and Fm(III) even after reduction with amalgamated zinc[2]. It is a striking difference from the 4f group that

---

[2]) Surprisingly enough, COHEN, ATEN and KOOI (1968) present evidence for syn-crystallization of californium(II) $(Z = 98)$ in $EuSO_4$. If ζ_{5f} is sufficiently large, this $5f^{10}$ system is much more stable than $5f^9$ according to eq. (4.32) but it needs strong effects of intermediate coupling to explain why $5f^{11}$ and $5f^{12}$ do not show comparable stabilities.

it is easier to produce Md(II) than Am(II) whereas it is distinctly much more difficult to prepare Tm(II) than Eu(II). Actually, EDELSTEIN et al. (1966) reduced Am(III) in CaF_2 to strongly coloured Am(II) showing broad bands at 16.1, 17.2, 22.0 and 23.8 kK. According to electron spin resonance measurements, this species has the ground configuration $5f^7$ (EDELSTEIN and EASLEY, 1968). MULLINS et al. (1968) present evidence for the equilibrium

$$3 \text{ Am} + 2 \text{ PuCl}_3 \rightleftharpoons 2 \text{ Pu} + 3 \text{ AmCl}_2 \tag{6.3}$$

when distributing americium between liquid plutonium and fused $NaCl - KCl$ at 700° C. However, Am(II) is certainly much more reducing than Eu(II). Metallic Am is not soluble with blue colour in liquid ammonia, as is Eu and Yb, and DAVID and BOUISSIÈRES (1968) present evidence from the formation of 4f and 5f group amalgams for aqueous citrate solutions that Am is a quite normal, trivalent metal.

DAINTON (1952) discussed absorption spectra of 3d group M(II) aqua ions having very broad, weak bands in the ultraviolet. Whereas the maximum 40.5 kK can be determined for $Fe(H_2O)_6^{++}$ the other bands are actually so uncharacteristic that a definite position cannot be given. DAINTON related approximate positions 33 kK for V(II), 31 kK for Cr(II), and above 48 kK for Mn(II), Co(II) and Ni(II), with the standard oxidation potentials for forming $M(H_2O)_6^{+3}$ also given in Table 4.3, and incidentally very similar to the values obtained assuming the absolute potential of the standard hydrogen electrode to be 36 kK. Hence, there is no doubt that one 3d electron is removed from the central atom. However, this is no certain evidence for the electron arriving at a great distance in the solvent. If the empty 4s shell has roughly the same energy in all the 3d-group M(II) aqua ions, it may equally well be argued that the excited configuration is $3d^{q-1}4s$. This configuration is also known from the gaseous ions, and the lowest term occurs at 44 kK in V^{++}, 50 in Cr^{++} and 63 in Mn^{++}. When more than five 3d electrons are present in the groundstate, the first transition is spin-forbidden such as ^7S at 30 kK (^5S at 41 kK) for Fe^{++} or ^3D at $21.9 - 24.0$ kK (^1D 26.3 kK) for Cu^+. There may be a kind of compensating influence from the strongly σ-antibonding character of the empty 4s orbital and from the general tendency toward lower $l \rightarrow l'$ excitation energy in M(II) than in M^{++} discussed above. The latter trend may be connected with the fractional charge of the central atom being considerably lower than the oxidation state in certain cases. Thus, Pu(IV) hexahalides in Table 6.2 are known from other evidence (such as the nephelauxetic effect) to be relatively covalent and it is seen that the variation of the $5f \rightarrow 6d$ excitation energy going from Pa(IV) over U(IV) and Np(IV) is far smaller than for the less covalent aqua ions. Unfortunately, the absorption spectrum of the colour-less copper (I) aqua ions is not very well known because of the ready disproportionation to Cu(II) and the metal. However, colourless $Cu(NH_3)_2^+$ has distinctly the first absorption band at higher energy than the $3d^94s$ levels of Cu^+. The intense bands close to 25 kK of crystalline CuCl, CuBr and CuI can be described (JØRGENSEN, 1962a) as transitions from a mole-cular orbital consisting of the higher (xy, yz, xz) sub-shell of Cu(I) mixed with ligand σ orbitals to the 4s orbital. FROMHERZ and MENSCHICK (1929) studied Cu(I) in octahedral chromophores. Thus, NaCl shows a 3d $\rightarrow$ 4s transition at 39.2 kK and NaBr at 38.6 kK. Aqueous solutions containing a large excess of Cl^- or Br^- show bands at 36.8 and 36.2 kK, respectively. Gaseous Ag^+ has the four levels of $4d^95s$ distributed between 39.16 and 46.05 kK. VOLBERT (1930) found relatively narrow bands of silver (I) aqua ions at 44.7, 47.5 and 51.9 kK which can most sensibly be described as 4d $\rightarrow$ 5s transitions (the writer confirmed the values 44.6, 47.7 and 51.7 kK for a 0.1 M $HClO_4$ solution). Similar results were obtained for $Ag(NH_3)_2^+$. On the other hand, it is quite certain that Zn(II) and Cd(II) aqua ions are far less absorbing at 50 kK. It is surprising that d^{10}-systems in very low oxidation states, such as $Co(CO)_4^-$ and $Co(PF_3)_4^-$, are not coloured in the visible region. One must conclude that the σ-anti-bonding character of the s orbital is far more important in these complexes. On the other hand, Fe(II) and Ag(I) aqua ions have wavenumbers only a few kK higher than the corresponding gaseous ions.

MILES (1965) discussed the general relations between the standard oxidation potentials and the wavenumbers of inter-shell transitions (generally increasing 8 kK per volt of the oxidation potential) and elec-tron transfer bands (generally decreasing 8 kK for each volt of increasing oxidizing character of the oxidized complex as measured by the oxidation potential of the corresponding reduced species). It is certainly true that the same principles determine the variations in the two cases unless entirely exceptional conditions take place. Thus, keeping the oxidation state constant, the oxidation potential and the $l \rightarrow l'$ excitation energies normally increase across a given transition group, and spin-pairing energy corrections may influence the two variations in a parallel manner. How-ever, it cannot be concluded that there is necessary quantitative connec-tion between the electrochemical and spectroscopic quantities. What is perhaps more important is that keeping the element constant, the $l \rightarrow l'$ excitations usually get more difficult and the electron transfer from filled M. O. easier as a function of increasing oxidation state.

The 4f $\rightarrow$ 5d transitions in Eu(II) aqua ions produce the photochemical result of hydrogen evolution (DOUGLAS and YOST, 1949)

$$Eu_{aq}^{++} + H_{aq}^+ + h\nu \ (27 \ kK) \ \rightarrow \ Eu_{aq}^{+3} + \tfrac{1}{2} H_2 \,. \tag{6.4}$$

Analogous reactions occur following $3d \rightarrow 4s$ excitation of $Fe(H_2O)_6^{++}$. In this connection, it is interesting to note that $FeSO_4, 7H_2O$ containing this ion is photo-conducting in the ultraviolet, the slightly delocalized 4s electron is rapidly liberated in the solid (SULZER and WAIDELICH, 1963).

There is no certain evidence known for Laporte-allowed $3d \rightarrow 4p$ transitions. Actually, since we study complexes in the condensed states, and not gaseous molecules, it is by no means certain that 4p orbitals have a lower energy than the ionization limit leading to "vertical" (and not adiabatic) photo-conductivity. Studies in the vacuo ultra-violet above 50 kK may be of some help in the future. The closest we are to observe this type of transition are the strong, relatively narrow, bands of $Pt(CN)_4^{--}$ where a 5d electron is excited to an orbital of the correct symmetry type a_{2u} in D_{4h} to combine characteristics of the central atom 6pz orbital and an appropriate linear combination of empty π-orbitals of the four cyanide ligands. In quadratic complexes, this p orbital has the exceptional property of not being σ-anti-bonding (GRAY and BALLHAUSEN, 1963). Salts of this anion frequently show fluorescent emission in the visible and extremely intense absorption bands above 20 kK not present in the colourless solutions. The dichroic properties of this band suggest the transition to involve the same, low-lying, empty M.O. of partial 6pz-character (MONCUIT and POULET, 1962).

The $5s^2$-family In(I), Sn(II) and Sb(III) have a very weak and a much stronger transition in the ultraviolet which, to a certain approximation, can be considered to have the excited levels $^3P_1$ and $^1P_1$ belonging to the configuration 5s5p. By the same token, the $6s^2$-family Tl(I), Pb(II) and Bi(III) have the first of the two transitions only some two to three times weaker than the second one. This is readily compatible with intermediate coupling between $^3P_1$ and $^1P_1$ of 6s6p. This assignment was first made by SEITZ (1938) in the case of Tl(I) in alkali metal halide crystals. KLEEMANN and FISCHER (1966) recently performed a very careful study of the chromophores $Tl(I)X_aY_{6-a}$ in mixed crystals of KCl—KBr; KCl—KI; and KBr—KI. Table 6.3 gives some data for s^2-complexes in crystals and solution. This subject was reviewed by McCLURE (1959), EPPLER (1961), JØRGENSEN (1962a and 1962b) and recently, more halide complexes were investigated by WALTON, MATTHEWS and JØRGENSEN (1967).

It is possible to derive a consistent set of moderately decreased Landé parameters ζ_{np} and strongly decreased $K_{av}(ns, np)$ from the absorption spectra observed. The only trouble for this treatment is that it is by no means certain that the symmetry of the chromophores is approximately spherical or at least cubic (in which case the degeneracy between the three p-orbitals is not removed). Actually, ORGEL (1959) collected extensive evidence that the stereochemistry of most s^2-compounds is highly distorted, the lone-pair being a mixture of comparable amounts of s and one of the p-orbitals. It is not easy to know whether crystals actually remaining cubic, such as TlCl, PbS, K_2SeBr_6 and Cs_2TeCl_6, only are cubic as a statistic time-average and show a strong tendency toward distortion on instantaneous pictures. There are many cases known where one ligand seems to be replaced by the lone-pair according to SIDGWICK and GILLESPIE's idea (GILLESPIE, 1963), e.g. in square-pyramidal IF_5 and XeF_5^+ (cf. also JØRGENSEN, 1967) or in NH_3 and ClO_3^- for that matter. Iodine (V) and xenon (VI) also form pyramidal tri-oxo complexes.

The s^2-chromophores tend toward having additional weak bands between the two main transitions. This tendency may be excused by $^3P_2$ but it may also sometimes be due to instantaneous deviations from cubic symmetry. If ORGEL's suggestion of considerable mixing of $p\sigma$ and s can be applied, the transitions from the occupied, bonding combination to the two $p\pi$ orbitals should have lower wavenumbers than the transitions to the higher, anti-bonding combination of $p\sigma$ and s. However, if the distortion from cubic symmetry is moderate, the triplet level of the higher M.O. configuration may occur at lower energy than the singlet level of the lower configuration. The dependence of the wavenumbers on the ligands for a given central atom

$$\text{gaseous ion} > H_2O > OH^- \sim \text{enta}^{-4} > Cl^- > Br^- > I^- \qquad (6.5)$$

is similar to the nephelauxetic series. However, the differences seen in Table 6.3 are not as large as they would be for electron transfer from the ligands to the central atom. Thus, the moderate shift from chloride to bromide complexes may be compared with the shift of $4f \to 5d$ and $5f \to 6d$ transitions.

Table 6.3. *Transitions (in kK) to $^3P_1$ and $^1P_1$ of the excited 5s 5p or 6s 6p configuration*

	gaseous	aqua ion	chloride	bromide
In(I)	43.35, 63.03	—	33, 36, 43	34, 39
Sn(II)	55.20, 79.91	—	34.7, 38.2, 45.0	32, 41
Sb(III)	66.70, 95.95	42.6?	34.8, 43.9	27.8, 32.3, 37.0
Te(IV)	78.02, 111.71	—	26.3, 37	—
Tl(I)	52.39, 75.66	46.7	40.5, 51.0	38.1, 47.2
Pb(II)	64.39, 95.34	48.0	36.8, 51.0	33.2, 44.9
Bi(III)	75.93, 114.60	45.0	30.5, 45.0	26.8, 38.5
Po(IV)	—	—	23.9	—

The halide anions isoelectronic with the noble gases present the opposite problem of the post-transition-group s^2-systems. There is no serious doubt that the prevailing symmetry is cubic or even quasi-spherical, but the nature of the excited orbital has been much discussed (cf. JØRGENSEN, 1962a and 1967). In alkali metal halide crystals, or in solvents without pronounced electron affinity such as water and alcohols, there is little doubt that the excited orbital is essentially spherical and hence is an s-orbital. However, the radial function in, say, iodide, may be rather different from 6s for the neutral iodine atom (the gaseous I$^-$ probably has no stable 6s orbital at all) and is adapted to the prevailing conditions, in particular by being orthogonalized on the filled orbitals of the surrounding atoms. The next Laporte-allowed transitions are expected to 5d-like orbitals of iodide. Actually, TEEGARDEN and BALDINI (1967) have recently presented evidence for the excited configuration 5p^{5}5d of iodide in RbI and 4p^{5}4d of bromide in RbBr.

By the same token as the 5s — 5p energy differences in Table 6.3 are comparable to those observed for the neutral cadmium atom (30.66 and 43.69 kK) and the 6s — 6p differences comparable with the neutral mercury atom (39.41 and 54.07 kK) if not somewhat smaller, it is possible to compare the lowest excited levels having $J = 1$ and odd parity of gaseous Cs$^+$ and xenon with iodide under various conditions:

	Cs$^+$	Xe	KI	RbI	I$^-$ in H$_2$O	CH$_3$OH	CH$_3$CN
5p^{5}6s	107.91	68.05	46.8	46.0	44.2	45.2	
	122.87	77.19	53.9	52.2	51.4		40.6
							47.7
5p^{5}5d	106.22	79.99	55.5	53.6			
	110.95	83.89	58.5	56.1			
	123.64	93.62	63	63			

$$(6.6)$$

When the ionization energy 24.8 kK of gaseous I$^-$ is compared with that, 97.83 kK, of xenon, it is seen that iodide takes an intermediate position between the two gaseous species. Actually, the similarity between crystals and the aqueous solution is even more pronounced than seen from eq. (6.6) since the shift toward higher energy by cooling the solution to the same low temperature as applied to the crystals can be extrapolated to 46.7 kK. It may be mentioned that solid xenon has strong absorption lines at 67.34 and 76.63 kK whereas a solid solution of Xe in krypton has higher wavenumbers, 71.68 and 78.44 kK (cf. JØRGENSEN, 1967). An interpolation in eq. (6.6) suggests that iodide behaves as if the effective charge were — 0.5. A similar comparison can be made between gaseous Rb$^+$, krypton and bromide:

	Rb⁺	Kr	KBr	RbBr	Br⁻ in H₂O	D₂O	CH₃CN
4p⁵5s	⎰ 134.88	80.92	54.6	53.2	51.1	51.5	46.0
	⎱ 143.47	85.85	58.6	57.2	53.2	53.6	49.1
4p⁵4d	⎧ 126.45	97.09	67	63.6			
	⎨ 140.62	99.65	69.5	66			
	⎩ 145.63	104.89					

$$(6.7)$$

again suggesting Br($-$I) having a kind of effective charge -0.6. By the same token as the wavenumbers of Br($-$I) are increased relative to gaseous Br⁻, it is expected that the opposite sign of the Madelung potential decreases the 4p excitation energy of rubidium (I). Actually, caesium (I) salts seem to have bands in the region 80 to 90 kK (not present in comparable sodium salts) which may be related to the 5p excitations of eq. (6.6). Crystalline LiF was studied by MILGRAM and GIVENS (1962) finding absorption bands at 104, 113 and 118 kK due to fluoride 2p excitation, at 182 and 196 kK due to fluoride 2s excitation, and at 500 kK probably originating in the 1s shell of lithium (I).

The general problem of $l \to l'$ excitations in halides has been obscured by photochemical arguments, the 6s electron of iodide liberating H₂ in solution in analogy to eq. (6.4) and by the tendency toward internal ionization (photo-conductivity) and photo-electric emission from the surface of alkali metal halide crystals. This is a far too complicated and frequently too unresolved a situation to discuss here, but is worth noting that the concept of localized primary inter-shell excitation is gaining more support than was the case in solid state physics around 1960.

We discuss the general topic of np-shell excitation of halides in complexes in Chapter 7 because it is intimately connected with the question of electron transfer to the central atom. However, there is some evidence that np $\to (n+1)$s excitations still can be detected, e.g. in PtX₆⁻⁻ (JØR-GENSEN and BRINEN, 1962) and ZnX₄⁻⁻ (BIRD and DAY, 1967 and 1968). There is some tendency in such cases to form a common Rydberg orbital having no angular node-plane for the whole complex anion, and it is not always possible to distinguish between electron transfer to a central atom s orbital and excitation to a symmetric linear combination of ligand s orbitals. This is also true for organic molecules such as CX₄.

Another type of physical experimentation is related to trapping isolated atoms in solid argon or other unreactive matrices at liquid helium temperature. It was recognized by REICHARDT and BONHOEFFER (1931) that the transition to ³P₁ at 39.41 kK in mercury atoms can be recognized as a broadened (or divided) and slightly shifted line in the (very dilute) solutions in water, methanol and hexane. PRENER, HANSON and WILLIAMS (1953) studied Hg atoms in zeolites, VINOGRADOV and GUNNING (1964) in various solvents and BREWER, MEYER and BRABSON (1965) in solid Xe, Kr, O_2, N_2, Ne and Ar (arranged according to increasing shift toward higher wavenumbers, amounting to 1.27 kK in argon). MCCARTY and ROBINSON (1959) initiated this technique studying Na and Hg atoms and reactive diatomic molecules such as NH and C_2 in solidified noble gases. SCHNEPP (1961) succeeded in studying isolated Mg and Mn atoms in such matrices, DULEY (1967) Zn and Cd showing some tendency toward formation of Zn_2 and Cd_2 at higher concentration, DULEY and GARTON (1967) studied transitions in In to ^4P and ^2S of the excited configuration $5s5p^2$ showing interesting effects of auto-ionization, and ANDREWS and PIMENTEL (1967) Li in solid Ar, Kr and Xe, where very pronounced dilution is necessary before the $2s \rightarrow 2p$ transition can be detected (in Xe at 14.68, 14.95 and 15.28 kK, whereas it occurs in the gaseous atom at 14.90 kK).

One interesting conclusion of these experiments is that one does not need extremely long distances to the nearest neighbour atoms before the spectral lines of a given atom having strongly allowed s →p or p →s transitions can be detected in absorption. This result, which would not be expected on the basis of group theory alone, is connected with the fact that one can perform atomic spectroscopy with electric discharges in moderately dilute gaseous samples. Actually, the diffuse character of d-energy levels observed under unfavourable conditions (for the atomic spectroscopist) of too concentrated vapour may be connected with the degeneracy; external perturbations can produce five Kramers doublet states of an alkali metal atom (cf. eq. 3.12) whereas s-levels cannot be split but only shifted to a small extent. By the same token as astrophysical conditions prevailing in the corona or in gaseous nebulae may be more favourable for the detection of highly forbidden emission lines than any laborabory experiment feasible, the stars having surface temperatures around 10000° C are also the best for showing the Balmer series of hydrogen $2 \rightarrow n$ with n from 3 to above 30 in absorption. It is remembered that the order of magnitude of average radius is n^2 bohr; hence, the effective diameter of hydrogen atoms having $n = 30$ is above 900 Å. However, spectral lamps in laboratories can be remarkably successful for bringing out high members of series spectra both in emission and absorption. Thus, KRATZ (1949) observed series up to 5s →77p in Rb and up to 6s →73p in Cs.

Another aspect of the experiments involving species trapped at very low temperature in transparent, inert solids is that spectra may be studied of constituents of comet tails, such as C_2, C_3 and CN, or species known from interstellar absorption such as CH and OH. Actually, the bright colours of the outer planets, which are remarkably heterogeneously distributed in the case of Jupiter, may be caused by highly reactive molecules produced by photochemical reactions and conserved at the low temperature of the surface. However, vibrational overtones of NH_3 and CH_4 also contribute to the colour, as do the overtones of H_2O in the blue water of the Earth's oceans.

We are not going here to discuss the opposite end of the problem, the influence of high concentrations of foreign gases on the position of atomic spectral lines. These effects are usually less spectacular than the influence of solid noble gases.

Gaseous diatomic molecules frequently have oxidation states quite unfamiliar in the chemistry of the condensed states. Thus, the atmospheres of cool stars which are so dark red that we just are able to observe them, have their absorption spectra strongly influenced by species such as TiO, ZrO, AlO, CN and MgH giving characteristic progressions showing vibrational and rotational structure and weak isotope shifts. The first volume of HERZBERG's three-volume treatise (1950) rationalized the many experimental details known about such species which also provided the fundament for MULLIKEN and HUND's M.O. theory. It is interesting that frequently, though by no means always, it is possible to recognize $l \rightarrow l'$ excitations of one of the constituent atoms. This approximation was used around 1930 for the diatomic hydrides MH of metallic elements, partly connected with somewhat exaggerated confidence in the extreme of the "united atom" model, but, later lost general public appeal. DUNN (in press) recently revived the idea in a brillant analysis of the gaseous ScO, YO and LaO having, to a remarkably good approximation, $s\sigma \rightarrow p\pi$ and $s\sigma \rightarrow p\sigma$ transitions in the visible. The opinions on electronic structure of these molecules have had a complicated pre-history, and the writer (1964) had to collect all his courage before suggesting that the ground terms $^4\sum$ assumed by molecular spectroscopists for ScO and LaO are indeed $^2\sum$ superposed the unusually large effects of hyperfine structure caused by the nuclear magnetic moments. The important conclusion is that though the oxidation states are M(II) and O($-$II) as discussed in Chapter 7, M(II) does not contain one $d\delta$ electron but needs the presence of an essentially non-bonding $s\sigma$ electron. CHEETHAM and BARROW (1967) compared data for all the transition group MX. The groundstate of ScF is $^1\sum$ corresponding to the filled orbital $(s\sigma)^2$ and low-lying excited states (~ 2 kK) $^3\Delta$ and $^1\Delta$ belonging to the M.O. configuration $(3d\delta)$ $(4s\sigma)$. Actually, $^1\Pi$ of $(3d\pi)$ $(4s\sigma)$ is observed at 10 kK and $^1\sum$ of $(3d\sigma)$ $(4s\sigma)$ at 16 kK. The unusual symmetry type $^3\Phi$ of the level at 15 kK corresponds to the first d^2-configuration $(3d\delta)$ $(3d\pi)$ since $^3\sum^-$ of $(3d\delta)^2$ has not been detected. However, the parameters of interelectronic repulsion may place $^3\Phi$ at lower energy (JØRGENSEN, 1964). The isoelectronic molecule TiO recognized to have the ground term $^3\Delta$ belonging to the M.O.

configuration $(3d\delta)$ $(4s\sigma)$ as also $^1\Delta$ following at 1.kK higher energy. $^1\sum$ of $(4s\sigma)^2$ is situated at 3 kK, $^1\Pi$ of $(3d\pi)$ $(4s\sigma)$ at 12 kK and $^3\Phi$ and $^1\Phi$ of $(3d\delta)$ $(3d\pi)$ at 14 and 18 kK. $^3\Delta$ of $(3d\delta)$ $(3d\sigma)$ is observed at 19 kK. The 4d and 5d groups have their $s\sigma$ orbitals slightly more stable in the diatomic oxides than the 3d group. Thus, the groundstate of ZrO is $^1\sum$ of $(5s\sigma)^2$ just below $^3\Delta$ and of NbO and TaO possibly $^2\Delta$ which most probably belongs to $(d\delta)$ $(s\sigma)^2$ rather than low-spin $(d\delta)^3$. On the other hand, it seems that the groundstate of VO is $^4\sum^-$ of $(3d\delta)^2$ $(4s\sigma)$, and DUNN kindly informed the writer that he believes the same to be true for NbO.

The conditions across the d groups are extremely complicated and not yet clarified. The value $S=3$ of the $^7\sum$ groundstate of the diatomic molecules MnH, MnF, MnCl, MnBr and MnI demonstrates beyond doubt the presence of the high-spin d5-core of the M.O. configuration $(3d\delta)^2$ $(3d\pi)^2$ $(3d\sigma)$ $(4s\sigma)$. The oxides show another type of complication at the end of the transition groups: it is nearly certain that CuO and AgO have $^2\Pi$ ground terms in which case they are electron transfer states M(I) O($-$I) in the sense discussed in Chapter 7.

Table 6.4 gives s $\rightarrow$ p excitations of effective one-electron systems M and M+ and of the alkaline earth atoms M having two levels $J=1$ of the excited configuration (ns) (np). The 0,0 energy differences of approximate $s\sigma \rightarrow p\pi$ and $s\sigma \rightarrow p\sigma$ transitions of MX are also given in kK. Since all three types of orbitals are roughly non-bonding, this list of origins (HERZBERG, 1950) approximately corresponds to the vertical transitions as well. The relativistic splitting between the levels of $^2\Pi$ having $\Omega = ^1/_2$ and $^3/_2$ can be shown to be compatible with the $p\pi$ orbitals being concentrated mainly on the M atom. The inverted $^2\Pi$ of BeF and MgF would be an argument for electron transfer character (such as the excited state $^2\Pi$ at 23.9 kK of BO having the oxidation states B(I)O($-$I) and the M.O. configuration $\sigma^2\sigma^2\pi^3$ $(\sigma^*)^2$ where σ^* is a mixture of boron 2s and 2pσ character on the rear side of the boron atom) if it was not for the negligible relativistic effects in the p shells of Be and Mg.

The separation between $^2\Pi$ and $^2\sum$ in MX varies in a rather unexpected way, and it is by no means certain that all of the cases of $^2\sum$ in Table 6.4 actually belong to the excited configuration $(p\sigma)^1$. The hydrides MH show $^2\Pi$ at a lower wavenumber than the heaviest halides, the excitation energies decreasing along the series MF > MCl > MBr > MI > MH. The moderate variation observed is quite typical for $l \rightarrow (l+1)$ transitions and not at all in agreement with any electron transfer character. Unfortunately, the high-energy pσ orbital has not yet been clearly located in the monohydrides.

Barium (I) halides have two sets of excited $^2\Pi$ and $^2\sum$ levels. Though Dr. DUNN undoubtedly is right in assuming that dλ orbitals will not be

Table 6.4. *Excited levels of alkali and alkaline-earth metal one- and two-valence-electron systems and the adiabatic (non-vertical) excitation energies of alkaline-earth monohalides and monohydrides in gaseous state*

$ns \to np$	$ns \to np$	$M = (ns)^2 \to (ns)(np)$			MH $^2\Pi$	MF $^2\Pi$	$^2\Sigma$	MCl $^2\Pi$	$^2\Sigma$	MBr $^2\Pi$
		M	$^3P_1$	$^1P_1$						
Li 14.904	Be⁺ 31.93, 31.94	Be	21.98	42.57	20.02	33.24, 33.22	—	27.97	—	—
Na 16.96, 16.97	Mg⁺ 35.67, 35.76	Mg	21.87	35.05	19.22	27.85, 27.81	37.15	26.46, 26.52	—	25.77, 25.88
K 12.99, 13.04	Ca⁺ 25.19, 25.41	Ca	15.21	23.65	14.37, 14.45	16.48, 16.56	18.89	16.09, 16.16	16.85	15.92, 15.98
Cu 30.54, 30.78	Zn⁺ 48.48, 49.35	Zn	32.50	46.75	23.2, 23.6	36.99, 37.36	—	33.59, 33.98	48	32.13, 32.54
Rb 12.58, 12.82	Sr⁺ 23.71, 24.52	Sr	14.50	21.70	13.3, 13.6	15.07, 15.35	17.30	14.82, 15.11	—	14.70, 15.00
Ag 29.55, 30.47	Cd⁺ 44.14, 46.62	Cd	30.66	43.69	22.2, 23.2	—	—	31.39, 32.51	45.40	—
Cs 11.18, 11.73	Ba⁺ 20.26, 21.95	Ba	12.64	18.06	{ ~10	11.65, 12.28	14.07	10.99, 11.88	—	—
					{ 14.6, 15.05	19.99, 20.20	24.15	19.06, 19.45	25.47	18.65, 19.19
	Yb⁺ 27.06, 30.39	Yb	17.99	25.07	—	—	—	17.88, 19.37	—	—
Au 37.36, 41.17	Hg⁺ 51.48, 60.61	Hg	39.41	54.07	24.6, 28.28	—	—	35.80, 39.70	23.42	~35, 38.57
	Ra⁺ 21.35, 26.21	Ra	14.00	20.72	—	—	—	14.78, 15.39	—	—

strongly mixed with other l-values, the writer still suspects that the second $^2\Pi$ is due to $5d\pi$ mixed with sufficient $6p\pi$ character to make transitions to and from $6s\sigma$ somewhat allowed. The only known $^2\Pi$ of RaCl is closer related to the 19 kK set of BaX than to the 11 kK set, among other characteristics by the moderate separation of $\Omega = {}^1/_2$ and $^3/_2$, and should be predominantly $6d\pi$. The $^2\Pi$ of YbCl should also be mainly $5d\pi$ along this argument.

By the same token as gaseous polyatomic molecules can show Rydberg orbitals collectively belonging to the positive core, it is clear that diatomic molecules have a way of situating most of the density of a non-bonding σ orbital on the rear side of the M atom, whereas such molecules would condense to metallic materials, or, in actual practice, disproportionate to MX_2 and elemental M. Since this concentration on the rear side may be described as a mixture of $p\sigma$ and s character, this would also contribute to the electric dipole moments of transitions to $d\pi$ orbitals. Actually, the chemical s^2-species may very well show a comparable admixture of $p\sigma$ with s. It is interesting that ONAKA, MABUCHI and YOSHIKAWA (1967) by a study of the circular dichroism induced by an external magnetic field (Faraday effect) demonstrated that the so-called $^3P_1$ excited levels of thallium (I) and lead (II) in alkali metal halide crystals has the gyromagnetic factor g between 0.6 and 0.8 whereas, in spherical symmetry, $^3P_1$ has $g = 1.5$ and $^1P_1$ $g = 1$.

Hence, the description of excited levels in terms of l-orbitals is much more a successful classification than it can be considered to be a good approximation to all aspects of the wavefunction.

X-ray absorption and emission spectra are among the clearest manifestations of nl-shells of individual atoms in chemical compounds. BONNELLE (1968) recently published an excellent review. The most general chemical effect is a shift of all the absorption bands as a monotonic function of the central atom oxidation state. One can show (JØRGENSEN, 1962c, chapter 12) that the ionization energies of all the inner shells change by an amount of the same order of magnitude, 90—110%, as that of the loosest bound electrons. Consequently, the shifts of emission lines involving inner shells of different energy are much smaller than of absorption processes going to low-lying empty orbitals. If one accepts the experimental difficulties of working in the soft X-ray region above 5Å (BONNELLE, 1966) one may study 2p excitation of 3d group compounds or 3d excitation of 5f group compounds, showing interesting chemical effects and the behaviour of the partly filled shell. It is also possible to detect interelectronic repulsion effects in excited states containing two partly filled shells such as $2p^5 3d^9$ of Ni(II) (BONNELLE and JØRGENSEN, 1964). Certain weak satellite absorption and emission lines of species such as MnO_4^- can be ascribed to electron transfer, e.g. from M. O. of p-like sym-

metry type mainly localized on the four oxygen ligands to the 1s hole on
manganese (BEST, 1966). Recently, *photo-electron spectroscopy* (HAG-
STRÖM, NORDLING and SIEGBAHN, 1965) has been a valuable alternative
for determining ionization energies of inner shells in solids. In Chapter
7, we return to the photo-electron spectroscopy of gaseous molecules
which has become an important tool for measuring ionization energies
of penultimate M.O. (TURNER, 1968). In all of these cases, the inner shells
show much more of an individual existence than is frequently allowed in
text-books.

Bibliography

ANDREWS, L., and G. C. PIMENTEL: J. Chem. Phys. **47**, 2905 (1967).

AXE, J. D., and P. P. SOROKIN: Phys. Rev. **130**, 945 (1963).

BAGNALL, K. W., and D. BROWN: J. Chem. Soc. (A) 275 (1967).

BARNARD, R., J. I. BULLOCK, and L. F. LARKWORTHY: Chem. Commun. 1270 (1967).

BEST, P. E.: J. Chem. Phys. **44**, 3248 (1966).

BIRD, B. D., and P. DAY: Chem. Commun. 741 (1967) and J. Chem. Phys., **49**, 392
(1968).

BONNELLE, C., and C. K. JØRGENSEN: J. Chim. Phys. **61**, 826 (1964).

— Ann. Phys. (Paris) [14] **1**, 439 (1966).

— In: Physical Methods in Advanced Inorganic Chemistry. (Editors: H. A. O. HILL
and P. DAY) p. 45. London: Interscience 1968.

BREWER, L., B. MEYER, and G. D. BRABSON: J. Chem. Phys. **43**, 3973 (1965).

BRON, W. E., and W. R. HELLER: Phys. Rev. **136**, A 1433 (1964).

BRYANT, B. W.: J. Opt. Soc. Am. **55**, 771 (1965).

BUTEMENT, F. D. S.: Trans. Faraday Soc. **44**, 617 (1948).

CHEETHAM, C. J., and R. F. BARROW: Advances in High Temperature Chemistry **1**,
7 (1967). New York: Academic Press.

COHEN, D., and W. T. CARNALL: J. Phys. Chem. **64**, 1933 (1960).

COHEN, L. H., A. H. W. ATEN, and J. KOOI: Inorg. Nucl. Chem. Letters **4**, 249 (1968).

DAINTON, F. S.: J. Chem. Soc. 1533 (1952).

DAVID, F., and G. BOUISSIÈRES: Inorg. Nucl. Chem. Letters **4**, 153 (1968).

DOUGLAS, D. L., and D. M. YOST: J. Chem. Phys. **17**, 1345 (1949).

DULEY, W. W.: Proc. Phys. Soc. **91**, 976 (1967).

—, and W. R. S. GARTON: Proc. Phys. Soc. **92**, 830 (1967).

DUNN, T. M.: private communication.

DUPONT, A.: J. Opt. Soc. Am. **57**, 867 (1967).

EDELSTEIN, N., W. EASLEY, and R. McLAUGHLIN: J. Chem. Phys. **44**, 3130 (1966).

— — J. Chem. Phys. **48**, 2110 (1968).

EPPLER, R. A.: Chem. Rev. **61**, 523 (1961).

FREED, S.: Phys. Rev. **38**, 2122 (1931).

FRIED, S., and J. C. HINDMAN: J. Am. Chem. Soc. **76**, 4863 (1954).

FROMHERZ, H., u. W. MENSCHICK: Z. Physik. Chem. **B 3**, 1 (1929).

GILLESPIE, R. J.: J. Chem. Educ. **40**, 295 (1963).

GRAY, H. B., and C. J. BALLHAUSEN: J. Am. Chem. Soc. **85**, 260 (1963).

HAGSTRÖM, S., C. NORDLING, and K. SIEGBAHN: Alpha-, Beta- and Gamma-ray
Spectroscopy, Vol. 1. Amsterdam: North-Holland Publishing Co. 1965.

HERZBERG, G.: Molecular Spectra and Molecular Structure, Vol. 1. Spectra of
Diatomic Molecules (2. Ed.). New York: D. Van Nostrand 1950.

HULET, E. K., R. W. LOUGHEED, J. D. BRADY, R. E. STONE, and M. S. COOPS: Science **158**, 486 (1967).
JØRGENSEN, C. K.: Mat. fys. Medd. Dan. Vid. Selskab **30**, no. 22 (1956a).
— Acta Chem. Scand. **10**, 1503 (1956b).
— Solid State Phys. **13**, 375 (1962a).
— Absorption Spectra and Chemical Bonding in Complexes. Oxford: Pergamon Press 1962b. U. S. distributor: Addison-Wesley.
— Orbitals in Atoms and Molecules. London: Academic Press 1962c.
—, and J. S. BRINEN: Mol. Phys. **5**, 535 (1962).
—, R. PAPPALARDO, and H. H. SCHMIDTKE: J. Chem. Phys. **39**, 1422 (1963).
—, and J. S. BRINEN: Mol. Phys. **6**, 629 (1963).
— Mol. Phys. **7**, 417 (1964).
— In: Halogen Chemistry (Ed. V. GUTMANN) **1**, 265 (1967). London: Academic Press.
KAPLYANSKII, A. A., V. N. MEDVEDEV, and P. P. FEOFILOV: Optics and Spectroscopy (English Transl.) **14**, 351 (1963).
—, and P. P. FEOFILOV: Optics and Spectroscopy (English Transl.) **16**, 144 (1964).
KLEEMANN, W., u. F. FISCHER: Z. Physik **197**, 75 (1966).
KLINKENBERG, P. F. A., and R. J. LANG: Physica **15**, 774 (1949).
KRATZ, H. R.: Phys. Rev. **75**, 1844 (1949).
LOH, E.: Phys. Rev. **147**, 332 (1966).
— Phys. Rev. **154**, 270 (1967).
MALÝ, J., and B. B. CUNNINGHAM: Inorg. Nucl. Chem. Letters **3**, 445 (1967).
McCARTY, M., and G. W. ROBINSON: Mol. Phys. **2**, 415 (1959).
McCLURE, D. S.: Solid State Phys. **9**, 399 (1959).
—, and Z. KISS: J. Chem. Phys. **39**, 3251 (1963).
McELANEY, J. H.: J. Opt. Soc. Am. **57**, 870 (1967).
MILES, J. H.: J. Inorg. Nucl. Chem. **27**, 1595 (1965).
MILGRAM, A., and M. P. GIVENS: Phys. Rev. **125**, 1506 (1962).
MONCUIT, C., et H. POULET: J. Phys. Radium **23**, 353 (1962).
MULLIKEN, R. S.: J. Am. Chem. Soc. **86**, 3183 (1964).
MULLINS, L. J., A. J. BEAUMONT, and J. A. LEARY: J. Inorg. Nucl. Chem. **30**, 147 (1968).
O'CONNOR, J. R., and J. II. CHEN: Phys. Rev. **130**, 1790 (1963).
ONAKA, R., T. MABUCHI, and A. YOSHIKAWA: J. Phys. Soc. Japan **23**, 1036 (1967).
ORGEL, L. E.: J. Chem. Soc. 3815 (1959).
PIPER, T. S., J. P. BROWN, and D. S. McCLURE: J. Chem. Phys. **46**, 1353 (1967).
PRENER, J. S., R. E. HANSON, and F. E. WILLIAMS: J. Chem. Phys. **21**, 759 (1953).
REICHARDT, H., u. K. F. BONHOEFFER: Z. Physik **67**, 780 (1931).
REISFELD, M. J., and G. A. CROSBY: Inorg. Chem. **4**, 65 (1965).
REISFELD, R., and A. GLASNER: J. Opt. Soc. Am. **54**, 331 (1964).
RYAN, J. L., and C. K. JØRGENSEN: Mol. Phys. **7**, 17 (1963).
— — J. Phys. Chem. **70**, 2845 (1966).
SCAIFE, D. E., and A. W. WYLIE: J. Chem. Soc. 5450 (1964).
SCHNEPP, O.: J. Phys. Chem. Solids **17**, 188 (1961).
SEITZ, F.: J. Chem. Phys. **6**, 150 (1938).
SPECTOR, N.: J. Opt. Soc. Am. **57**, 312 (1967).
STEWART, D. C.: Light Absorption ... III, ANL-4812. Illinois: Argonne 1952.
STRUCK, C. W., and F. HERZFELD: J. Chem. Phys. **44**, 464 (1966).
SUGAR, J.: J. Opt. Soc. Am. **53**, 831 (1963).
SULZER, J., and W. WAIDELICH: Physica Status Solidi **3**, 209 (1963).
TEEGARDEN, K., and G. BALDINI: Phys. Rev. **155**, 896 (1967).

TITLE, R. S., P. P. SOROKIN, M. J. STEVENSON, G. D. PETTIT, J. E. SCARDEFIELD, and J. R. LANKARD: Phys. Rev. **128**, 62 (1962).

TURNER, D. W.: In: Physical Methods in Advanced Inorganic Chemistry (Editors H. A. O. HILL and P. DAY), p. **74**. London: Interscience 1968.

VINOGRADOV, S. N., and H. E. GRUNNING: J. Phys. Chem. **68**, 1962 (1964).

VOLBERT, F.: Z. Physik. Chem. **A 149**, 382 (1930).

WAGNER, M., and W. E. BRON: Phys. Rev. **139**, A 223 (1965).

WALTON, R. A., R. W. MATTHEWS, and C. K. JØRGENSEN: Inorg. Chim. Acta **1**, 355 (1967).

YOUNG, J. P.: Inorg. Chem. **6**, 1486 (1967).

7. Electron Transfer Spectra and Collectively Oxidized Ligands

It was frequently argued in the period 1950—60 that a fully ionic description of many, though not all, transition group complexes must be a fairly good approximation since the electrostatic model of the ligand field apparently worked as well. Only slowly was it realized that it is the integral number of electrons in a partly filled shell which produces the useful results as a matter of classifying the excited levels, and that not much was lost in the strength of the argumentation if the partly filled shell is delocalized to some extent on the ligands as long it retains many of the essential characteristics of d and f orbitals, and in particular the corresponding 2 or 3 angular node-planes (node-cones counting for two). The apology for the ligand field theory went entirely wild then it was argued by some authors that the approximate validity must have something to do with the variation principle. Not only does no variation principle exist for energy differences, but the principle itself has entirely lost its backbone for any system containing more than ten electrons. Thus, the correlation energy, i.e. the difference between the full non-relativistic energy and the best Hartree-Fock solution, is more negative than the energy of the loosest bound electron in any atom heavier than neon. This is another way to say that an infinite number of excited levels having the same symmetry type as the groundstate occur in the energy interval between the Hartree-Fock approximation and the actual groundstate.

No, the application of the ligand field theory is connected with symmetry properties and group-theoretical engineering, exploiting the mathematical structures in a chemically meaningful way. This is very similar to the situation in spherical symmetry; the classification of excited levels having definite parity and J and fairly well-defined S and L into configurations where each nl-shell contains zero, one, two,... $(4l+2)$ electrons, also works in nearly all cases though the total wavefunctions Ψ *cannot* be anti-symmetrized Slater determinants, and though the correlation energies in most cases are larger than the energy differences of chemical and spectroscopic interest (JØRGENSEN, 1962d). Actually, if the idea of Ψ is abandoned for the benefit of first-order and second-order density matrices, there may be the positive result valid that the Hartree-Fock electronic density in our three-dimensional space may be remarkably close to the actual electronic density, whereas the average reciprocal interelectronic distance $\langle 1/r_{12} \rangle$ distinctly is smaller for the actual situation than for the Hartree-Fock approximation. However, it is quite clear that even the discrepancies remaining on a six-dimensional level (three spatial variables of each of the two electrons considered at the time when additive interelectronic repulsion is evaluated) are so serious that the utility of the one-electron classification must reside in some more general principle. This question is not yet completely clarified. In poly-atomic entities, the most important consequence is the *relevant and irrelevant symmetry components*. M. O. configurations based on a different symmetry from that of the nuclear positions may give a *better* classification of the energy levels observed than the strict symmetry. In particular, weakly interacting systems at a large separation would not have non-closed-shell states compatible with an integral number of electrons in the fully delocalized M. O. This is why the energy band description must be looked upon with outmost suspicion in non-metallic materials (JØRGENSEN, 1962 a). In solids, we have another, related problem: once it is admitted that they have an over-all dense continuum of excited levels above a given lower limit (being the groundstate in the case of metals) we may be interesting in a weighting according to the transition probability from the groundstate. It is obvious that for a spectroscopist, it is much more important to know the position of narrow intervals of excited levels having transitions to the groundstate connected with a large non-diagonal element of electric dipole moment, than to be informed about adjacent levels hardly having any transition probability at all. Thus, the narrow absorption bands and even narrower emission lines studied by X-ray spectroscopy correspond to excited levels far up in the continuum; but experimentally, they may be as informative as atomic spectral lines which also, sometimes, show auto-ionization when excited levels in the continuum are involved.

Electron transfer bands occur at low energy if one of the constituents of a polyatomic entity is strongly reducing and another strongly oxidizing. It has now been argued by some authors that in order to be validly classified as electron transfer transitions, the reducing and the oxidizing region must be spatially separated. Thus, in a halide complex MX_N^{+z-N} it would be argued that only if the N halide ligands X^- had electronic densities well separated from the central ion M^{+z} would the transition be true electron transfer. Actually, the definition of electron transfer bands would to some extent be a corollary of the full ionicity assumed in the electrostatic model of the ligand field. However, a much more plausible definition is based on M. O. theory. If the filled M. O. are localized mainly on the ligands X, though they may be localized to a minor extent on the central atom M, and if the empty or partly filled M. O. receiving the electron constitutes an *l*-orbital of M delocalized to some extent on the X atoms, we talk about electron transfer in the excited state. As a matter of fact, very few ligands show low-lying empty orbitals with the exception of conjugated carbon-systems (such as dipyridyl, acetylaceto-nate,...; cf. JØRGENSEN, 1962c) and in the case where an electron is transferred from the partly or completely filled *l*-shell mainly concentrated on M to such low-lying, empty M.O., we talk about *inverted electron transfer*.

Because of the delocalized character of all the M.O. involved, it has been argued that one should rather speak about "charge transfer" bands. It is true that the electronic densities in our three-dimensional space are changed in such a way that the fractional charges of the constituent atoms normally are changed by only a fractional amount. Actually, an optical transition with a perceptible dipole moment can only occur if the excited and the depleted M.O. co-exist somewhere in space. However, in the writer's opinion, it is very uncharacteristic to call something charge transfer. Besides the fact that the separation of fractional charges in the groundstate of a heteronuclear molecule frequently is described by this word, a much more serious disadvantage is that internal transitions in a partly filled d shell going from a less to a more anti-bonding sub-shell generally involve charge transfer form the central atom to the ligands. It might then be argued that this is exactly the reason why a complex has to be fairly electrovalent in order to distinguish between internal transitions in the partly filled shell and charge transfer transitions. However, this is a far too conservative point of view. Our discussion proceeds rather differently according to whether the chromophore considered has a centre of inversion or not. In the former case, the electronic levels of even and odd parity cannot be mixed. That means that internal d-transitions between levels of even parity cannot be confused with Laporte-allowed electron transfer between the even groundstate and odd excited levels. The presence of a centre of inversion does not protect against mixing with Laporte-forbidden electron transfer bands, but this is not very serious in actual practice because the distance between the first, Laporte-forbidden, and the following Laporte-allowed electron transfer band usually is rather small. What has an influence, however, is the absence of a centre of inversion when the chromophore performs vibrations corresponding to odd normal modes. If the internal d-transition is only separated 2 or 3 kK from the first electron transfer band, it may achieve considerable intensity and sometimes a molar extinction ε above 300 which normally can be considered the dividing line in d-group complexes. However, the vibronic coupling inducing intensity to Laporte-forbidden transitions is only strongly active over small energy differences, and JØRGENSEN (1962b, p. 103), ENGLMAN (1963) and FENSKE (1967)

have discussed why the ratio between the oscillator strength P_f of Laporte-forbidden and P_a of relatively adjacent Laporte-allowed transitions
follow the approximate rule

$$\frac{P_f}{P_a} = \frac{(W_{af})^2}{(\nu_a - \nu_f)^2} \cdot \frac{\nu_f}{\nu_a} \tag{7.1}$$

where the effective non-diagonal element W_{af} nearly always is between
1.5 and 2.5 kK, about the same value as 2δ, the half-width of a typical
absorption band toward both smaller and larger wavenumbers.

The situation is quite different in chromophores lacking a centre of
inversion of their time-average symmetry. In some cases, such as in
$Co(NH_3)_5Cl^{++}$ or $Cr(H_2O)_5F^{++}$, there is hardly any difference from regular
octahedral complexes, and one may even argue (JØRGENSEN, 1967a) that
this fact might be taken as an indication of a Gillespie-type (dynamic
second-order Jahn-Teller effect consisting of vibronic coupling with
excited electronic levels of opposite parity) of distortion occurring in
the instantaneous picture of $Co(NH_3)_6^{+3}$ and $Cr(H_2O)_6^{+3}$ as well. On the
other hand, the internal 3d-transitions in tetrahedral CoX_4^{--} have been
known for a long time to have ε between 500 and 1000 around 15 kK,
far removed from any other, intense bands. Though the instantaneous
symmetry of these complexes does not seem to be as high as T_d since the
inter-sub-shell transitions split perceptibly like in all known tetrahedral
d-group complexes (JØRGENSEN, 1967b), this fact is still not very readily
explained. From the point of view of group theory, the available mech-
anisms for high intensities of transitions in the partly filled shell are
mixtures between p- and d-orbitals having the same symmetry, and
mixtures between d-orbitals and ligand orbitals. In view of the very high
intensity of transitions between σ-bonding and σ-anti-bonding orbitals,
it is probable that minor admixtures of the latter type are the main
reason for high intensities in tetrahedral complexes. In trigonal-bipyramidal chromophores MX_5 where the X-atoms originate in phosphines,
arsines or selenium-containing ligands, the bands in the visible may have
the very high ε-values between 5000 and 10000. However, *energy-wise*
these transitions behave exactly as internal d-transitions, and the excited
levels can be treated by the angular overlap model or any other ligand
field concept. This is particularly true for the iron (II) arsine complexes
studied by HALFPENNY, HARTLEY and VENANZI (1967) where it is incon-
ceivable that the bands in the visible are due to electron transfer. In
chromophores distinctly lacking a centre of inversion, the one-electron
operator can mix even and odd orbitals having highly differing diagonal
elements of energy, and it is quite conceivable that the non-diagonal
element of one-electron energy for such a mixing is some 10 to 20 kK,
much larger than the quantity W_{af} of eq. (7.1).

Our reason for using the category "electron transfer" is that the oxidation state of the central atom in the excited level is one unit smaller than in the groundstate; or, in the case of inverted electron transfer, that the central atom oxidation state is increased one unit in the excited level. Hence, there is an intimate connection between the possibility of determining a spectroscopic oxidation state for the central atom both in the excited and ground states and the classification of the transition being electron transfer (± 1 unit) or of other types, such as internal p-, d-, f-transitions or $f \rightarrow d$ transitions keeping the central atom oxidation state constant. In the M.O. configuration, it is exactly one electron which jumps from one M.O. to another in the case of electron transfer, but it is true that the fractional charge of a number of atoms in the molecule or complex ion may change at the same time. It is also true that the sum of the changes of fractional charges may be smaller than one because both M.O. have co-existing squared amplitudes on the same atom. The one electron is jumping somewhere else than in our three-dimensional space. In the following, we defend and explicit the statement

If an atom in a polyatomic entity has an oxidation state, it is determined by the preponderant electron configuration correctly classifying the symmetry types of the manifold consisting of the groundstate and the lowest excited states. If one-electron substitutions are made in the preponderant configuration, a large number of the excited states should be described.

Hence, one has to decide in each individual case whether one can define a reasonable spectroscopic oxidation state or not. A well behaved set of examples are the transition group hexahalide complexes MX_6^{+z-6} which have been discussed at length (JØRGENSEN, 1959, 1967b, 1968b; JØRGENSEN and PREETZ, 1967). One of the interesting features of the hexahalides is that the positions of the electron transfer bands vary in a regular way as a function of the halide ligand X^- and as a function of the central atom M in a given oxidation state $+z$. It has been possible to introduce *optical electronegativities* x_{opt} by the comparison of the wavenumber of the first Laporte-allowed electron transfer band corrected for spin-pairing energy of eq. (4.31) σ_{corr} in the case of transitions to the lower d-like sub-shell:

$$\sigma_{corr} = [x_{opt}(X) - x_{opt}(M)] \cdot 30 \text{ kK}. \tag{7.2}$$

In the case of transitions to the upper, σ-anti-bonding, sub-shell, $\sigma_{corr} - \Delta$ is considered. The necessary corrections for spin-pairing energy and the sub-shell energy difference Δ are known from internal d-shell transitions or can be readily extrapolated from analogous complexes. Table 7.1 gives the optical electronegativities for d-group central atoms derived from hexahalide complexes. Of various reasons, partly discussed below, other complexes containing less than six halide ligands tend to indicate slightly lower values of x_{opt} for the central atom.

Once a large number of experimental data were connected, it was possible to derive optical electronegativities for other ligands than the four halides. Table 7.1 includes several such values, but many are less certain than for the halides. No conjugated carbon-containing ligands are included; the loosest bound orbital of acetylacetonate is known to have $x_{opt} = 2.7$ (JØRGENSEN, 1962c). NH_3 and amines RNH_2 containing only one lone-pair have very high optical electronegativities, though not as high as the σ-orbital of chloride ligands. Nitrogen-containing ligands having two lone-pairs show much lower x_{opt} as also suggested by the much lower wavenumber of electron transfer bands of NH_2^- in $Pt(NH_3)_5$ NH_2^{+3} compared with $Pt(NH_3)_6^{+4}$ (JØRGENSEN, 1956). Pt(IV) complexes of NCl_2^- obtained by chlorinating ammonia complexes were studied by KUKUSHKIN (1965).

Table 7.1. *Optical electronegativities of d-group central atoms in hexahalide complexes and of various ligands. Values for π and σ orbitals relative to the M–X axis are given.* $dtp^- = (C_2H_5O)_2PS_2^-$. $dsep^- = (C_2H_5O)_2PSe_2^-$

						π	σ
3d⁰	—	Ti(IV) 2.05	—	—	F⁻	3.9	(4.4)
3d³	Cr(III) (1.9)	Mn(IV) (2.7)	—	—	Cl⁻	3.0	3.4
3d⁶	—	Ni(IV) 3.4	—	—	Br⁻	2.8	3.3
					I⁻	2.5	3.0
					NH_3	—	3.3
4d⁰	—	Zr(IV) 1.6	Nb(V) 1.85	Mo(VI) 2.1	RNH_2	—	3.2
4d³	Mo(III) 1.7	Tc(IV) 2.25	—	—	N_3^-	2.8	—
4d⁴	—	Ru(IV) 2.45	—	—	NCO^-	3.0	—
4d⁵	Ru(III) 2.1	Rh(IV) 2.65	—	—	H_2O	3.5	—
4d⁶	Rh(III) 2.3	Pd(IV) 2.75	—	—	ROH	3.1	—
					O^{--}	(3.2)	—
					SO_4^{--}	3.2	—
5d⁰	—	—	Ta(V) 1.8	W(VI) 2.0	R_3P	—	(2.6)
5d²	—	—	—	Os(VI) 2.6	S^{--}	2.5	—
5d³	—	Re(IV) 2.1	—	Ir(VI) 2.9	dtp^-	2.7	—
5d⁴	—	Os(IV) 2.2	—	Pt(VI) 3.2	R_2S	2.9	—
5d⁵	Os(III) 1.95	Ir(IV) 2.4	Pt(V) 3.0	—	R_3As	—	(2.5)
5d⁶	Ir(III) 2.25	Pt(IV) 2.6	—	—	$dsep^-$	2.6	—

SCHMIDTKE and GARTHOFF (1967) found that azide N_3^- is as reducing as bromide. The behaviour of the triatomic pseudo-halides is not yet completely understood. Thus, SCHMIDTKE (1967) presents evidence for nitrogen-bound NCS^- having $x_{opt} = 2.6$ whereas the sulphur-bound form of this *ambidentate* ligand should have $x_{opt} = 2.85$. However, it is conceivable that the very intense bands corresponding to the latter value are not produced by the loosest bound orbital of thiocyanate, but rather a σ-like orbital of higher ionization energy. Oxide ligands present other problems (JØRGENSEN, 1962d and 1968b) mainly because of the very strong π-anti-bonding effect on the partly filled d-shell.

LEMPKA, PASSMORE and PRICE (1968) measured the ionization energy of the two components ($\omega = {}^3/_2$ and ${}^1/_2$) of the π orbitals and the highest filled σ orbital of gaseous HX molecules by photo-electron spectroscopy. In the case of σ ionization, much vibrational structure is co-excited, and one can make a distinction between the adiabatic ionization energy (allowing the internuclear distance to re-arrange) and the vertical ionization energy (like the maximum of an absorption band corresponding to the same internuclear distance as the groundstate). The ionization energies are in kK:

	π		σ		
	$\omega = {}^3/_2$	$\omega = {}^1/_2$	adiabatic	vertical	
HF		129.5	150	161	(7.3)
HCl	102.9	103.4	131	134	
HBr	94.1	96.8	123	126	
HI	83.7	89.1	112	115	

For our purposes, it is interesting that the π ($\omega = {}^3/_2$) ionization energy rather closely is $(11 + 30\, x_{opt})$ kK. Further on, the distance to the σ ionization energy is some 32 kK, much higher than the difference 12 to 15 kK indicated by the electron transfer spectra of hexahalides. There may be two main reasons for this difference. The proton in HX may be a stronger perturbation on X^- than the central atom is in a hexahalide. Another important fact is that the excitation energy tends to show a smaller variation than the ionization energy $I(\lambda)$. To a first approximation (JØRGENSEN, 1962d) the electron transfer band occurs at

$$I(\lambda) - I(d) + J(d, d) - J(\lambda, d) \tag{7.4}$$

where the J-integrals are parameters of interelectronic repulsion comparable to $A*$ of eq. (3.36). Since the σ-orbital is concentrated closer to the central atom than are the π-orbitals, $J(\sigma, d)$ is expected to be larger than $J(\pi, d)$ and the difference in excitation energy (7.4) consequently smaller than $I(\sigma) - I(\pi)$. For comparison, it may be mentioned that Dr. D. R. LLOYD measured by photo-electron spectroscopy of WF_6 the adiabatic ionization energies of the four sets of three degenerate π orbitals 125.4, 131.8, 135.7 and 142.4 kK, and two of the σ sets 152 and 159.5 kK.

The hexahalides of the 4f group (RYAN and JØRGENSEN, 1966) can be treated along similar lines again producing higher values of x_{opt} for the central atoms than found for other lanthanide complexes. Table 7.2 gives the results for the chemically very reactive hexahalides measured in acetonitrile solution, and for Nd(IV) and Dy(IV) from the reflection spectra of Cs_3MF_7 kindly communicated by Dr. L. B. ASPREY and Dr. R. HOPPE. The results for 5f group complexes have previously been reviewed (JØRGENSEN, 1967b). Table 7.2 gives the uncorrected optical electronegativity x_{uncorr} obtained by the simple technique of inserting the wavenumber of the first electron transfer band in eq. (7.2) and the value x_{opt} obtained by corrections, not only for the spin-pairing energy, but also for the other quantities E^3 and ζ_{nf} described in eq. (4.32). The reason why it is interesting to compare x_{uncorr} in the f groups is that electrode standard oxidation potentials such as those given in Table 4.3, and the general chemical tendency for a given central atom to decrease its oxidation number by one unit, is much closer connected with x_{uncorr} than would be the case in the d groups. For M(III) in the 4f group, x_{opt} smoothly increases 0.11 unit per increasing unit of atomic number, and the values for M(IV) are 1.5 to 2 units higher than for the isoelectronic M(III) in perfect agreement with the chemical behaviour of the lanthanides. On the other hand, the 5f group shows a much less dramatic variation as a function of the oxidation state and is closer comparable to the 5d group in Table 7.1. Again, the oxide complexes produces difficulties for a consistent treatment in terms of optical electronegativities. The main reason seems to be the *anisotropic* behaviour of MO_2^{++} (JØRGENSEN, 1965). The distance $U - X$ is so much longer than $U - O$ in complexes such as $UO_2X_4^-$ that the equatorial plane containing the X ligands seems to have x_{uncorr} of U(VI) as low as 1.8, whereas the behaviour of the firmly bound linear group OUO is compatible with $x_{uncorr} = 2.3$ and x_{opt} of oxide 3.1. Incidentally, the luminescence of uranyl compounds is caused by this electron transfer band.

Table 7.2. *Uncorrected and corrected optical electronegativities for f group hexahalides (and Cs_3NdF_7 and Cs_3DyF_7)*

		x_{uncorr}	x_{opt}			x_{uncorr}	x_{opt}
$4f^5$	Sm(III)	1.6	1.05	$5f^1$	U(V)	2.2	1.95
$4f^6$	Eu(III)	1.9	1.2		Np(VI)	2.6	2.35
$4f^{12}$	Tm(III)	1.5	1.85	$5f^2$	U(IV)	1.8	1.5
$4f^{13}$	Yb(III)	1.8	1.95		Pu(VI)	2.85	2.5
$4f^0$	Ce(IV)	2.1	2.05	$5f^3$	Np(IV)	1.95	1.75
$4f^2$	Nd(IV)	3.0	2.5	$5f^4$	Pu(IV)	2.1	2.0
$4f^8$	Dy(IV)	3.05	3.6				

Returning to a specific example of a d group hexahalide, let us consider $IrCl_6^{--}$. The oxidation state of the central atom in the groundstate is Ir(IV) because magnetic and many other properties of the complex show it has the lower 5d-sub-shell nearly filled, containing five of the six possible electrons. When iridium contains five d-like electrons, it is *ipso facto* Ir(IV) as can be seen from the comparison

$$
\begin{array}{cccccc}
d^3 & d^4 & d^5 & d^6 & d^8 & d^{10} \\
Ir(VI) & Ir(V) & Ir(IV) & Ir(III) & Ir(I) & Ir(-I) .
\end{array}
\tag{7.5}
$$

Whereas there is no doubt that the groundstate of gaseous Ir^{+4} would contain 68 electrons in the closed shells characterizing Hf^{+4} and $5d^5$, there is some doubt whether Ir^- would not spontaneously loose an electron, and no doubt what so ever that both Ir^0 and Ir^- would contain one or two 6s electrons replacing some of the 5d electrons. The reason why $Ir(-I)$ is mentioned in eq. (7.5) is the existence of complexes such as $Ir(PF_3)_4^-$.

The *spectroscopic oxidation state* Ir(IV) connected with the presence of five electrons in the partly filled shell is an entirely different concept from the valence-bond structures such as $Ir^0(Cl)_4(Cl^-)_2$ and $Ir^+(Cl)_3(Cl^-)_3$ one might write in an attempt to describe the charge distribution inside the complex. Actually, the fundamental difficulty for the valence-bond description is that there is no intrinsic difference between the groundstate and literally thousands of excited states, all consisting of an intermixing of thousands of "resonance structures" (JØRGENSEN, 1962b). This problem is already serious enough in cases such as SO_4^{--} but it has catastrophic consequences if one attempts a description of d group complexes, mainly because each alternative Ir^0, Ir^+, Ir^{+2},... corresponds to a large number of individual states. The superiority of the M.O. theory is clearly shown by the fact that the groundstate and any excited state of $IrCl_6^{--}$ accessible by internal 5d-transitions have the oxidation state Ir(IV) with five 5d-like electrons according to (7.5) and $Cl(-I)$ with a closed-shell configuration terminating with six 3p-like electrons. The electron transfer bands in the visible and near ultraviolet have excited levels where the central atom has the oxidation state Ir(III) with six 5d-like electrons and the six chloride ligands are lacking one delocalized electron. In valence-bond language, this would correspond to the resonance structure $Cl(Cl^-)_5$. It is worth remembering that each Cl atom has six readily available 3p-states (eight if 3s is taken into account) and that the permutation of the "hole" on six chloride ions produces 36 states. This particular result is exactly the same one obtains in M.O. theory considering one hole in the 18 highest filled M.O. Since an odd number of electrons is involved, the states must be degenerate two and two, forming 18 Kramers doublets.

In the high symmetry of octahedral $IrCl_6^{--}$, further degeneracies are required by group theory. Thus, the relatively weak, Laporte-forbidden electron transfer band at 17.0 kK is produced by three almost (the separation is due to relativistic effects) coinciding Kramers doublets, and the strong band at 20.5 kK by three other almost coinciding Kramers doublets. The first excited level of Ir(III) having one electron in the higher sub-shell (the first triplet of this M.O. configuration occurs in $IrCl_6^{-3}$ at 16.3 kK and the two singlets at 24.1 and 28.1 kK) corresponds to a broad electron transfer band of $IrCl_6^{--}$ at 43.1 kK, where the electron is transferred from the set of three orbitals having odd parity producing the relatively narrow band at 20.5 kK. The energy difference between 20.5 and 43.1 kK corresponds to the sub-shell energy difference Δ corrected for spin-pairing energy and other effects of interelectronic repulsion. Ir(IV) is obviously more oxidizing then Ir(III) and has a higher optical electronegativity in Table 7.1. Correspondingly, the broad electron transfer band of $IrCl_6^{-3}$ produced by the transfer to the higher sub-shell occurs at higher energy, 48.5 kK. Since the lower sub-shell is completely occupied in $IrCl_6^{-3}$, there are no electron transfer bands at lower wavenumbers.

The excited levels of electron transfer bands of $IrCl_6^{--}$ where the oxidation state of the central atom has been decreased from Ir(IV) to Ir(III) are said to involve the *collectively oxidized ligands* Cl_6^{-5}. We are not ascribing the fractional oxidation number $-5/6$ to each chloride ligand but rather accepting the delocalized character of the depleted M.O. by proposing a solidaric treatment of the six equivalent ligands. This is again an entirely different situation from valence-bond treatment where Cl and Cl^- undoubtedly would contribute roughly the same probabilities to even the groundstate of $IrCl_6^{--}$. The collectively oxidized set of ligands have lost their closed-shell characteristics which is not the case in any normal chloride, especially not outside the transition groups, such as CCl_4, Al_2Cl_6 or $LaCl_3$.

One may ask the pertinent question whether the groundstate of $IrCl_6^{--}$ may not involve collectively oxidized Cl_6^{-5}. That this is not the case can be seen from the regular shift of electron transfer bands toward higher energy in the series of decreasing Z-value of the central atom $IrCl_6^{--} <$ $OsCl_6^{--} < ReCl_6^{--} < ... < HfCl_6^{--}$ and correspondingly increasing optical electronegativities Hf(IV) $< ... <$ Re(IV) $<$ Os(IV) $<$ Ir(IV) in Table 7.1. The coherent treatment of all electron transfer bands of hexahalide complexes makes it highly unprobable that any of the hexahalides contain Cl_6^{-5} in their groundstate. Actually, U(VI) was at a time suspected of having $x_{opt} = 3.0$ rather than 2.3 extrapolated from Table 7.2 (JØRGENSEN, 1965; a corollary of this story is that UF_6 cannot have an electron transfer band in the near ultra-violet) in which case the dark coloured

UCl_6 might have been a uranium (V) complex of collectively oxidized ligands. Quite generally, a complex showing intense absorption bands at low energy, say in the near infra-red, is much more liable to contain collectively oxidized ligands or not having any oxidation state at all, as we shall see. In the case of $IrCl_6^{--}$, OWEN and GRIFFITHS' pioneer work on iridium *and* chlorine nuclear hyper-fine structure on the electron spin paragenetic resonance curve of $IrCl_6^{--}$ (for a theoretical discussion, see STEVENS, 1953) demonstrated that the electrons in the lower sub-shell spend roughly 70% of their time on the iridium atom and 30% on the ligands, i. e. some 5% on each chloride. It is somewhat an accident that the squared amplitudes evaluated in this fashion from the hyper-fine structure is so close to normalize according to eq. (5.8) because it would be quite conceivable that L. C. A. O. was not a perfect approximation for electronic densities of the partly filled shell close to the nuclei. Anyhow, it is an interesting question what one would have commented if the distribution had been 40% on Ir and 10% on each Cl. The order of two sub-shells would be the same from $HfCl_6^{--}$ to $IrCl_6^{--}$, but somewhere along this series, the lower sub-shell crossed the filled M. O. and was completely filled, now behaving as an inner shell. It is of course a tautology to say that in a transition group complex, the electron affinity of the partly filled shell is necessarily smaller than the ionization energy of any of the filled M. O.; otherwise, an electron would be transferred spontaneously. However, it is still a question whether it would not be more sensible to call our hypothetical 40% case Ir(IV) rather than Ir(III). This problem is of minor practical importance in $IrCl_6^{--}$ but intrusive in complexes of strongly oxidizing central atoms where the lowest empty orbital is strongly σ-anti-bonding and not only moderately π-anti-bonding. The situation has not yet deteriorated in $PtCl_6^{--}$. The excited triplet at 22.1 kK and the singlet at 28.3 kK can still be recognized as conventional "ligand field" transitions where an electron is moved from the lower to the higher sub-shell, whereas the first intense electron transfer band occurs at 38.2 kK.

The centre of inversion of $PtCl_6^{--}$ is an obvious advantage; it is quite clear that the 5d-transitions are Laporte-forbidden even → even and the 38.2 kK band even → odd. However, troubles start in $PtBr_6^{--}$. The weak shoulders (JØRGENSEN, 1956) in the 19—23 kK region are too weak to be Laporte-allowed electron transfer bands; but they have too low energy to be 5d-inter-sub-shell transitions. The characteristics of the latter category may have been mixed up with the Laporte-forbidden electron transfer bands. The situation is far more serious in quadratic $AuCl_4^-$. It is beyond dispute that the intense bands at 31.8 and 44.1 kK are due to electron transfer (from π and σ orbitals, respectively) to the empty, σ-anti-bonding dδc-orbital having the central atom part proportional to

$(x^2 - y^2)/r^2$. There is a doubtful shoulder at 26 kK which may either be
the transition of one of the eight 5d-like electrons to this empty orbital,
and hence be a subject for ligand field theory, or it may be an extraordinarily weak electron transfer band. In the writer's opinion, there is little
doubt that the first internal 5d-transitions would occur in $AuCl_4^-$ at higher
energy than the isoelectronic $PtCl_4^{--}$ having the first triplet level at
18.2 kK and the corresponding singlet at 25.7 kK. MASON and GRAY
(1968) argue that the energy differences between the sub-shells may
not always increase from M(II) to M(III) in quadratic chromophores
as they do in octahedral symmetry. However, the important fact is
that $AuCl_4^-$ has no bands *before* 25 kK, and it can be concluded that
the internal 5d-transitions roughly coincide with the electron transfer
bands and hence get intensified via eq. (7.1). The situation is rather
different in gold (III) complexes lacking a centre of inversion such
as the diethylenetriamine complexes Au den X^{++} studied by BADDLEY
et al. (1963). The moderately intense bands ($\varepsilon \sim 1000$) in the 26—33 kK
region have lower wavenumbers than the weaker bands of the corresponding Pt den X^+ containing $Pt(II)N_3X$ chromophores. In these
cases, it cannot be decided whether the excited states correspond
to the oxidation state Au(III) or Au(II) with collectively oxidizing
ligands. However, the excited levels have to be very low-lying before
any serious doubt would arise whether the groundstate is Au(III).
One would probably argue that any quadratic gold complex having
the $(x^2 - y^2)$ like orbital empty is Au(III). The danger is that a filled
M.O. at lower energy might consist of so much $5d\delta c$ that one would
have to decide for Au(I) which would be highly surprising in view of the
quadratic stereochemistry.

This problem is clearly actualized in quadratic copper complexes
(JØRGENSEN, 1966b). Normally, they have nine 3d-like electrons, the
last situated in a $(x^2 - y^2)$ like M.O. with somewhat larger squared
amplitude on Cu(II) than on the ligands. Now is the question what to
say if this σ-anti-bonding orbital was 40% on Cu and 60% on the ligands.
In view of the quadratic stereochemistry, one would be reluctant to
describe the complex as Cu(I) and collectively oxidized ligands, though
it is obvious that there must be a lower limit somewhere; Cu(II) would
be incompatible with 10% on Cu and 90% on the ligands. The situation
is entirely different if the half-filled M.O. has *another symmetry-type* than
appropriate for $(x^2 - y^2)$. One might go so far as to say that any quadratic
copper complex not having the $(x^2 - y^2)$ behaviour of the half-filled
M.O. cannot be reasonably described as Cu(II). Since such situations are
most liable to occur for strongly coloured substances, electron paramagnetic resonance can sometimes be of considerable help establishing
the symmetry type and determining the nuclear effects of different nuclei.

The four halides are a particularly fortunate class of ligands. The difficulties of establishing a set of oxidation states in a compound usually occur either because the ligands are strongly reducing; or because they readily change between different forms having different numbers of electrons in the free ligand. The conduction electrons in metals to be discussed in Chapter 8 in many ways belong to the latter category. In this chapter we are not going to concentrate on the case of a *single* ligand not being *innocent*, i.e. allowing the oxidation states of the other constituent atoms to be determined. We just mention typical cases such as O_2^{--}, O_2^{-}, O_2 or NO^-, NO, NO^+ and return in Chapter 9 to the question whether there is any sense in which BH_3 or PF_3 when forming complexes can be suspected for being $\overline{BH_3}^-$ and $\overline{PF_3}^-$. Actually, certain of the simplest ligands are particularly difficult to ascribe oxidation states; the fact that H^- has only one electron-pair more than H^+ will be shown in Chapter 9 to have rather disastrous effects on adjacent atoms. In this chapter, we rather concentrate on cases where two or more ligands possibly are collectively oxidized or reduced (i.e. one or more electrons have been added to low-lying empty M.O. delocalized on the ligands).

The absorption spectra of complexes of sulphur-containing ligands are highly interesting. In the case of non-conjugated systems such as dithiophosphates $(RO)_2PS_2^-$, dithiocarbamates $R_2NCS_2^-$ and xanthates $ROCS_2^-$ (R indicating σ-bonded alkyl or aryl groups) the internal d-transitions can be recognized in octahedral chromophores such as $Cr(III)S_6$, $Co(III)S_6$, $Rh(III)S_6$, $Ir(III)S_6$ and quadratic chromophores such as $Ni(II)S_4$ and $Pd(II)S_4$ (KIDA and YONEDA, 1955; SCHÄFFER, 1961; JØRGENSEN, 1962c and 1968a). Though electron transfer bands are observed in the ultraviolet of these species, and in the visible of $Fe(III)S_6$ and $Ru(III)S_6$, there is not the slightest doubt about the assignment of oxidation states to the central atoms, though the strongly pronounced nephelauxetic effect (see Tables 5.6 and 5.8) clearly demonstrates rather covalent bonding (JØRGENSEN, 1963b; LIVINGSTONE, 1965). It is worth to mention two characteristic features of sulphur-containing ligands. There is a frequent tendency toward three-coordination of sulphur. This is true not only for sulphonium cations R_3S^+ and for thioether complexes MSR_2 where three carbon atoms, or two carbon and a metal atom is bound to sulphur; but also for the general situation that RS^- may bridge two different metal atoms. Thus, xanthates are known to polymerize of this reason, and JICHA and BUSCH (1962) studied several interesting cases of mercaptide-bridging. In all of these cases, no impairment is made on the oxidation states though it may be recalled that binary sulphides most frequently are dark coloured, either due to deviations from stoichiometry or due to other collective effects to be discussed in Chapter 8. Actually, the dithiophosphate ligands are particularly useful because they protect the brightly coloured chromophores against polymerization and make the compounds soluble in organic solvents, and because the first internal transitions in the electronic system of the ligand occur at unusually high energy in the ultra-violet.

Another interesting feature is that most sulphur-containing ligands can be considered as *pseudo-halogens* and can be reversibly oxidized to a dimer:

$$2 \, RS^- \rightleftharpoons RSSR + 2 \, e^- . \qquad (7.6)$$

The definition of a pseudo-halogen involves this dimer (with exception of N_3^- where three N_2 are formed with a loud bang) and the formation of white or yellow precipitates with Ag^+ (this is the reason why F^- was not always called a halide last century). Whereas the "vertical" (invariant internuclear distances) electron transfer transitions of Fe(III) or Cu(II) oxidize RS^- to RS in the complex, the adiabatic chemical reaction producing stable Fe(II) or Cu(I) includes the dimerization (7.6).

However, the occurrence of more than one sulphur-containing ligand in a given complex can sometimes produce unexcepted complications. Thiuramdisulphide $[(C_2H_5)_2NCS]_2S_2$ (the compound "antabus" interfering with the metabolism of ethanol) is the oxidized dimer dtc_2 of diethyldithiocarbamate dtc^- and Professor K. A. JENSEN pointed out that when the olive-green $Nidtc_2$ is oxidized under certain conditions to dark red $Ni \, dtc_2^{++}$ which can be obtained also from Ni^{++} and dtc_2, it is not easy to know whether a nickel (IV) complex of dtc^- is formed or whether the CSSC bridge of (7.6) is retained. VÄNNGARD and ÅKERSTRÖM (1959) and ÅKERSTRÖM (1959) studied blue $Ag \, dtc_2$ having a band at 16.3 kK and the corresponding gold (II) complex $Au \, dtc_2$.

Quite recently, there has been a great interest in a certain class of conjugated, bidentate sulphur-containing ligands which in the complexes may be dimercapto anions, dithioketones or something intermediate:

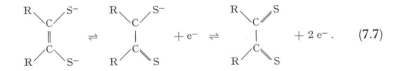

The pale yellow sodium salt of maleonitriledithiolate mnt^{--} (R $=$ CN) was prepared by BÄHR and SCHLEITZER (1957) by condensation of CN^- and CS_2 in dimethylformamide. Since 1962, three independent groups of workers started intensive studies of complexes of such ligands (denoted $\mathscr{L}^{--}$ or $\mathscr{L}$ in the following) and terminal groups such as R $=$ H, CF$_3$,

C_6H_5 are known. However, besides direct reaction between $\mathscr{L}^{--}$ and M^{+z}, other routes of preparation exist. When KING (1963) reacted the molybdenum (0) carbonyl with a heterocyclic starting material

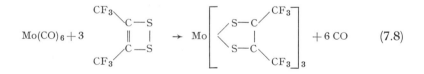

$$Mo(CO)_6 + 3 \quad \begin{array}{c} CF_3 \\ \diagdown \\ C-S \\ \| \quad | \\ C-S \\ \diagup \\ CF_3 \end{array} \quad \rightarrow \quad Mo \left[\begin{array}{c} CF_3 \\ S-C \diagup \\ < \quad | \\ S-C \\ \diagdown \\ CF_3 \end{array} \right]_3 \quad + 6\,CO \qquad (7.8)$$

one can understand why it was suggested by some authors that the dark violet compound is a complex of Mo(0) with $\mathscr{L}$ rather than of Mo(VI) (as proposed by KING) with $\mathscr{L}^{--}$. SCHRAUZER and MAYWEG (1964) prepared $Ni\mathscr{L}_2$ ($R=C_6H_5$) from solid NiS and diphenylacetylene $C_6H_5CCC_6H_5$.

All authors agreed that moderately coloured, red or orange $Ni\mathscr{L}_2^-$ and green $Pd\mathscr{L}_2^-$ contain quadratic chromophores $Ni(II)S_4$ and $Pd(II)S_4$ much in the same way as the non-conjugated complexes such as $Ni\,dtc_2$ and $Pd\,dtp_2$ (JØRGENSEN, 1962e). However, the intensely coloured $Ni\mathscr{L}_2$ and $Ni\mathscr{L}_2$ soluble as monomeric species in unreactive organic solvents posed many problems. SCHRAUZER and MAYWEG (1965) argued from the characteristic, intense absorption bands that essentially aromatic systems occur, and that $Ni\mathscr{L}_2$ is an indefinite intermediate between $Ni(0)\mathscr{L}$ and $Ni(IV)\mathscr{L}^{--}$ though probably closer to the first alternative. GRAY (1965) reviewed the M.O. calculations of such species and suggested essentially Ni(II) $3d^8$ remaining a well-defined oxidation state with a collectively oxidized set of ligands $\mathscr{L}_2^{-3}$ or $\mathscr{L}_2^-$. Thus, the Wolfsberg-Helmholz approximation applied to $Ni\,mnt_2^-$ (SHUPACK et al., 1964) suggested that the half-filled orbital ($S=1/2$) remains about 26% on the nickel atom and has such a symmetry type (a_g) as to consist mainly of sulphur π orbitals in the molecular plane (these orbitals are called π_h orbitals by the group at Columbia University) whereas the central atom contribution corresponds to a filled orbital (mainly of 3d character) at lower energy. Hence, the central atom has eight d-like electrons and is Ni(II). Professor HARRY GRAY has a great confidence in numerical M.O. calculations; JØRGENSEN (1966a) suggested another alternative where the collectively oxidized set of ligands $\mathscr{L}_2^{-3}$ has lost one electron and $\mathscr{L}_2^{-2}$ both electrons from a delocalized M.O. mainly consisting of "π" orbitals perpendicular on the molecular plane, what the Columbia people call π_v orbitals. We have to mention the group-theoretical technicality that the Cartesian coordinates x and y normally are directed through the four unidentate ligands in quadratic chromophores MX_4 having the tetragonal symmetry D_{4h}. Complexes of two bidentate ligands no longer have a four-fold axis, and it is necessary to turn the x and y axes $45°$ in the molecular plane with the result that the strongly σ-anti-bonding d orbital is written $(xy)/r^2$ and not $(x^2 - y^2)/r^2$. We let the y-axis bisect the $C - C$ bonds and the x-axis separate the two ligands. It is noted that the Columbia scientists make the opposite choice of x and y, in which case the symmetry types b_2 and b_3 in orthorhombic symmetry D_2 are interchanged. From the number of angular node-planes and from the binding properties of the central atom, it is expected that the order of π_v orbitals is

$$1 b_{1u} < 1 b_{3g} < 2 b_{1u} < 2 b_{3g} < 1 b_{2g} < 1a_u \leqslant 2 b_{2g} < 2 b_{2g} < 2 a_u \qquad (7.9)$$

where the double inequality sign separates the two empty orbitals of highest energy (in $\mathscr{L}_2^{-4}$) from the orbitals filled in the dimercaptide anion case. It is suggested that the collective oxidation removes one or both electrons from the orbital $1a_u$. This orbital is similar to an f orbital $(x^2 - y^2)z/r^3$ by having three mutually orthogonal node-planes.

SCHMITT and MAKI (1968) argue that the unpaired spin $(S=\frac{1}{2})$ of Co mnt_2^{-2} and Ni mnt_2^- resides in a b_{3g} orbital being a mixture of $3d(yz)$ and π_v orbitals, whereas Rh mnt_2^{-2} (MAKI et al., 1964) has a half-filled a_g orbital partly consisting of $(3z^2 - r^2)/r^2$. In this case, the central atom undoubtedly has seven 4d-like electrons and has the oxidation number Rh(II). On the other hand, if the experimental results can be interpreted as the unpaired electron residing in the a_u orbital just mentioned, we have essentially a Rh(I) complex with eight 4d-like electrons and the collectively oxidized set of ligands mnt_2^{-3}. If this should be the case, the complex is one of the rare d group ions with well-defined odd parity; compounds containing an odd number of f electrons also have odd parity of Ψ. In this connection, it may be mentioned that the only known monomeric rhodium (II) complexes are blue-green $Cl_2Rh(P(o - C_6H_4CH_3)_3)_2$ (BENNETT et al., 1966) and the red hexamethylbenzene cation $Rh(C_6(CH_3)_6)_2^{++}$ (FISCHER and LINDNER, 1964). Most previously reported Rh(II) complexes have turned out to be Rh(III) hydrides.

BALCH (1967) demonstrated that ligands L such as $X(C_6H_5)_3$ ($X = P$, As, Sb, Bi) cleave dimeric $[Co\mathscr{L}_2]_2^-$ and $[Fe\mathscr{L}_2]_2^-$ (antiferromagnetic coupling, HAMILTON and BERNAL, 1967) to monomeric $LCo\mathscr{L}_2$ ($S=0$) and $LFe\mathscr{L}_2$ ($S=\frac{3}{2}$) whereas the isoelectronic $LCo\mathscr{L}_2$ has $S=\frac{1}{2}$. It is plausible that the two latter categories have defined oxidation states Fe(III) and Co(IV) whereas the diamagnetic Co(III) complex only is unusual by being five-coordinated and not octahedral. According to WILLIAMS et al. (1965) Co mnt_2^- is diamagnetic in cyclohexanone and in pyridine but has $S=1$ in dimethylsulphoxide. Salts of Co mnt_2py^- have been isolated.

The behaviour of six-coordinated complexes $M\mathscr{L}_3$ is extraordinarily complicated. STIEFEL et al. (1966) reported that the diamagnetic $Mo\mathscr{L}_3$ and $W\mathscr{L}_3$ and the paramagnetic $(S=1/2)$ $Mo\mathscr{L}_3^-$, $W\mathscr{L}_3^-$ and $Re\mathscr{L}_3$ all are almost perfectly trigonal-prismatic D_{3h} for a variety of R-substituents. There is no absolute arguments against these species being d^0- and d^1-systems with strong covalent bonding of the $\mathscr{L}^{--}$ ligands and low-lying electron transfer bands. However, when $V\mathscr{L}_3$ was studied by EISENBERG et al. (1966) the dimercapto formulation would involve the chemically unacceptable vanadium (VI). This compound is isomorphous with $Re\mathscr{L}_3$ and apparently involves a vanadium (IV) complex of the collectively oxidized $\mathscr{L}_3^{-4}$, whereas the trigonal-prismatic $V\mathscr{L}_3^-$ may be a diamagnetic V(V) complex of $\mathscr{L}_3^{-6}$. The orbital depleted in $\mathscr{L}_3^{-4}$ may be a g-like orbital of symmetry a_1'' having four nodal planes, one bisecting all three C—C bonds and the three others producing alternating hexagons of positive and negative sign on the three sulphur atoms above the first node-plane and on the three sulphur atoms below in such a way as approximately to have even parity (JØRGENSEN, 1966a). In their M.O. treatment, SCHRAUZER and MAYWEG (1966) assume Cr(IV) in $Cr\mathscr{L}_3$ and V(III) in $V\mathscr{L}_3^-$. It is interesting that $Cr\mathscr{L}_3^-$ has $S-1$ and $Mo\mathscr{L}_3^-$ $S=0$. However, GRAY and colleagues rather consider $Re\mathscr{L}_3$ to be a border-line case between Re(VI) and Re(II) according to whether a filled sub-shell of two e' orbitals are ascribed to the central atom or not. It is quite conceivable that the squared amplitude is slightly higher than 0.5 on rhenium; however, the description as a Re(VI) complex has the advantage that the order of the 5d-like sub-shells is that predicted by the angular overlap model for σ-bonding alone (JØRGENSEN, 1966a) and in particular, that one electron occurs in the lowest orbital a_1' having 5dσ as the main constituent.

Quite recently, $V\mathscr{L}_3^-$ was shown by STIEFEL, DORI and GRAY (1967) to have a distorted structure intermediate between that of a trigonal anti-prism (or octahedron as a special case) and that of a trigonal prism. Nothing is known about the stereochemistry of pale brown $Cr\mathscr{L}_3^{-3}$ which may be the first non-octahedral six-coordinated chromium (III) complex. It is interesting that all 3d group M mnt_3^{--} behave as low-spin M(IV) complexes (S values in parentheses): Ti(0), V(1/2), Cr(1), Mn(3/2), Fe(1). McCLEVERTY et al. (1968) studied magnetic properties and spectra of maleonitrile dithiolates, and Table 7.3 gives some of the results. There is no compelling reason not to consider all the examples as M(IV) and M(III) complexes of mnt^{--}. The transitions in the visible are less intense than of oxidized quadratic species such as $M\mathscr{L}_2^-$ and $M\mathscr{L}_2$ and might serve as an argument for the loosest bound ligand orbital essentially having even parity. The $3d^2$-transitions of Cr(IV) occur at low energy and with high intensity, as expected; it is not easy to distinguish between internal sub-shell and electron transfer transitions in such systems. The magnetism and the position of the first band would be compatible with octahedral $Cr(III)S_6$ and $Co(III)S_6$; however, the M.O. configurations $(a_1')^1(e')^2$ and $(a_1')^2(e')^4$ in a trigonal prism D_{3h} would be equally compatible, as well as a large number of less symmetric alternatives.

Table 7.3. *Main absorption bands of maleonitriledithiolate complexes measured by* McCleverty, Locke, Wharton *and* Gerloch *(1968). Wavenumbers of maxima in kK, molar extinction coefficients* ε *in parentheses*

$V mnt_3^{-2}$ $(S=1/2)$	10.3 (2800), 17.2 (4700), 23.4 (6600), 32.4 (21600)
$Cr\, mnt_3^{-2}$ $(S=1)$	4.8 (1200), 9.9 (200), 14.4 (2900), 20.3 (4800), 28.7 (12700)
$Cr\, mnt_3^{-3}$ $(S=3/2)$	14.1 (420), 17.7 (1700), 21.1 (6000), 23.5 (8500)
$Mn\, mnt_3^{-2}$ $(S=3/2)$	12.1 (2400), 15.8 (2800), 31.9 (29800)
$Fe\, mnt_3^{-2}$ $(S=1)$	12.4 (3100), 24.4 (7100)
$Fe\, mnt_3^{-3}$ $(S=1/2)$	12.5 (2200), 24.4 (8900), 27.9 (16300)
$Co\, mnt_3^{-3}$ $(S=0)$	14.9 (1100), 21.5 (6800)
$Mo\, mnt_3^{-2}$ $(S=0)$	15.0 (5600), 20.0 (2700), 25.6 (10000)
$W\, mnt_3^{-2}$ $(S=0)$	17.5 (4000), 20.2 (6300), 23.4 (3500), 26.3 (7400)

It is interesting that even H_2 tdt $= 3, 4$-dimercaptotoluene ("dithiol") behaves as $H_2\mathscr{L}$ (Williams et al., 1965) though it is only conjugated in the sense of (7.7) via one of the two Kekulé structures. Actually, green Co tdt_2^- is a remarkably stable, quadratic Co(III) complex having $S=1$, probably because of adjacent energy of the (x^2-y^2) and xz-like orbitals. However, the ready one-electron redox reactions of most $\mathscr{L}$-complexes certainly show a pronounced delocalized character of the loosest bound electrons, and one may still ask the question whether the ligands really are innocent, allowing a central atom oxidation state to be defined, or whether the description in terms of collectively oxidized sets of ligands $\mathscr{L}_2^{-2}$ in quadratic or $\mathscr{L}_3^{-4}$ in trigonally prismatic chromophores is not oversimplified.

One might have hoped that very accurate crystallographic measurements might have thrown some light on the problem. Sartain and Truter (1966, 1967) have reported the Ni$-$S distance 2.10 Å in Ni$\mathscr{L}_2$, $\mathscr{L}$ being $S_2C_2(C_6H_5)_2$. This is just slightly shorter than 2.146 Å in Ni mnt_2^- and 2.165 Å in Ni mnt_2^{--}. In the three examples, the mean C$-$S distance is 1.71, 1.714 and 1.75 Å and the C$-$C (olefinic) bond length 1.37, 1.356 and 1.33 Å. Hence, there seems to be gradual change from an oxidized form of $\mathscr{L}^{-\delta}$ to $\mathscr{L}^{--}$.

On the other hand, if the olefinic bonds are not conjugated with the ligating sulphur atoms, the ligands seem to be quite innocent and to behave like dithiocarbamates. Thus, iso-maleonitrilodithiolate i-mnt^{--}

$(NC)_2C=CS_2^{--}$ forms quite normal quadratic $M(II)S_4$ with $M = Ni$, Cu, Pd and Pt and octahedral $Fe(III)S_6$ $(S = ^5/_2)$ and $Co(III)S_6$ $(S = 0)$ in contrast to the conjugated dithiolates (WERDEN, BILLIG and GRAY, 1966). The quadratic $M(II)S_4$ are also formed by N-cyanodithiocarbimate $NCN = CS_2^{--}$ studied by FACKLER and COUCOUVANIS (1966) and COTTON and McCLEVERTY (1967). See also the writer (1968a).

LOCKE and McCLEVERTY (1966) showed that mnt^{--} can occur in mixed complexes also containing CO and $C_5H_5^-$. This is a case of symbiosis of PEARSON's soft ligands (JØRGENSEN, 1964a).

The ligand glyoxal bis (o-mercaptoanil) H_2 gma forming gma^{--}

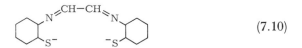

(7.10)

was recently studied by HOLM et al. (1967) in black Ni gma and the reduction products Ni gma^- and Ni gma^{--}, and in the corresponding Zn gma and Zn gma^-. If the analytical results are correct, one rather would speak about a ligand reduced from gma^{-2} to gma^{-3}. A closely related molecule is $H_2\mathscr{L} = $ 2-mercaptoaniline with $H\mathscr{L}^-$ being NH_2C_6 H_4S^- (a better name than aminothiophenol). Under anaerobic conditions, nickel (II) forms a pale yellow, quite normal looking $Ni(H\mathscr{L})_2$. With oxygen, a dark blue oxidation product is formed, which previously was formulated as a binuclear Ni(IV) complex. However, today it is thought of as the deprotonated $\mathscr{L}^{--}$ $NHC_6H_4S^{--}$ being oxidized according to (7.7) to the collectively oxidized $\mathscr{L}_2^{--}$ in the dark blue nickel (II) complex $Ni\mathscr{L}_2$. The reason why we say that $\mathscr{L}_2'$ occurs rather than two radical ligands $\mathscr{L}^-$ is that the collective influence is needed for forming delocalized M.O. besides the fact that $Ni\mathscr{L}_2$ is diamagnetic. In GRAY's description, there is a logical connection between conjugated ligands $\mathscr{L}$ containing two S groups, the intermediate case 2-mercaptoaniline (or half of gma) containing one S and one NH group, and the other extreme of two NH groups.

Actually, BALCH and HOLM (1966) studied oxidized o-phenylenediamine (1,2-diaminobenzene) complexes. It had been thought for a long time (cf. JØRGENSEN, 1963b) that the normal ligand $H_2\mathscr{L}$ or $C_6H_4(NH_2)_2$ rarely forms bidentate complexes as do aliphatic diamines. However, the complexes of the Schiff base bis (salicylidene) o-phenylenediimine have been known for a long time to be perfectly stable, and MARKS, PHILLIPS and REDFERN (1967) recently reported blue-violet nickel (II) complexes $Ni(H_2\mathscr{L})_3^{++}$ having absorption spectra corresponding to octahedral $Ni(II)N_6$ with $S = 1$. DUFF (1968) confirmed these findings and described $Ni(H_2\mathscr{L})_2X_2$ and octahedral $Ni(H_2\mathscr{L})X_2$ undoubtedly involving halide bridges like many other nickel (II) amine-halides. BALCH and HOLM's deprotonated complexes $Ni\mathscr{L}_2^-$ having NH groups bound to Ni(II) are very readily oxidized stepwise to $Ni\mathscr{L}_2^-$, $Ni\mathscr{L}_2$, $Ni\mathscr{L}_2^+$ and $Ni\mathscr{L}_2^{++}$. In accordance with GRAY's arguments, all these species must be considered as Ni(II) complexes of collectively oxidized ligands, the limiting form $\mathscr{L}$ being o-benzoquinonediimine $C_6H_4(=NH)_2$. The dark blue $Ni\mathscr{L}_2$ has a very complicated absorption spectrum in solution with some six narrow bands in the visible, the strongest at 12.65 kK having $\varepsilon = 55000$. Very similar spectra are observed of the dark blue $Pd\mathscr{L}_2$ and $Pt\mathscr{L}_2$. It may be worth the warning that not all strong colours produced by organic molecules and d group ions are due to metallic complexes; PERKIN's discovery of the oxidation of aniline to the multi-annular dyestuff mauveïne well known from Great Britain's earlier stamps; or the very complicated behaviour of phenols in the company of $FeCl_3$ may serve as useful examples.

RÖHRSCHEID, BALCH and HOLM (1966) studied complexes of $H_2\mathscr{L}$ pyrocatechol$=1,2$-dihydroxobenzene with many elements. The sulphonic acid of this molecule is a well-known analytical reagent "tiron" and gives bright solution colours with oxidizing central atoms such as Fe(III), Ce(IV) and Ti(IV) obviously caused by electron transfer bands. What is less obvious is that their wavenumbers vary with the number of ligands; thus, $Fe\mathscr{L}(H_2O)_4^+$ is blue, $Fe\mathscr{L}_2(H_2O)_2^-$ red and $Fe\mathscr{L}_3^{-3}$ yellow (cf. JØRGENSEN, 1962b, p. 196 and 1963b). RÖHRSCHEID et al. also studied 3,4,5,6-tetrachloropyrocatecholates. White $Zn\mathscr{L}_2^-$ and olive-green $Cu\mathscr{L}_2^-$ contain quite innocent ligands; but blue-green $Ni\mathscr{L}_2$ and dark brown $Ni\mathscr{L}_2^-$ presumably contain collectively oxidized sets of ligands. The limiting form is $\mathscr{L}$ being o-benzoquinone which is very unstable as a free molecule. CROWLEY and HAENDLER (1962) studied dark coloured complexes of transition group ions with o-phenanthrenequinone and chrysenequinone. CALDERAZZO and HENZI (1967) give arguments for blue-black $Mo(OC_6H_4O)_3$ and $MoX(OC_6H_4O)_3^-$ being Mo(VI) complexes of the neutral p-quinone. They do not seem monomeric.

Quite generally, the 1,3-diphenols and p-quinones are known to be related by intermediate steps of oxidation. Though the dark coloured compound quinhydrone is known to be a loose molecular adduct of hydroquinone HOC_6H_4OH and OC_6H_4O, other quinones sich as 2,3,5,6-tetramethyl-quinone ("duroquinone") are known to be reduced by one electron to the semiquinone. KIMURA et al. (1967) studied the near ultra-violet spectra of the deprotonated forms $HOC_6H_4O^-$ and $OC_6H_4O^{--}$ and the corresponding free radicals HOC_6H_4O and $OC_6H_4O^-$. It is seen that the deprotonated hydroquinone $OC_6H_4O^{--}$ can be related to the quinone via two steps of one-electron oxidation. Thus, VLČEK and HANZLIK

(1967) reacted the strongly reducing, olive-green, square-pyramidal $Co(CN)_5^{-3}$ with quinone obtaining a red (strong band, $\varepsilon = 8000$ at 23.0 kK) complex

$$(NC)_5CoOC_6H_4OCo(CN)_5^{-6} \qquad (7.11)$$

which slowly decomposes to the cobalt (III) complexes $Co(CN)_5(H_2O)^{--}$ and $(NC)_5CoOC_6H_4OH^{-3}$. The latter hydroquinone complex can be oxidized to an unstable $(NC)_5CoOC_6H_4O^{--}$. The ligand in (7.11) is not necessarily innocent; further investigation is necessary.

In this connection, it may be mentioned that a monomeric CO^- has not yet been isolated though it is isoelectronic with NO. The so-called alkali metal carbonyls analyzing as MCO have turned out (WEISS and BÜCHNER, 1963; BÜCHNER and WEISS, 1964; BÜCHNER, 1966) to consist of a mixture of two dimers and one hexamer:

$$K^+ (OC \equiv CO)^{--} K^+$$

$$(7.12)$$

The salt from acetylenediol is known to contain a linear dianion. The second, organo-metallic, constituent makes the mixture pyrophoric in air. In the case of alkali metal salts, the formulation of the hexakis-anion of fully deprotonated hexahydroxobenzene raises no serious doubt; but if complexes of such an anion was formed, one might expect some tendency toward oxidation of the ligand toward quinonic forms, and the metallic elements might not have an oxidation state. Actually, yellow compounds containing a series of adjacent carbonyl groups are known, such as the three CO groups of the lactone formed by dehydrogenation of ascorbic acid. However, the cyclic squarates $C_4O_4^{-2}$ and croconates $C_5O_5^{-2}$ seem to form perfectly normal complexes (WEST and HSIEN YING NIU, 1963). The squarates have exactly the colours expected for 3d group octahedral $M(II)O_6$ and $M(III)O_6$. The croconates of colourless cations such as Ca(II) and Al(III) are yellow, and the deep purple Fe(II) croconate suggests inverted electron transfer to low-lying empty M.O. like in dipyridyl and picolinate (cf. JØRGENSEN, 1963b). The crystal structure of $MnC_5O_5, 3H_2O$ shows $Mn(II)O_6$ polymerized in such a way that each Mn is connected to three water molecules, to two oxygen atoms of one croconate anion and to one oxygen of another (GLICK and DAHL, 1966). KAUFMAN (1964) performed M.O. calculations of various $C_nO_n^{-z}$. Regular hexagonal $C_6O_6^{-4}$ was calculated to have $S = 1$. Since the observed anion is diamagnetic, it is probably distorted in some way.

Many colours in Nature or produced by industry are caused by phenolate-quinone tendencies in oxygen-containing aromatic compounds. BAYER (1962, 1966) and STEVENSON (1965) have reviewed the pigments in flowers, which sometimes can be influenced by complex formation with metallic elements. The red Al(III) and purple Be(II) and Sn(IV) lakes formed with alizarine and its derivates are well known both from qualitative analysis and from flags and uniforms. Since this book is about oxidation numbers, we are not discussing here the properties of quinoid structures giving rise to low-lying excited levels in such molecules and complex ions; the groundstate is rarely perturbed to such an extent that any serious doubt arises about the oxidation state of connected central atoms.

In certain cases, the oxidized forms of quinoid structures can be isolated as entities having positive S. If we write $\mathscr{L}$ for N, N, N', N' tetramethyl-p-phenylenediamine $(CH_3)_2NC_6H_4N(CH_3)_2$, Wurster's blue cation is $\mathscr{L}^+$, and ALBRECHT and SIMPSON (1955) described the intense absorption bands in the 16—18 kK region of this species. THOMAS, KEELER and McCONNELL (1963) studied magnetic coupling in crystals of $\mathscr{L}^+ClO_4^-$ and CHOPOORIAN et al. (1964) the blue oxidized form produced photochemically from p-phenylenediaminetetraacetic acid. We are not here discussing the charge-transfer complexes formed between organic molecules of high electron affinity and comparatively low ionization energy, as reviewed by BRIEGLEB (1961) and MURRELL (1961). However, we may mention that FORSTER and THOMSON (1963) studied the visible absorption bands produced by mixtures of organic electron acceptors and $\mathscr{L}$ in solution.

Complexes of such radical ligands have not been very much studied until now. EHRENBERG et al. (1966) studied reduced solutions of flavine (isoalloxazine) containing zinc and cadmium, and concluded from the electron paramagnetic resonance spectrum that the 4s and 5s central atom orbital contains a great proportion of the un-paired spin density and may, to a certain approximation, be described as Zn(I) and Cd(I) complexes. Further d-group flavosemiquinone complexes were studied by MÜLLER et al. (1968).

Even aliphatic tertiary amines R_3N are rather reducing, and fragmentary evidence from electron transfer spectra is available that the optical electronegativity of the lone-pair is lower than the value 3.2 given in Table 7.1 for primary amines RNH_2. Whereas phosphines R_3P and arsines R_3As are readily oxidized to R_3PO and R_3AsO and frequently extract oxygen atoms from other compounds, the chemical oxidation of R_3N seems to go through R_3N^+ and produces various decomposition products in a higher yield than R_3NO. The latter amineoxides are the most stable with aryl R-groups. However, the behaviour of nitrogen atoms are rather different when incorporated in the heterocyclic skeleton of aromatic compounds.

The bidentate ligands 2,2'-dipyridyl and 1,10-phenanthroline form complexes of three rather different types. The absorption spectra of the pink nickel (II) complexes Ni dip_3^{++} and Ni phen_3^{++} are extremely similar to those of other diamines, such as purple Ni en_3^{++}, or violet $\text{Ni(NH}_3)_6^{++}$ for that matter (JØRGENSEN, 1955). There is no indication of anything but σ-bonds formed by the six nitrogen atoms in these complexes. Quite many dip- and phen-complexes of central atoms in normal oxidation states show a similar behaviour such as nearly colourless Mn $\text{dip}_3^{++}(S=^5/_2)$ and yellow Co $\text{dip}_3^{++}(S=^3/_2)$. Unidentate pyridine shows essentially the same type of behaviour. However, there has been rather much conflicting evidence as to whether Ni py_6^{++} exists; there seems to be steric hindrance preventing more than four pyridine molecules from coordinating most central atoms. (NORBURY et al., 1967) but Fe py_6^{++} occurs in a crystal structure (DOEDENS and DAHL, 1966). In this connection, it may also be mentioned that nearly all $\text{MX}_2 \text{phen}_2^+$ (M = Cr, Co, Rh, Ir,...) seem to have the two halides X in cis-position. The repulsion between the two planar phenanthroline molecules seems to prevent trans-$\text{Mphen}_2\text{X}_2^+$ from existing; and Pd phen_2^{++} seems to be somewhat distorted away from the purely planar stereochemistry.

The second type of complexes of aromatic heterocycles have strong colours due to inverted electron transfer spectra where an electron is transferred from the partly or completely filled d-shell of the central atom to a low-lying empty M.O. mainly concentrated on the ligands. However, these transitions do not remove the possibility of assigning a definite oxidation state to the central atom. WILLIAMS (1955) demonstrated the character of inverted electron transfer of the strong transitions in octahedral Fe dip_3^{++} and Fe phen_3^{++} and in tetrahedral Cu dip_2^+ and Cu phen_2^+ by studying the influence of substituting groups in the dip- and phen-molecules. The latest theoretical treatment of this problem was given by DAY and SANDERS (1967). For other studies of Fe(II) complexes of this kind see JØRGENSEN (1957) and KRUMHOLZ (1965 and 1967).

The sub-shell energy difference Δ which is only insignificantly larger for nickel (II) complexes of aromatic diimines than of aliphatic diamines, is somewhat larger in the iron (II) complexes. However, this seems to be connected mainly with the contraction of the M − X distances in low-spin d^6-complexes containing no σ-anti-bonding electrons. In other cases, either is Δ unusually large, or the spin-pairing energy parameter D unusually small. Thus, the intensely purple chromium (II) complexes Cr dip_3^{++} and cis-Cr dip_2X_2 have $S = 1$, as is true for Cr(CN)_6^{-4}. It may be remarked that strongly coloured high-spin $(S=2)$ iron (II) complexes are known, such as dark blue cis-Fe phen$_2$X$_2$ studied by KÖNIG and MADEJA (1967). In the case of X = NCS$^-$ and NCSe$^-$, the complex (also with dip) changes over from $S=2$ to $S=0$ at a rather sharp temperature by cooling (KÖNIG et al., 1968). An even more surprising observation was made by KÖNIG and MADEJA (1966) finding $S=1$ for the malonate Fe phen$_2$ mal. However, according to private communication from Dr. E. KÖNIG, this compound is perhaps not monomeric.

There exist ligands such as acetylacetonates (JØRGENSEN, 1962c) and picolinates (cf. JØRGENSEN, 1963b and MERCIER and PÂRIS, 1965) where reducing central atoms such as Ti(III), V(II) or Fe(II) produce inverted electron transfer bands, but where the reduced form of the ligand has not been isolated in definite compounds. This is not the case for 2,2'-dipyridyl. The deep purple dip$^-$($S=^1/_2$) and dark green dip^{--} $(S=0)$ are known in non-aqueous solutions, and hence, it is not clear whether apparently very low oxidation states of the central atom actually correspond to collectively reduced ligands. Professor S. HERZOG has prepared rather unexpected compounds under anaerobic conditions. The most surprising about these compounds is perhaps their low magnetic moments. HERZOG and BERGER (1966) report purple Ca dip$_4$ (2.27 Bohr magnetons) and Sr dip$_4$ (2.34 B. M.) suggesting some antiferromagnetic coupling decreasing the susceptibility from that expected for $S=1$ (or even a statistical mixture of 75% triplet and 25% diamagnetic molecules). HERZOG and GUSTAV (1966) find $S=1$ for deep red-brown Y dip$_4^-$, $S=^3/_2$ for wine-red Y dip$_3$ and $S=^1/_2$ for dark red La dip$_4$. PAPPALARDO (1968) studied absorption spectra of such species formed by the 3d and 4f groups. HERZOG and ZÜHLKE (1966) report olive-green Zr dip$_3^-$ (1.10 B. M.), blue-violet Zr dip$_3^-$ (1.70 B. M.) and green Zr dip$_3$ (0.31 B. M.). It is easier to understand that magenta Nb dip$_3^-$ is diamagnetic and blue Nb dip$_3$ has $S=^1/_2$ (cf. HERZOG and WULF, 1966) since the two complexes $might$ be low-spin Nb ($-$I) and Nb (0). Actually, the isoelectronic series blue-violet Ti dip$_3^-$, blue V dip$_3$, blue Cr dip$_3^+$ and blue (but this time due to normal electron transfer) Fe dip$_3^{+3}$ has such an electron spin resonance spectrum (WEBER, 1961) that it may be argued that all the species are essentially 3d^5-systems with $S=^1/_2$. The missing link is Mn dip$_3^{++}$ known to have $S=^5/_2$. The interpretation involving Ti($-$I) and V(0) would suggest that the blue-green Ti dip$_3^{--}$ and dark red V dip$_3^-$ are diamagnetic 3d^6-systems with Ti($-$II) and V($-$I). If this is true, it represents the lowest oxidation states known at the beginning of the 3d group. Trigonal-bipyramidal 3d^8-systems such as Cr(CO)$_5^{--}$, Mn(CO)$_5^-$ and Fe(CO)$_5$ and tetrahedral 3d^{10}-systems such as Fe(CO)$_4^{--}$, Co(CO)$_4^-$ and Ni(CO)$_4$ take over the rôle of making low oxidation states. However, there is a profound difference between these colourless complexes of CO (and analogous complexes of PF_3) where no doubt can be emitted regarding the oxidation state of the central atom, and the strongly coloured dipyridyl complexes where the ligands may not be innocent at all. According to private communication from Dr. ROMANO PAPPALARDO, Cr dip_3 which has many reasons for being considered as a normal $3d^6$-system like the red isonitrile complexes $Cr(CNR)_6$ (MALATESTA, 1959) and colourless $Cr(CO)_6$, actually has strong bands already in the near infra-red which is not a good sign for being Cr(0) with well-defined oxidation state.

On the other hand, when BENNETT and TAUBE (1968) find strong bands of V dip_3^{++} at 15.4 and 24.9 kK and of $Vphen_3^{++}$ at 15.5 kK, they can be considered as inverted electron transfer bands of V(II) quite analogous to Fe(II). MAHON and REYNOLDS (1967) pointed out that the spectrum of Fe dip_3^- is remarkably similar to that of dip^-, whereas these authors consider blue-violet Fe dip_3 to be a genuine $3d^8$-system implying Fe(0). The iso-electronic Mn dip_3^- has the high magnetic moment 3.7 B.M. according to HERZOG and SCHMIDT (1963). According to ALBRECHT (1963) the chelating angle NMN is only 74° in Ti dip_3, V dip_3 and Cr dip_3.

Organo-metallic dipyridyl complexes are not particularly coloured, unless the dipyridyl ligand is reduced. COATES and GREEN (1962) report, not unexpectedly, that Be $dipCl_2$ is colourless, whereas Be $dip(CH_3)_2$ is yellow (band at 25.3 kK) and Be $dip(C_2H_5)_2$ red (21.7 kK). This is in striking contrast to dark green Be dip_2 and deep violet-blue Be dip_2^-. It is also interesting that Zn $dip(CH_3)_2$ is only yellow, whereas the $4s^2$ Ga(I) complex Ga dip_2^+ is red (cf. JØRGENSEN, 1963b). The boron (III) complexes $B(C_6H_5)_2$ dip^+ are normally pale yellow (BANFORD and COATES, 1964) with exception of the orange anhydrous iodide. It is a quite common tendency for anhydrous iodides of dipyridyl complexes to show electron transfer bands from the 5p shell of iodide to the low empty M.O., and such orange colours frequently disappear by the introduction of crystal water (HARRIS and MCKENZIE, 1963). On the other hand, the apparent boron (I) complex $(CH_3)_2NB$ dip is deep red (KUCK and URRY, 1966), these authors prepared also the apparent boron (II) complex $[(CH_3)_2N]_2B$ dip, the apparent boron (0) B dip_2, and the deep violet Si dip_3. FOWLES et al. (1968) believe to have a brown, polymeric material Na_2TiCl_4 from which they make black titanium (II) complexes such as Ti $dipCl_2$, Ti $phenCl_2$, Ti py_2Cl_2 and Ti $(CH_3CN)_2Cl_2$. FARVER and GWYNETH NORD (1967) give polarographic evidence for Tl dip_2^{++} as an intermediate by the reduction of Tl dip_2^{+3} to Tl dip_2^+. In view of all what we have said, it is possible either that this species is a thallium (II) complex, or that dipyridyl has been reduced.

For a long time, it was believed that phenanthroline would not allow the same apparently low oxidation states as dipyridyl. This is mainly due to the tendency of the former molecule to be hydrogenated by the solvent under the influence of the reducing agent (alkali metals etc.). There is little doubt that phen can do the same things as dip; blue-violet $V \text{ phen}_3^-$ and $Cr \text{ phen}_3$ are known today. By the way, colourless $Cr(CO)_6$ can be substituted by quite normal ligands such as $Cr(CO)_5X^-$ $(X = Cl, Br, I)$, $Cr(CO)_5NH_3$ etc., and the yellow colour observed may be due to tetragonal splitting of the first spin-allowed $3d^6$-transition. $Cr(CO)_4$ dip is deep orange and $Mo(CO)_4$ dip red; these colours are probably caused by inverted electron transfer bands.

Many aromatic hydrocarbons can accept electrons from alkali metals in non-aqueous solvents such as tetrahydrofuran or liquid ammonia. The anions formed have quite characteristic absorption spectra with relatively narrow bands in the visible. If an odd number of electrons is added, S is usually $1/2$. If an even number of electrons is added, S may be 1 under favorable conditions. Thus, the deep blue pentaphenyl cyclopentadienyl cation $(C_6H_5C)_5^+$ studied by BRESLOW et al. (1963) has $S = 1$ for the groundstate. The cyclopentadienide anion $C_5H_5^-$ is best known by inorganic chemists for containing six "π" electrons (in the sense of planar molecules) like benzene C_6H_6 and the tropylium cation $C_7H_7^+$. These species are able to form "sandwich" complexes with all carbon atoms bound at equivalent positions relative to the central atom, like the Fe(II) ferrocene $Fe(C_5H_5)_2$ and the Co(III) cobalticinium cation $Co(C_5H_5)_2^+$. Thus, the molecule $Mo(C_5H_5)(C_7H_7)$ is a molybdenum (0) complex of the ligands $C_5H_5^-$ and $C_7H_7^+$. In this connection, it may be mentioned that cyclopentadiene can be fully mercurated to $C_5(HgCl)_6$ (WATT and BAYE, 1964).

The bis-methylene (the two C-bridges in m-position) $C_6H_5CC_6H_4$ CC_6H_5 has $S = 2$ according to ITOH (1967). Simple molecules having $S = 1$ include diazomethylene CNN, cyanonitrene NCN and the weakly bent dicyanomethylene NCCCN (WASSERMAN et al., 1965). Complexes of these simple molecules have not been reported; however, it is worth remembering the potential existence of organic molecules having positive S values. We are returning to this problem in chapter 9, especially in connection with ligands such as O_2 and NO.

The phthalocyanine anion pc^{--} is mainly known in complexes of divalent metals Mpc as well as the metal-free form H_2pc. However, M(III) may form MpcCl and M(IV) species such as MpcO (MOSER and THOMAS, 1963; LEVER, 1965). DODD and HUSH (1964), TAUBE (1966) and CLACK et al. (1967) studied the reduction of Mpc to Mpc^-, Mpc^{-2}, $Mpc^{-3}, \ldots$ There is general agreement that in most cases, the added electrons are distributed in the heterocyclic ligand and that the central atom conserves the oxidation state M(II). However, $Copc^{-2}$ and $Copc^{-4}$ seem derived from Co(I), quite reasonably for a quadratic low-spin d^8-system. CLACK and HUSH (1965) studied the related one-electron reductions of Mg(II) and Zn(II) porphin complexes.

VEILLARD and PULLMAN (1965) calculated M.O. configurations for the biologically important Fe(II) porphyrins and vitamin B 12. It was the general feeling that the ionization energies of aromatic systems and of the central atom 3d shell are so closely comparable that there is a great probability that many such ligands are not innocent. This may very well be the case when O_2 is connected with hemoglobin; but vitamin B 12 is more and more considered as an octahedral $Co(III)N_4X_2$ complex, X referring to the groups (H_2O, CN^-, CH_3^-,...) which may be bound perpendicular to the plane of the cobalamine (DAY, 1967a and 1967b). It is very interesting that certain ligands such as dimethylglyoximate dmg^- may be used as model substances for vitamin B 12 and perform comparable reactions (SCHRAUZER, 1966; SCHRAUZER and WINDGASSEN, 1966; COSTA et al., 1968). Besides the cobalt (III) complexes such as $Co\,dmg_3$ and $Co\,dmg_2X_2^-$, the cobalt (II) complexes $Co\,dmg_2X_2^{-2}$ ($X = SeCN,SCN,I,Br,Cl$) were studied by BURGER and PINTÉR (1967). In this connection, it may be mentioned that DAY, HILL and PRICE (1968) studied the reduction and methylation of cobalt phthalocyanines:

$$Co^{II}pc \xrightarrow{\ N_2H_4\ } Co^{I}pc^- \xrightarrow{\ CH_3I\ } CH_3Co^{III}pc\,. \qquad (7.13)$$

It is perhaps more surprising that the Schiff base (acting as an acid when forming complexes) bis (salicylidene) ethylenediimine H_2salen may show comparable properties. CALDERAZZO and FLORIANI (1967) and FLORIANI et al. (1968) showed that red Co^{II} salen ($S=^1/_2$) can be reduced to green Co salen$^-$ and to a dark green Co salen^{--}. One may prepare olive-green CH_3Co^{III} salen, red-orange CH_3Co^{III} salen (H_2O) and red CH_3Co^{III} salen (py) via Grignard reagents acting on the cobalt (II) complex or alkylhalides reacting with the green mono-anion:

$$Co^{II}\ salen\ +CH_3MgX \rightarrow CH_3Co^{III}\ salen+MgX^+$$
$$Co\ \ salen^-+CH_3X\ \ \ \rightarrow CH_3Co^{III}\ salen+X^-\,. \qquad (7.14)$$

The magnetic properties of some of the compounds prepared are rather surprising. Black diamagnetic Co^{III} salen I was also prepared.

CALDERAZZO and FLORIANI (1968) also reduced the well-known orange Ni^{II} salen (low-spin d^8 $Ni(II)N_2O_2$) to dark blue-green Ni salen$^-$ and olive-green Cu^{II} salen to bright red Cu salen^{--}. In these cases, it is felt that the reduction takes place on the ligand. The sodium salt of Fe salen$^-$ was shown to react with benzyl chloride; and the iron (III) iodide with phenyl Grignard reagent:

$$Fe\ salen^-\ +C_6H_5CH_2Cl \rightarrow C_6H_5CH_2Fe^{III}\ salen+Cl^-$$
$$Fe\ salen\ I+C_6H_5MgBr \rightarrow C_6H_5Fe^{III}\ salen\ \ \ \ +MgBrI\,. \qquad (7.15)$$

The deep brown iron (III) organo-metallic compounds formed seem to have $S = \frac{5}{2}$. It may be recalled that the stereochemistry of iron (III) Schiff base-complexes is fairly complicated. GERLOCH and MABBS (1967) described how recrystallization of dimeric $Fe_2\,salen_2Cl_2$ from nitromethane produces a square-pyramidal Fe salen Cl weakly solvated in the crystal by a CH_3NO_2 molecule. EARNSHAW et al. (1968) discussed how dark brown Fe^{II} salen is readily oxidized to orange $(Fe^{III}\,salen)_2O$ and also the magnetic properties of dimeric $Mn_2\,salen_2$. Each Mn(II) has $S = \frac{5}{2}$ but there is an antiferromagnetic coupling.

Returning to blue Fe dip_3^{3+} and Fe $phen_3^{+3}$ (both having $S = \frac{1}{2}$) it may be mentioned that these species are only formed by removing one electron (with strong oxidants) from the red Fe(II) complexes. If $Fe(H_2O)_6^{+3}$ is treated with dip or phen, oxo-bridged complexes are obtained. KHEDEKAR et al. (1967) point out that species such as (Fe phen$_2$ Cl)$_2$O^{++} containing a linear bridge $Fe^{III}O\,Fe^{III}$ are not necessarily low-spin; it is quite conceivable that they have $S = \frac{5}{2}$ with strong antiferromagnetic coupling.

After this discussion of nitrogen-containing ligands, it may be asked whether phosphines and arsines involve similar difficulties. Actually, a large number of multidentate ligands containing P and As are known to form 3d group complexes with strong colours and absorption bands in the visible having ε in the range 1000 to 10000. However, in most cases, the oxidation state of the central atom remains defined. It is somewhat more surprising that *energy-wise* these transitions behave like inter-3d-sub-shell transitions. NORGETT, THORNLEY and VENANZI (1967a and 1967b) have demonstrated beyond doubt that the low-spin trigonal-bipyramidal chromophores M(II) As$_4$X [with M = Fe$(S = 1)$, Co$(S = \frac{1}{2})$, Ni$(S = 0)$ and the corresponding Pd and Pt] can be described by ligand field theory assuming the order of orbitals in three sub-shells (dπc and dπs; dδc and dδs; dσ) predicted by σ-anti-bonding effects and neglecting π-bonding. NORGETT and VENANZI (1968) applied the angular overlap model to the slightly distorted case Co(II)P$_4$X. It is interesting to compare with the high-spin complexes M(II)N$_4$X formed by the quadridentate amine N{CH$_2$CH$_2$N(CH$_3$)$_2$}$_3$ (CIAMPOLINI, 1966 and 1969; cf. JØRGENSEN, 1966a) and having ε between 10 and 100. The exorbitant intensity of the inter-sub-shell transitions in the phosphine and arsine complexes must be connected with the low energy of the electron transfer bands corresponding to transitions from rather diffuse (because of low electronegativity of the ligands) σ-bonding orbitals mainly consisting of ligand lone-pairs to the σ-anti-bonding sub-shell of the central atom. Since both the σ-bonding and anti-bonding orbitals are highly delocalized, they tend to co-exist in our space of three dimensions, and the oscillator strength of the absorption band is correspondingly large. In arsine complexes having a centre of inversion, the distinction between Laporte-forbidden inter-sub-shell transitions and those of the electron transfer transitions going from even to odd states is rather sharp. There is a certain mixing due to the vibronic coupling according to eq. (7.1) but it diminuishes rapidly when the difference between ν_a and ν_f is more than a few kK.

Actually, F. G. MANN invented the bidentate diarsine das o-C_6H_4 $\{As(CH_3)_2\}_2$ which was used by NYHOLM (1949) and collaborators for making many interesting complexes (for references, see also JØRGENSEN, 1963b). In many cases, unusually high oxidation states were reported; halogens X_2 such as Cl_2 and Br_2 form Fe das$_2$X$_2^{+2}$ and Ni das$_2$X$_2^{+2}$ apparently containing Fe(IV) and Ni(IV) which is surprising in view of the reducing character of the diarsine. YAMADA (1967) discusses the absorption spectra of cobalt (III) complexes such as Co das$_3^{+3}$ and *trans* (and *cis*) Co das$_2$X$_2^+$ of which the moderately intense internal $3d^6$-transitions have nearly the same positions as in the corresponding ethylenediamine complexes, whereas an electron transfer band in the near ultraviolet indicates $x_{opt} = 2.45$ for das (i.e. the odd linear combination of the σ-bonding lone-pairs), cf. Table 7.1. JENSEN and JØRGENSEN (1965) studied spectra of the intensely coloured d^6-systems ($S = 1$) Co(III)P$_2$X$_3$ having trigonal-bipyramidal structure with the three chloride or bromide ligands in the equatorial plane. Their conclusion of $x_{opt} = 2.6$ for the phosphine $P(C_2H_5)_3$ is less certain when considering the absence of a centre of inversion in the complexes. Actually, the one-electron operator may have non-diagonal elements mixing σ-bonding lone-pairs with the d-sub-shells in a way perhaps 10 or 50 times more effectively than W_{af} in eq. (7.1). On the other hand, it is only a necessary and not a sufficient condition for exceedingly intense inter-sub-shell transitions to lack a centre of inversion. Thus, Co(NH$_3$)$_5$Cl^{+2} has hardly more intense bands than Co(NH$_3$)$_6^{+3}$ though the chromophores are usually considered to have the symmetry C_{4v} and O_h without and with a centre of inversion. Only the low-electronegativity ligands containing P, As, S, Se, ... show the extremely intense bands. Though most such ligands contain aryl groups, it is known from the work of DYER and MEEK (1967) that there is no essential difference between entirely aliphatic ligands in such cases and the corresponding ligands containing benzene rings, and the conceivable conjugation with the aromatic systems seems to have no influence on the absorption spectra. The tendency toward high intensity increases for the multidentate ligands S < Se < P < As.

KREISMAN et al. (1968) presented electron spin resonance evidence that *trans*-Ni das$_2$Cl$_2^+$ is not a conventional Ni(III) complex but lacks one electron in a linear combination of bonding σ-lone pairs. The authors found a striking similarity with the e.s.r. of das$^+$ produced by far ultraviolet irradiation of das in frozen ethanol at $88°$ K. It would be tempting to conclude with KREISMAN et al. that the depleted orbital has two node-planes perpendicular on the $NiAs_4$ plane. However, this orbital might be more stabilized by bonding with the nickel 3d orbital of same symmetry (the one half-filled in Cu(II) complexes) and the depletion might take place in the odd combination having only one node-plane bisecting both diarsine ligands. The 4p bonding of this orbital may be insignificant. It would be worth looking more into this situation; the frequent high oxidation numbers such as Fe(IV) and Ni(IV) found in crystalline phosphides and arsenides reviewed by F. HULLIGER and discussed in Chapter 8 might suggest a quite conventional Ni(III) formulation of Ni $das_2Cl_2^+$ in what case the e.s.r. similarity is spurious.

	^4E	$^4B_2$		
Cr das$_2$ Cl$_2^+$	17.0	(20.8)		
Cr das$_2$ Br$_2^+$	16.3	(20.0)		
Cr das$_2$ I$_2^+$	15.4	—		
	^1E	$^1A_2$	$^1T_2$ (O$_h$)	
Fe das$_2$ Cl$_2$	13.5	22.4	25.5	
Fe das$_2$ Br$_2$	12.9	22.5	26.0	
Fe das$_2$ I$_2$	12.7	22.5	25.5	
Fe das$_3^{++}$		21.4	26.6	
Co das$_3^{+3}$		23.2	—	

$$(7.16)$$

FELTHAM and SILVERTHORN (1968) studied recently absorption spectra of Fe(II), Cr(III) and Co(III) complexes and report the excited levels in kK (symmetry types for $trans$-complexes having D_{4h}) eq. (7.16) showing a perfectly normal behaviour of tetragonal and octahedral complexes (cf. also SCHÄFFER and JØRGENSEN, 1965). The only slightly surprising aspect is that Fe(II)As$_6$ has nearly as high a sub-shell energy difference Δ as Co(III)As$_6$ reminding one about Fe(CN)$_6^{-4}$ and Co(CN)$_6^{-3}$ in Table 5.8. FELTHAM and SILVERTHORN also report the parameters for Fe das$_3^{++}$ (not correcting for sub-shell configuration intermixing)

$$\Delta = 22.7 \text{ kK} \qquad B_{35} = 325 \text{ K} \qquad \beta_{35} = 0.36 \qquad (7.17)$$

showing a very extensive nephelauxetic effect (cf. also eq. 5.21) explaining why this iron (II) complex as well as $trans$-Fedas$_2$X$_2$ are low-spin.

Of course, it cannot be excluded that some oxidation states, such as Cr(III), Fe(II), Co(II) and Ni(II) (cf. also RODLEY and SMITH, 1967) and Co(III) form diarsine complexes having normal inter-sub-shell transitions, whereas other, apparently high oxidation states, have strong absorption bands in the visible and may contain non-innocent ligands being oxidized, rather than as the dubious dipyridyl cases, reduced. This problem needs further clarification.

The diarsine ligands also involve a variety of unresolved stereochemical problems. One would wish that NYHOLM's diamagnetic red Ni das_3^{++} was not regular octahedral, but say, square-pyramidal Ni(II)As$_5$ as many other d^8-complexes now known. COLLINGE, NYHOLM and TOBE (1964) prepared H_2Cdas^{+2} and $Cdas_2^{+4}$ and pointed out that these species can be considered as carbon (IV) complexes. The point of view of an inorganic chemist looking at organic compounds, in particular simpler molecules, may seem unfamiliar to the organic chemist. $CH_3NH_3^+$ *may* be thought of as an ammonia-trihydrido complex of carbon (IV) which is deprotonated to methylamine CH_3NH_2 slightly more difficultly than NH_4^+ to NH_3, whereas $Pt(NH_3)_6^{+4}$ is a stronger acid than NH_4^+ when forming $Pt(NH_2)(NH_3)_5^{+3}$. In many ways, carbon has the same chemical properties as one of the platinum group elements; the aqua ions of aliphatic species such as $CH_3OH_2^+$ are strong acids readily forming the hydroxo complexes CH_3OH (this is less true for the anhydrous $C(C_6H_5)_3^+$) and *simple* di-hydroxo compounds are known such as $OC(OH)_2$ and $Cl_3C\,CH(OH)_2$ in solution. The main difference from the 5d group is that carbon is so small that it most frequently has the coordination number 4 or 3. However, this situation can change if the ligands also are sufficiently small. In the tetramer $Li_4(CH_3)_4$ each carbon is six-coordinated to three lithium atoms of the central Li_4 tetrahedron and to three hydrogen atoms (WEISS and LUCKEN, 1964). Colourless Be_2C crystallizes in the CaF_2 type, and hence the coordinations are $Be(II)C_4$ and $C(-IV)Be_8$ in a cube. It is customary to call all carbon compounds having the coordination number higher than four electron-deficient, such as numerous cases of five-coordination where a methyl group bridges two metal atoms, say Be(II) or Al(III). This name is rather unsuitable for colourless compounds. It is, of course, true that diborane B_2H_6 can be reduced to $B_2H_6^2$ iso-electronic and isosteric with ethane; but the description as electron deficient compounds comes from a belief in the octet rule which cannot be shared by the inorganic chemist interested in elements with higher atomic number than silicon. Most transition group complexes have too many electrons, relative to the octet rule; but $Cr(NH_3)_6^{+3}$ is again "electron-deficient" without any conspicuous observable effects.

From the point of view of oxidation states, the phosphine and arsine complexes sofar studied have not presented the same problems as the conjugated sulphur- and nitrogen-containing ligands discussed above. This may be due to the accident that conjugated phosphine complexes have not yet attracted attention. After all, the whole class of conjugated sulphur-containing ligands (as contrasted to the innocent ones) was not recognized before 1962. Obviously, chemistry is at its very beginnings at many essential points.

The *interaction between ligands* is more important than was believed around 1960. Direct evidence comes from the absorption spectra of hexa-halide complexes to be discussed below, but crystallographic studies of complexes of conjugated sulphur-containing ligands have also revealed interesting facts. GRAY has legitimately emphasized that the planar system MS_4 of many bis-complexes $M\mathscr{L}_2$ is close to a square rather than an elongated rectangle. What is perhaps even more striking is that the three faces MS_4 of trigonal-prismatic $M\mathscr{L}_3$ also are quadratic with a very small tolerance. This suggests considerable bonding between the ligands; and it may be extremely pertinent when McCLEVERTY writes $M\text{ mnt}_3^{-z}$ as the combined formula $MS_6C_6(CN)_6^{-z}$ disregarding the individuality of the three ligands. We return to this question when discussing chemical bonding in solid d-group compounds.

Normally, the absorption spectra rather indicate repulsions between the ligands. This stems from the fact that the low-lying excited levels correspond to the excitation of ligand-ligand anti-bonding orbitals. The energetic effects observed on the individual orbitals are far larger than the total energy due to inter-ligand repulsion because other, deeper, orbitals are ligand-ligand bonding. Our suggestion for the collectively oxidized set of ligands $\mathscr{L}_2^{-2}$ of quadratic or $\mathscr{L}_3^{-4}$ of trigonal-prismatic structure is indeed that one orbital of strongly inter-ligand anti-bonding character has lost its two electrons. Direct evidence can be presented in green ytterbium (III) cyclopentadienide $Yb(C_5H_5)_3$ having a weak electron transfer band close to 15 kK (PAPPALARDO and JØRGENSEN, 1967). The three tightly compressed ligands are expected to present a high M.O. with three node-planes and, accidentally, having very close to well-defined odd parity. It is not easy to have a clear idea about the reducing character of $C_5H_5^-$ and in particular, about its optical electro-negativity. However, there are many reasons to believe that the electron transfer to the available hole in the central atom $4f^{13}$ has anomalous low wavenumber. Thus, $Yb(C_5H_5)_2Cl$ and related compounds with less squeezing of the ligands are only orange; and the larger central ion europium (III) forms a dark red $Eu(C_5H_5)_3$ (as contrasted to yellow $Eu(C_5H_5)_2$) apparently having the first electron transfer band at a slightly higher wavenumber than Yb(III) in disagreement with the generally valid series of x_{uncorr} in Table 7.2.

In the hexahalides MX_6^{+z-6}, JØRGENSEN (1959) suggested the order of orbitals mainly concentrated on the ligands

$$\pi\ t_{1g} > \pi\ t_{2u} > (\pi + \sigma)\ t_{1u} > \ldots > (\sigma + \pi)\ t_{1u} > \ldots \qquad (7.18)$$

partly based on the argument of MCCLURE (1959) that the number of angular node-planes is $4, 3, 1, \ldots, 1, \ldots$ among the observed excitable orbitals. However, this argument neglects the common symmetry type t_{1u} for three π and three σ orbitals (each oriented along one of the Cartesian axes). The higher eigenvalue hence corresponds to an additional node-cone, and the effective number of angular node-planes in (7.18) is $4, 3, 3, \ldots, 1, \ldots$ Consequently, no certain conclusion can be drawn relative to the order of the two sets having three node-planes (this number may be called "l" in quotation marks) except that the empty p orbitals of the central atom were expected to stabilize the two sets t_{1u} whereas t_{2u} can only combine with central atom f-orbitals and higher l-values, Actually, HENNING, MCCAFFERY, SCHATZ and STEPHENS (1968) determined the *Faraday effect*, i.e. the optical activity induced by an external magnetic field giving rise to circular dichroism similar to that observed for absorption bands of optically active complexes, for $IrCl_6^{--}$. Since the signs of the Faraday effect expected for the two excitations are opposite, and known, it became clear that the order of orbitals (7.18) actually is

$$\pi\, t_{1g} > (\pi + \sigma)\, t_{1u} > \pi\, t_{2u} > \ldots > (\sigma + \pi)\, t_{1u} > \ldots \qquad (7.19)$$

which colloquially is called the "Virginian Revolution" since the measurements were made at the University of Virginia in Charlottesville. There is little doubt that (7.19) obtains for all usual hexahalide complexes, and actually, certain effects in hexabromides are now easier to understand. If (7.19) had been realized earlier, the optical electronegativities x_{opt} of Table 7.1 would probably have been based on the second Laporte-allowed transition due to the excitation of the approximately non-bonding πt_{2u} orbitals and the values would have 0.1 unit lower for the central atoms.

Anyhow, the separation between πt_{1g} and πt_{2u} is remarkably constant 6 to 8 kK and *entirely* due to ligand-ligand anti-bonding effects. Correspondingly, the 4f group hexahalide complexes studied by RYAN and JØRGENSEN (1966) have unusually high optical electronegativities in Table 7.2 because the set of πt_{1g} orbitals with four angular node-planes produces the first electron transfer band. The strong inter-ligand repulsion also explains the rather paradoxical fact noted by BALCHAN and DRICKAMER (1961) that electron transfer to the lower 5d-sub-shell in hexahalide anions decreases their wavenumber as a function of increasing pressure up to the region of some 100000 atm., whereas Δ increases some 20—40% and the inter-sub-shell transitions shift towards higher wavenumbers. This is another way of saying that the minimum of the potential surface of electron transfer states occurs for slightly smaller internuclear distances than for the groundstate, whereas the excited states of inter-sub-shell transitions rather are expanded. The red-shift cannot be understood on basis of an electrostatic model alone, because the electron transfer annihilates a part of the Madelung potential. However, if the first electron transfer bands have ligand-ligand anti-bonding effects in the interval 6 to 10 kK, it is easy to imagine that these effects become more important for shorter internuclear distances $M-X$. Actually, crystals with smaller $M-X$ distance, say A_2MX_6 arranged according to $A=$

$$\text{As}(C_6H_5)_4^+ \sim N(C_4H_9)_4^+ > N(CH_3)_4^+ > Cs^+ > Rb^+ > K^+ > \ldots \qquad (7.20)$$

show red-shifts of $IrCl_6^{--}$ (JØRGENSEN, 1963a) also varying with the main constituent of Cs_2MCl_6 substituted with a few % Ir:

$$CeCl_6^{--} > SnCl_6^{--} > PbCl_6^{--} > PtCl_6^{--} \sim IrCl_6^{--} \qquad (7.21)$$

and organic solvents have a remarkable effect on $IrBr_6^{--}$, $OsBr_6^{--}$ and OsI_6^{--} (JØRGENSEN, 1962f) where the red-shift

$$\begin{aligned} H_2O > HCF_2CF_2CH_2OH > CH_3OH \sim C_2H_5OH > CH_3NO_2 \sim CH_3CN \\ > CHCl_3 > CHCl_2CHCl_2 > (CH_3)_2CO \sim (CH_3)_2SO \end{aligned} \qquad (7.22)$$

probably indicates decreasing $M-X$ distances. The advantage of a complex such as $IrBr_6^{--}$ is that it exchanges its bromide ligands with the solvent extremely slowly. The fine structure of the narrow band groups also change in a characteristic way with the ligands.

The importance of ligand-ligand anti-bonding was also confirmed by a study of the mixed complexes $IrCl_xBr_{6-x}^{--}$ and $OsCl_xBr_{6-x}^{--}$ (including the *cis*- and *trans*-isomers) by JØRGENSEN and PREETZ (1967). The difference 6 kK between the intrinsic excitation energies of the chloride and bromide ligands is not sufficient to compensate for the inter-ligand effects. The effects of intermediate coupling are also important in bromide and iodide complexes.

SCHMIDTKE (1968) has investigated the topological problem what would be the order of one-electron energies in the cluster X_6 without considering the central atom M. This argument separates the orbitals strongly according to their number of node-planes, except that the higher eigen-value of symmetry type t_{1u} has unusually high energy. This treatment throws a certain light on (7.19) but it is imperative *also* to consider the influence of chemical bonding with definite l-orbitals of the central atom. What is perhaps even more important is the different diagonal elements of energy of π and σ orbitals, as discussed in connection with eq. (7.3). When observing electron transfer spectra, excitation energies rather than ionization energies are determined. However, the effects of interelectronic repulsion such as eq. (7.4) are expected mainly to compress the differences between the orbital energies (7.19) relative to what they would be for ionization. SCHMIDTKE's ideas are more directly applicable to actual clusters such as Si_4^{-4} and P_4.

The general influence of ligand-ligand anti-bonding is also perceptible in the absorption spectra of d^8-phosphine-halide complexes discussed by CANADINE (in press). The quadratic chromophores $trans$-Pd(II)P_2X_2 have their intense electron transfer band at about 10 kK lower wavenumbers than otherwise comparable cis-Pd(II)P_2X_2. There may be a variety of contributions to this behaviour, partly the lower ionicity in the cis-isomer because of the different Madelung potential; but the main reason seems to be the odd combination of the two σ-bonding lone-pairs of the cis-isomer being needed for obtaining Laporte-allowed transitions to the $(x^2 - y^2)$-like empty orbital of the central atom, whereas the node-plane separating the two lone-pairs in the trans-isomer has much weaker energetic effect, the inter-ligand distance being approximately $\sqrt{2}$ times larger. Mixed phosphine- or arsine-halide complexes show quite complicated behaviour when the optical electronegativity of, say, the π orbitals of iodide approach that of the σ orbitals of the phosphine or arsine. Weak deviations (D_{2d} like $CuCl_4^{--}$ or C_{2v} as might be suggested from the stability and formation constants of cis-Pd(NH$_3$)$_2$(H$_2$O)$_2^{++}$; cf. RASMUSSEN and JØRGENSEN, (1968)) from quadratic structure are also suspected.

The absorption spectra of hexahalide complexes have been reviewed by JØRGENSEN (1967b and 1968b). Since the former review appeared, important data have been published by BRISDON, LESTER and WALTON (1967) on TiX_6^{--} and ZrX_6^{--} and by FURLANI and ZINATO (1967) on NbX_6^{-}. These species are much more difficult to study than the d^5-and d^6-systems because the tendency towards hydrolysis is much larger. As seen from Table 7.1, there is less difference between Nb(V) and Ta(V) than between Ru(IV) and Os(IV). This corresponds well with other chemical and crystallographic arguments, and it is expected that Zr(IV) and Hf(IV) will be rather similar.

Whereas ARISTOTLE was interested both in formal logics and in botany, there has not in recent times been a very strong interaction between chemistry and formal logics. One sporadic feature is that the British philosopher JEVONS pointed out that the natural sciences in actual practice frequently use definitions of the type "metals are materials having the properties common to sodium, iron and gold". This is very attractive alternative to the class-inclusion definitions normally used; the examples are chosen as different as possible and yet compatible with the intention. The specific density and the chemical reactivity achieve rather extreme variations going from sodium to gold. Iron was probably included by JEVONS in order to allow ferromagnetic materials, if it was not a tribute paid to the existence of the transition groups. The trouble with such definitions is a certain historical evolution. When radioactivity was discovered, uranium might have been included in order to have a radio-active metal included, serving as well other purposes such as combining strong chemical reactivity with high density and having a variety of oxidation states. However close JEVONS' type of definition may be to chemists' actual thinking, it can simply not hold water in the strictest sense (like predictions of the future, by the way) because no other metal is member of the class only consisting of sodium, iron and gold. Both ARISTOTLE and the reflecting chemist immediately start wondering about essential and accidental properties. This is a most complicated subject; at least the difference is not between negative (e. g. non-radioactive), and positive properties; an extensional classification of categorical proposi-tions can be made perfectly symmetric allowing biunique conversions (JØRGENSEN, 1964b and 1967c). The question is rather one connected with modal logics; the metallicity is known from actual materials and is not a construction of the same kind as differentiable functions. At one time, the class of non-metallic ferromagnets was thought to be as empty as earlier the class of aerostats heavier than air.

We meet a comparable problem when defining oxidation states from preponderant electron configurations. We are not very accustomed to think about *properties of classes* though BERTRAND RUSSELL elaborated his theory of types for treating this problem. That it is the class itself, and not its members individually, which may be numerous, has been clear for many years. It was more surprising for the chemist to meet the states of a neon atom classified in such a way that $1s^2 2s^2 2p^5 nl$ applies to all known excited levels before the first ionization limit. The low-lying excited levels of IrF_6 belong together with the groundstate to a situation where three 5d-like electrons and hence the oxidation state Ir(VI) describe satisfactorily the symmetry types and the relative order of energy. We have not committed us to any statement on the fractional charge δ of the central atom though it probably is in the interval between 2 and 3. The broad electron transfer bands of IrF_6 beginning in the near ultra-violet have excited levels belonging to a preponderant electron configuration containing four 5d-like electrons and *hence* Ir(V) and a collectively oxi-dized set of ligands F_6^{-5}. The latter statement is not meant to discuss the fractional charge of fluorine which anyhow is close to 0.5 both in the groundstate and the excited levels, but refer to the non-closed-shell behaviour of the six ligands. The class of a given set of members may have interesting or relevant properties, and other properties which obviously (to the chemist) are without interest. Thus, the class of the halogens F, Cl, Br, I, At have many characteristic properties though most people fall in JEVONS' trap if they try to explain them explicitly. For instance, it may be a chemically interesting statement that all halogens have odd atomic numbers. We cannot imagine, after MOSELEY, a situation where a new element would be called a halogen without having odd Z. On the other hand, it is considered an accident (only of interest to somebody taking only the first half of the alphabetic index of "Chemical Abstracts" from the shelf) that in English, the names of the halogens start with one of the ten first letters of the alphabet. If $Z = 85$ had been called syntomine (from συντομία = brevity) or taxyboline (from ταχυβόλον = rapid fire-gun), the rule would not have been valid.

Now, we are having new surprises all the time. The external world escapes our control in many ways, but in particular because we cannot know exactly how is the future. Hence, it is always useful to be prepared by considering how new facts will be combined with old ones.

In the writer's opinion, the proposition to assign the oxidation state Cr(III) to the central atom in $Cr(H_2O)_6^{+3}$ or Sm(II) to its dispersion in CaF_2 is much closer to determining Z of a halogen than to choosing the name, though there are admittedly both attitudes involved. One fundamental problem is that of isolability of constituents. Organic chemistry has been much more fortunate at this point than inorganic chemistry. The typical organic molecule is volatile and can, if necessary, be studied by electron diffraction or microwave rotational spectroscopy in the gaseous state. The separation by distilling columns have been further refined into techniques such as gas chromatography. This is why many school books state that chemical compounds (always) consist of molecules. Not an easy position to defend in the case of NaCl, GaAs, ruby, non-stoichiometric iron sulphide or, in general, for a great part of inorganic compounds. Solid state chemistry and non-Daltonian descriptions progressed very slowly relative to organic chemistry which, until the dawn of the polymer industry, always threw contemptuously down the drain the brown tars of black residues not being decent compounds.

Inorganic chemistry has another prototype of isolated entity, the complex ion. Certain cases such as $Co(NH_3)_6^{+3}$ or $PtCl_6^{--}$ have as much individuality as SF_6 or the typical organic molecules; others, such as $Ni(H_2O)_6^{++}$ may be stable constituents in crystalline salts and they definitely exist in aqueous solution as seen from the absorption spectrum but they are known to exchange with the water molecules of the solvent in somewhat less than a millisecond at room temperature. The inorganic chromophores $Cr(III)O_6$ or $Ni(II)O_6$ dispersed in crystalline or vitreous materials have an equal individuality; but how make a clear-cut limit to the surroundings? In anhydrous $NiCl_2$ or $CsNiCl_3$ we have the situation of octahedral $Ni(II)Cl_6$ but each chloride is connected with three or with two nickel atoms. In NiO, the octahedral $Ni(II)O_6$ has each oxygen bound to six nickel atoms. In all of these cases, we have given arguments for the oxidation state being Cr(III) and Ni(II) in an absolute sense. We are not isolating three or eight d-like electrons in a spatial way, but by assigning this integer to a preponderant electron configuration. We know already that if we had the computers available to give a total wavefunction Ψ even of mediocre quality for $Ni(H_2O)_6^{++}$, it would not be a well-defined electron configuration. Since we have already lost this battle in advance, it may be worth looking on what is liable to remain of present-day quantum mechanics in future descriptions.

ARISTOTLE and medieval scholars believed two things can be essentially identical; what is meant in daily life with two coins being identical. Later philosophers argued convincingly (cf. JØRGENSEN, 1964b) that two things cannot be identical in all properties; this would involve spatial relations of such a character as not only superposing the two things but just retaining one. The essential similarities which are so striking in the reproduction of animals or in the crystallization process were thought to be regularities at a much higher level, having nothing to do with identity in the logical sense. Quantum mechanics made a radical change of this opinion. The anti-symmetrization of Ψ is made in such a way that not only are the electrons in a kangaroo in Australia and the electrons in the Eiffel tower indiscernible; but in any situation where we consider both classes in our system, they are the *same*. This is one of the best arguments for considering electronic densities as something much more fundamental than the electrons which have been deprived of their individuality in such a fastidious way as the anti-symmetrization. The closed systems have among other quantum numbers an integer indicating the number of electrons. Since the singularity in the electrostatic potential $U(x,y,z)$ close to a nucleus is so strong, it may even be argued that this integer rather indicates the invarient difference between electron and positron countings at repeated measurements. This is a technicality without importance for us.

However, the questions of isolability achieve a new form. A story about NEWTON which seems to be less apocryphic than that about the apple says that he was once speculating about the tea he had stopped turning with a spoon. The leaves were collecting at the centre of the bottom of the cup, and the surface of the tea was still curved. Did the rest of the universe inform the tea that it was *still* rotating? MACH and EINSTEIN made long elaborations of this question; it is highly probable that we cannot isolate a system from a general background influence. It is quite clear that the atomic constituents of normal compounds are not even isolated from their nearest neighbours in this sense though Na^+Cl^- crystals seem to have fractional charges close to $+1$ and -1 i.e. electron numbers close to the values 10 or 18 one would find also for neon or argon atoms under normal conditions. But this is an extraordinarily sharp determination of an oxidation number, actually similar to the integer for complex ions or zero for neutral molecules. The spectroscopic oxidation state is something else: that the 3d-shell *contains* eight electrons in gaseous Ni^{++}, or in gaseous Ni^0 (with two 4s electrons added) or in $Ni(H_2O)_6^{++}$ is not something happening in our three-dimensional space. Suppose for a moment that we knew very detailed Ψ for a variety of excited or ionized states. We might then look for the electronic densities of these states and form the differences ∂ by each transition:

$$
\begin{aligned}
3d^8 &\rightarrow 3d^8 & \partial &\sim 0 \\
3d^8 &\rightarrow 3d^7 & \partial &\sim -\psi_{3d}^2 \\
3d^8 &\rightarrow 3d^7 \, x & \partial &\sim \psi_x^2 - \psi_{3d}^2 \, .
\end{aligned}
\tag{7.23}
$$

The internal $3d^8$ transitions differ mainly by higher values of the average reciprocal interelectronic distance $\langle r_{12}^{-1} \rangle$ in the excited states; the effect on the electronic density is due to weak second-order effects. The ionization might seem the obvious way of determining the distribution of 3d-density; but here, unfortunately, the second-order effects tend to be more important. The wisest choice is probably to try excited states with one electron in the orbital x, or y, or z,... and it is remarked that at best, we obtain each time the difference ∂ between the density of x (or y,...) and 3d. The only thing one can hope for is a fairly consistent convergence of the set of somewhat indeterminate results (7.23). The statement then is that 8 electrons occur in the somewhat indefinite 3d-shell thus determined. However, this is a very attenuated fashion of expressing the situation. It might already be argued that in complexes having positive S, the spatial distribution of uncompensated spin-density is in principle an observable (and has indeed been inferred from extremely delicate experiments on the neutron diffraction of $3d^5$ manganese (II) and $4f^{11}$ erbium (III) compounds) giving a much better estimate of ψ_{3d}^2 than (7.23). However, complexes may have $S = 0$ and still have a perfect preponderant d^6-configuration such as octahedral low-spin Co(III), Ru(II), Rh(III), Ir(III) and Pt(IV) complexes. In the case of $S = 0$, every single point of space can be shown to have vanishing density of uncompensated spin. In the case of Russell-Saunders coupling (this is for our purposes rather a technicality), the integral of the uncompensated spin-density over all space is proportional to S. It may be interesting to note that whereas the delocalized spin density of d-group complexes on the ligand atoms has the same sign as on the central atom (as would result from any conventional L.C.A.O. description) the very small density of uncompensated spin on the ligands of 4f group complexes, at least in the case of Gd(III), has the *opposite sign* of the central atom (FREEMAN and WATSON, 1962). This feature can only be incorporated in a Hartree-Fock description of the electron configuration $4f^q$ if the orbitals of filled nl-

shells are allowed to have slightly different radial functions according to whether they have the same or the opposite sign of m_s as the majority of the 4f electrons in a given state considered, having a definite value of M_S.

In the discussion of (7.23) it is already clear that a *manifold of states* is compared. Quantum-mechanical calculations have perhaps concentrated too much interest on a single state, such as the groundstate ^1S of the helium atom [again one of these identical entities which can be procured in any set of external conditions]. It might have been fruitful also to compare the class of low-lying, excited levels together with the groundstate. Thus, it is striking that all the excited states of helium below the first ionization energy can be *classified as if* they belong to configurations $(1s)^1 (nl)^1$. They have exactly the symmetry types ^3L and ^1L expected; and one of the aspects of the preponderant electron configuration being $1s^2$ is actually that one-electron substitutions describe the low-lying levels (which, of course, are less low-lying in helium than in all other neutral atoms). The integration of $|\partial|$ of eq. (7.23) over all space may not always correspond to values very close to 0 or to 1; but once the approximate one-electron densities have been established, it becomes clear that most optical excitations are either approximately zero-electron jumps (such as internal l^q-transitions) or approximately one-electron jumps (such as nearly all strong transitions observed). If a transition is approximately a two-electron jump, the energy difference may be known of other reasons, such as between levels of $3d^8$ and $3d^64s4p$ of an iron atom, but the oscillator strength is usually small, even when the two levels have opposite parity, and originates in the very fact that Ψ must be described by configuration intermixing (cf. JØRGENSEN, 1962d). This is a fortiori true for many-electron jumps.

The fortunate situation that spectroscopic oxidation states so frequently, though not always, can be defined for transition group compounds corresponds to the somewhat unexpected case that one can find shells, in particular the partly filled shell, which are so constituted that the "vertical" optical transitions (not modifying the internuclear distances of the groundstate) approximately correspond to one-electron excitations of these shells. In the case of closed-shell systems, it can frequently be discussed which of several alternatives is the better description. In Chapter 9, it is shown that CCl_4 undoubtedly involves C(IV) and Cl($-$I) whereas CH_4 may be C(IV)H($-$I) to be preferred of chemical reasons *or* C($-$IV)H(I). This is not the question of fractional charges δ as much as the question of spectroscopic properties and applied group theory. In the case of partly filled shells, such alternatives rarely are present, and the completely extreme case are the 4f-group M(III) with atomic spectra hardly dependent on the ligands.

It is becoming clear that chemistry realizes far fewer alternatives than an atomic spectroscopist might have thought of. From the logician's point of view, this is exactly what regularities in the natural sciences consist of: among the 2^n conceivable combinations of n properties in divalent logic $(ab, a\bar{b}, \bar{a}b, \bar{a}\bar{b}$ for $n = 2$ where $\bar{a}$ indicates non $-a)$ only a very few are represented in reality. Thus, until recently, the class of ferromagnetic materials was included in the class of metals. This illustrates the inherent surprises always possible; we know now that red EuS or green $CrBr_3$ by sufficient cooling become ferromagnets and remain isolators. It is much easier to be certain that something exists than its impossibility; the okapi or the non-conservation of parity in β-decay always tend to show up. This is why the difficult concept of essential similarities is so important for the natural sciences, not just any unimportant similarity. Actually, one can readily show that two things *cannot* be entirely different, having all properties different. If one defines the property c as being present when $a\bar{b}$ or $\bar{a}b$; and $\bar{c}$ when ab or $\bar{a}\bar{b}$, the two things having definite signs for all properties including a, b and c have at least one of the three properties in common. This is a logical type of omnipresent background property; but physics and chemistry probably present other indications of an inherent connection between everything.

There is little doubt that progress in chemistry and the other natural sciences most frequently take place via logically unsatisfactory conclusions by analogy. Thus, the octahedral chromophores $Cr(III)X_6$ are so frequent that the existence of a black semi-conducting $NaCrSe_2$ known from the crystal structure to have an octahedron of selenium atoms around each chromium atom makes one intimate the occurrence of $Cr(III)Se_6$ though the evidence is less direct than for the monomeric green $Cr(Se_2P(OC_2H_5)_2)_3$ of which the spectrum can be studied in organic solvents. Some of the concepts created by analogy may show a historic evolution; thus the idea of side-group behaviour has different connotations when discussing the chemical behaviour of scandium and gallium and when considering the preponderant configurations $[Ar]3d4s^2$ and $[Ar]3d^{10}4s^24p$ of the groundstate of the neutral atoms. Certain incomplete concepts are far more universal than others restricted to one professional group such as inorganic chemists. Thus, the idea of causality had a great impact on classical mechanics and obviously has much to be said for it. But there is one aspect which seems to be gone forever. Quantum events may have macroscopic consequencies of any size. Supposing we connect a Geiger counter with a gun and let a cat stand before the gun in so long a time that a radioactive preparation has 8% probability of triggering the Geiger counter and the gun. Sofar we can tell today, we have absolutely no way of predicting the outcome of this unpleasant experiment, though it obviously is of great interest for the cat, not to speak about the possible kittens subsequently born by the cat if the radioactive event did not take place. We have much less the feeling today that science can be made complete, and the extrapolation proposed by LAPLACE when writing about the celestial mechanics that the Universe is a perfectly predictable clock seems less plausible than ever.

It is psychologically interesting that the young quantum mechanics already has got a rather dogmatic, so-called Copenhagen, formulation. Fundamentally, we cannot know what physicists think about this subject in the future. Chemists are now running into problems they did not care about previously. The chemical bonding and the atomic spectroscopy as it developed since 1860 were so obviously difficult to adapt in the classical mechanics and electrodynamics that chemists lived a happy, isolated existence as exclusive experimentalists. It may be true that SCHRÖDINGER's equation (3.16) has reduced chemistry to applied mathematics; but it is of no help to anybody as long we cannot produce the interesting predictions. Contrary to the case of modern physics, chemistry has been in the fortunate situation that the sharp specialization and division of work between theorists and experimentalists have not yet appeared. It may even be argued that most important new ideas in chemistry since 1950 have been introduced by experimentalists interested in theory rather than the other way round. Much lip-service is paid to the importance of team-work; but for most people, it is easier to speak to themselves than to anybody else. Exclusive theoretical work tends to be degrading and sterile. By the same token as physics once was called natural philosophy (the name of "Philosophical Magazine" and of doctor's degrees in many countries are reminiscences of this notion) and specialized scientists later looked down on general philosophy, time may come when chemists look on deteriorated forms of theoretical chemistry with the same healthy suspicion. Chemistry is interesting in another way than computer technique.

Bibliography

ALBRECHT, A. C., and W. T. SIMPSON: J. Am. Chem. Soc. 77, 4454 (1955).
ALBRECHT, G.: Z. Chem. 3, 182 (1963).
BADDLEY, W. H., F. BASOLO, H. B. GRAY, C. NÖLTING, and A. J. POË: Inorg. Chem. 2, 921 (1963).
BÄHR, G., and G. SCHLEITZER: Chem. Ber. 90, 438 (1957).
BALCH, A. L., and R. H. HOLM: J. Am. Chem. Soc. 88, 5201 (1966).
— Inorg. Chem. 6, 2158 (1967).
BALCHAN, A. S., and H. G. DRICKAMER: J. Chem. Phys. 35, 356 (1961).
BANFORD, L., and G. E. COATES: J. Chem. Soc. 3564 (1964).

BAYER, E.: Chimia 16, 333 (1962).
— Angew. Chem. 78, 834 (1966).
BENNETT, L. E., and H. TAUBE: Inorg. Chem. 7, 254 (1968).
BENNETT, M. A., R. BRAMLEY, and P. A. LONGSTAFF: Chem. Commun. 806 (1966).
BRESLOW, R., HAI WONG CHANG, and W. A. YAGER: J. Am. Chem. Soc. 85, 2033 (1963).
BRIEGLEB, G.: Elektronen-Donator-Acceptor-Komplexe. Berlin–Göttingen–Heidelberg: Springer 1961.
BRISDON, B. J., T. E. LESTER, and R. A. WALTON: Spectrochim. Acta 23 A, 1969 (1967).
BÜCHNER, W., and E. WEISS: Helv. Chim. Acta 47, 1415 (1964).
— Helv. Chim. Acta 49, 907 (1966).
BURGER, K., and B. PINTÉR: J. Inorg. Nucl. Chem. 29, 1717 (1967).
CALDERAZZO, F., and C. FLORIANI: Chem. Commun. 139 (1967).
—, and R. HENZI: J. Organometall. Chem. 10, 483 (1967).
—, and C. FLORIANI: Chem. Commun. 417 (1968).
CANADINE, R. M.: J. Chem. Soc., in preparation.
CHOPOORIAN, J. A., K. O. LOEFFLER, W. F. MARZLUFF, and G. H. DORION: Nature 204, 180 (1964).
CIAMPOLINI, M.: Inorg. Chem. 5, 35 (1966).
— Struct. Bonding, in preparation.
CLACK, D. W., and N. S. HUSH: J. Am. Chem. Soc. 87, 4238 (1965).
— —, and J. R. YANDLE: Chem. Phys. Letters 1, 157 (1967).
COATES, G. E., and S. I. E. GREEN: J. Chem. Soc. 3340 (1962).
COLLINGE, R. N., R. S. NYHOLM, and M. L. TOBE: Nature 201, 1322 (1964).
COSTA, G., A. PUXEDDU, and G. TAUZHER: Inorg. Nucl. Chem. Letters 4, 319 (1968).
COTTON, F. A., and J. A. McCLEVERTY: Inorg. Chem. 6, 229 (1967).
CROWLEY, P. J., and H. M. HAENDLER: Inorg. Chem. 1, 904 (1962).
DAY, P., and N. SANDERS: J. Chem. Soc. (A) 1530 and 1536 (1967).
— Theoret. Chim. Acta 7, 328 (1967a).
— Coordin. Chem. Rev. 2, 109 (1967b).
—, H. A. O. HILL, and M. G. PRICE: J. Chem. Soc. (A) 90 (1968).
DODD, J. W., and N. S. HUSH: J. Chem. Soc. 4607 (1964).
DOEDENS, R. J., and L. F. DAHL: J. Am. Chem. Soc. 88, 4817 (1966).
DUFF, E. J.: J. Chem. Soc. (A) 434 (1968).
DYER, G., and D. W. MEEK: Inorg. Chem. 6, 149 (1967).
EARNSHAW, A., E. A. KING, and L. F. LARKWORTHY: J. Chem. Soc. (A) 1048 (1968).
EHRENBERG, A., L. E. G. ERIKSSON, and F. MÜLLER: Nature 212, 503 (1966).
EISENBERG, R., E. I. STIEFEL, R. C. ROSENBERG, and H. B. GRAY: J. Am. Chem. Soc. 88, 2874 (1966).
ENGLMAN, R.: Mol. Phys. 6, 345 (1963).
FACKLER, J. P., and D. COUCOUVANIS: J. Am. Chem. Soc. 88, 3913 (1966).
FARVER, O., and GWYNETH NORD: Chem. Commun. 736 (1967).
FELTHAM, R. D., and W. SILVERTHORN: Inorg. Chem. 7, 1154 (1968).
FENSKE, R. F.: J. Am. Chem. Soc. 89, 252 (1967).
FISCHER, E. O., and H. H. LINDNER: J. Organometall. Chem. 1, 307 (1964).
FLORIANI, C., M. PUPPIS, and F. CALDERAZZO: J. Organometall. Chem. 12, 209 (1968).
FORSTER, R., and T. J. THOMSON: Trans. Faraday Soc. 59, 296 (1963).
FOWLES, G. W. A., T. E. LESTER, and R. A. WALTON: J. Chem. Soc. (A) 1081 (1968).
FREEMAN, A. J., and R. E. WATSON: Phys. Rev. 127, 2058 (1962).
FURLANI, C., u. E. ZINATO: Z. Anorg. Allgem. Chem. 351, 210 (1967).

184   Electron Transfer Spectra and Collectively Oxidized Ligands

GERLOCH, M., and F. E. MABBS: J. Chem. Soc. (A) 1598 (1967).
GLICK, M. D., and L. F. DAHL: Inorg. Chem. 5, 289 (1966).
GRAY, H. B.: Transition Metal Chem. 1, 239 (1965).
HALFPENNY, M. T., J. G. HARTLEY, and L. M. VENANZI: J. Chem. Soc. (A) 627
    (1967).
HAMILTON, W. C., and I. BERNAL: Inorg. Chem. 6, 2003 (1967).
HARRIS, C. M., and E. D. McKENZIE: J. Inorg. Nucl. Chem. 25, 171 (1963).
HENNING, G. N., A. J. McCAFFERY, P. N. SCHATZ, and P. J. STEPHENS: J. Chem.
    Phys. 48, 5656 (1968).
HERZOG, S., u. M. SCHMIDT: Z. Chem. 3, 392 (1963).
—, u. K. GUSTAV: Z. Anorg. Allgem. Chem. 346, 150 and 162 (1966).
—, u. H. ZÜHLKE: Z. Chem. 6, 382 (1966).
—, u. R. BERGER: Z. Chem. 6, 434 (1966).
—, u. E. WULF: Z. Chem. 6, 434 (1966).
HOLM, R. H., A. L. BALCH, A. DAVISON, A. H. MAKI, and T. E. BERRY: J. Am.
    Chem. Soc. 89, 2866 (1967).
ITOH, K.: Chem. Phys. Letters 1, 235 (1967).
JENSEN, K. A., and C. K. JØRGENSEN: Acta Chem. Scand. 19, 451 (1965).
JICHA, D. C., and D. H. BUSCH: Inorg. Chem. 1, 872 and 878 (1962).
JØRGENSEN, C. K.: Acta Chem. Scand. 9, 1362 (1955).
— Acta Chem. Scand. 10, 518 (1956).
— Acta Chem. Scand. 11, 166 (1957).
— Mol. Phys. 2, 309 (1959).
— Solid State Phys. 13, 375 (1962a).
— Absorption Spectra and Chemical Bonding in Complexes. Oxford: Pergamon Press
    1962b. U. S. distributor: Addison-Wesley.
— Acta Chem. Scand. 16, 2406 (1962c).
— Orbitals in Atoms and Molecules. London: Academic Press 1962d.
— J. Inorg. Nucl. Chem. 24, 1571 (1962e).
— J. Inorg. Nucl. Chem. 24, 1587 (1962f).
— Acta Chem. Scand. 17, 1034 (1963a).
— Inorganic Complexes. London: Academic Press 1963b.
— Inorg. Chem. 3, 1201 (1964a).
— Logique et Analyse (Louvain) 7, 233 (1964b).
— Proceedings of the Symposium on Coordination Chemistry on Coordination
    Chemistry in Tihany 1964, p. 11. Budapest: Hungarian Academy of Sciences
    1965.
— Coordin. Chem. Rev. 1, 164 (1966a).
— The Biochemistry of Copper (Editors: J. PEISACH, P. AISEN, and W. E. BLUM-
    BERG), p. 1. New York: Academic Press 1966b.
— Advances in Chemistry Series 62, 161. Washington D. C.: American Chemical
    Society 1967a.
— Halogen Chemistry 1, 265 (1967b). London: Academic Press.
— Logique et Analyse (Louvain) 10, 141 (1967c).
—, u. W. PREETZ: Z. Naturforsch. 22a, 945 (1967).
— Inorg. Chim. Acta Rev. 2, in press (1968a).
— Chemical Applications of Spectroscopy. (Editor: B. G. WYBOURNE) New York:
    Interscience 1968b.
KAUFMAN, J. J.: J. Phys. Chem. 68, 2648 (1964).
KHEDEKAR, A. V., J. LEWIS, F. E. MABBS, and H. WEINGOLD: J. Chem. Soc.
    (A) 1561 (1967).
KIDA, S., and H. YONEDA: Nippon Kagaku Zasshi 76, 1059 (1955).

KIMURA, K., K. YOSHINAGA, and H. TSUBOMURA: J. Phys. Chem. **71**, 4485 (1967).

KING, R. B.: Inorg. Chem. **2**, 641 (1963).

KÖNIG, E., and K. MADEJA: J. Am. Chem. Soc. **88**, 4528 (1966).

— — Inorg. Chem. **6**, 48 (1967).

— —, and K. J. WATSON: J. Am. Chem. Soc. **90**, 1146 (1968).

KREISMAN, P., R. MARSH, J. R. PREER, and H. B. GRAY: J. Am. Chem. Soc. **90**, 1067 (1968).

KRUMHOLZ, P.: Inorg. Chem. **4**, 609 and 612 (1965).

— Inorg. Chim. Acta **1**, 27 (1967).

KUCK, M. A., and G. URRY: J. Am. Chem. Soc. **88**, 426 (1966).

KUKUSHKIN, Y. K.: Russ. J. Inorg. Chem. (English Transl.) **10**, 325 (1965).

LEMPKA, H. J., T. R. PASSMORE, and W. C. PRICE: Proc. Roy. Soc. (London) **A 304**, 53 (1968).

LEVER, A. B. P.: Advan. Inorg. Chem. Radiochem. **7**, 27 (1965).

LIVINGSTONE, S. E.: Quart. Rev. (London) **19**, 386 (1965).

LLOYD, D. R.: private communication.

LOCKE, J., and J. A. McCLEVERTY: Inorg. Chem. **5**, 1157 (1966).

MAHON, C., and W. L. REYNOLDS: Inorg. Chem. **6**, 1927 (1967).

MAKI, A. H., N. EDELSTEIN, A. DAVISON, and R. H. HOLM: J. Am. Chem. Soc. **86**, 4580 (1964).

MALATESTA, L.: Progr. Inorg. Chem. **1**, 284 (1959).

MARKS, D. R., D. J. PHILLIPS, and J. P. REDFERN: J. Chem. Soc. (A) 1464 (1967).

MASON, W. R., and H. B. GRAY: Inorg. Chem. **7**, 55 (1968).

McCLEVERTY, J. A., J. LOCKE, E. J. WHARTON, and M. GERLOCH: J. Chem. Soc. (A) 816 (1968).

McCLURE, D. S.: Solid State Phys. **9**, 399 (1959).

MERCIER, R. C., and M. R. PÃRIS: Bull. Soc. Chim. France 3577 (1965).

MOSER, F. H., and A. L. THOMAS: Phthalocyanine Compounds, Monograph No. 157. Washington D. C.: American Chemical Society 1963.

MÜLLER, F., P. HEMMERICH, and A. EHRENBERG: European J. Biochem. **5**, 158 (1968).

MURRELL, J. N.: Quart. Rev. (London) **15**, 191 (1961).

NORBURY, A. H., E. A. RYDER, and R. F. WILLIAMS: J. Chem. Soc. (A) 1439 (1967).

NORGETT, M. J., J. H. M. THORNLEY, and L. M. VENANZI: J. Chem. Soc. (A) 540 (1967a).

— — — Coordin. Chem. Rev. **2**, 99 (1967b).

—, and L. M. VENANZI: Inorg. Chim. Acta **2**, 107 (1968).

NYHOLM, R. S.: Quart. Rev. (London) **3**, 321 (1949).

PAPPALARDO, R., and C. K. JØRGENSEN: J. Chem. Phys. **46**, 632 (1967).

— Inorg. Chim. Acta **2**, 209 (1968).

RASMUSSEN, L., and C. K. JØRGENSEN: Acta Chem. Scand. **22**, 2313 (1968).

RODLEY, G. A., and P. W. SMITH: J. Chem. Soc. (A) 1580 (1967).

RÖHRSCHEID, F., A. L. BALCH, and R. H. HOLM: Inorg. Chem. **5**, 1542 (1966).

RYAN, J. L., and C. K. JØRGENSEN: J. Phys. Chem. **70**, 2845 (1966).

SARTAIN, D. R., and M. R. TRUTER: Chem. Commun. 382 (1966).

— — J. Chem. Soc. (A) 1264 (1967).

SCHÄFFER, C. E.: Abstracts of Papers for 140. Meeting of American Chemical Society, Chicago, September 1961, page 24 N.

—, and C. K. JØRGENSEN: Mat. fys. Medd. Dan. Vid. Selskab **34**, no. 13 (1965).

SCHMIDTKE, H. H.: Ber. Bunsenges. Physik. Chem. **71**, 1138 (1967).

—, and D. GARTHOFF: J. Am. Chem. Soc. **89**, 1317 (1967).

— J. Chem. Phys. **48**, 970 (1968).

SCHMITT, R. D., and A. H. MAKI: J. Am. Chem. Soc. 90, 2288 (1968).
SCHRAUZER, G. N., u. V. MAYWEG: Z. Naturforsch. 19b, 192 (1964).
— — J. Am. Chem. Soc. 87, 3585 (1965).
— Naturwissenschaften 53, 459 (1966).
—, and V. P. MAYWEG: J. Am. Chem. Soc. 88, 3235 (1966).
—, and R. J. WINDGASSEN: J. Am. Chem. Soc. 88, 3738 (1966).
SHUPACK, S. I., E. BILLIG, R. J. H. CLARK, R. WILLIAMS, and H. B. GRAY: J. Am. Chem. Soc. 86, 4594 (1964).
STEVENS, K. W. H.: Proc. Roy. Soc. (London) A 219, 542 (1953).
STEVENSON, P. E.: J. Mol. Spectr. 18, 51 (1965).
STIEFEL, E. I., Z. DORI, and H. B. GRAY: J. Am. Chem. Soc. 89, 3353 (1967).
TAUBE, R.: Z. Chem. 6, 8 (1966).
THOMAS, D. D., H. KEELER, and H. M. MCCONNELL: J. Chem. Phys. 39, 2321 (1963).
VÄNNGARD, T., and S. ÅKERSTRÖM: Nature 184, 183 (1959).
VEILLARD, A., and B. PULLMAN: J. Theoret. Biol. 8, 307 and 317 (1965).
VLČEK, A. A., and J. HANZLIK: Inorg. Chem. 6, 2053 (1967).
WASSERMANN, E., L. BARASH, and W. A. YAGER: J. Am. Chem. Soc. 87, 2075 (1965).
WATT, G. W., and L. J. BAYE: J. Inorg. Nucl. Chem. 26, 1531 (1964).
WEBER, G.: Z. Physik. Chem. (Leipzig) 218, 204 and 217 (1961).
WEISS, E., and W. BÜCHNER: Helv. Chim. Acta 46, 1121 (1963).
—, and E. A. C. LUCKEN: J. Organometall. Chem. 2, 197 (1964).
WERDEN, B. G., E. BILLIG, and H. B. GRAY: Inorg. Chem. 5, 78 (1966).
WEST, R., and HSIEN YING NIU: J. Am. Chem. Soc. 85, 2586 and 2589 (1963).
WILLIAMS, R., E. BILLIG, J. H. WATERS, and H. B. GRAY: J. Am. Chem. Soc. 88, 43 (1965).
WILLIAMS, R. J. P.: J. Chem. Soc. 137 (1955).
YAMADA, S.: Coordin. Chem. Rev. 2, 83 (1967).
ÅKERSTRÖM, S.: Arkiv Kemi 14, 403 (1959).

8. Oxidation States in Metals and Black Semi-Conductors

Even in cases where the optical absorption spectra cannot be measured of *4f-group compounds*, measurements of the macroscopic magnetic susceptibility or of the electron paramagnetic resonance spectrum nearly always allow the determination of the S, L, J ground level characterizing either the oxidation state M(II) or M(III). Table 6.1 described the variation of the position of the first excited level belonging to the configuration $4f^{q-1}5d$ of M(II) and of M(III) in CaF_2 crystals. It is seen that even in the case of Ce(III) or Tb(III), this configuration is extremely excited in the fluoride (or in the aqua ion, by the way). It is quite conceivable that the decreased excitation energy known from relatively covalent complexes of cerium (III) in solution (JØRGENSEN, 1956) may contribute to yellow or brown colours of highly covalent solid compounds; but there is no indication that the groundstate deviates from $4f^q$, i.e. M(III). The 4f group compounds MP, MAs, MSb and MBi crystallize in NaCl lattice and have magnetic moments expected for octahedral chromophores $M(III)X_6$ (HULLIGER, 1968b). These materials are black and behave as metals; but it is not completely certain whether they would be low-energy-gap semi-conductors if they were perfectly stoichiometric. It may be worth noting that the phenomenological criterion for a metallic sample is that its electric conductivity decreases or at most stays roughly constant as a function of increasing temperature, whereas the typical semi-conductor has a strongly increasing conductivity. If the results are reproducible and are not influenced by chemical rearrangement, the energy-gap is given empirically as the activation energy for the conduction from the Arrhenius equation as a function of the reciprocal absolute temperature. Certain metals are "atypical" or "accidental" by having low conductivities; elemental bismuth is such a case, and mercury and gallium show somewhat this tendency.

The 4f-group nitrides MN also crystallize in NaCl type; their magnetic moments are those expected for M(III) (DIDCHENKO and GORTSEMA, 1963) except CeN containing Ce[IV] and SmN, EuN and YbN apparently containing some M[II], and the metallicity observed is probably due to small deviations from stoichiometry, such as vacancies among the $N(-III)$ anions. The reflection and transmission spectra suggest that the intrinsic energy-gap is as large as 20 kK in the stoichiometrically pure materials (SCLAR, 1964). Cf. also SCHUMACHER and WALLACE (1966).

On the other hand, already M(II) in CaF_2 have groundstates $4f^{q-1}5d$ for M $=$ La, Ce and Gd according to Table 6.1. When materials having simultaneously a partly filled 4f shell and a 5d electron are undiluted, forming definite compounds, metallicity is observed. JØRGENSEN (1964) and HULLIGER (1968b) discussed how the $4f^{q-1}5d$ may be a little lower for covalent compounds than for M(II) in CaF_2. Thus, J. W. McCLURE (1963) found that SmS, EuS and YbS have the magnetic moments expected for M(II) and are semi-conductors, whereas all the other 4f-group monosulphides, though also crystallizing in NaCl type, are metallic and have the magnetic moments expected for M(III). The same is true for selenides and tellurides with the single exception that TmTe may perhaps be a semi-conducting thulium(II) compound. It is interesting that the mixed crystals $Nd_xSm_{1-x}Se$ studied by REID et al. (1964) have a very sharp limit of metallicity; they are semi-conductors for x < 0.12.

DRUDING and CORBETT (1959) prepared solid compounds (crystallizing in $PbCl_2$ lattice) such as dark green $NdCl_2$ and purple NdI_2 which are normal neodymium (II) compounds having intense $4f^4 \rightarrow 4f^35d$ absorption bands in the visible. The corresponding SmX_2, EuX_2, TmX_2 and YbX_2 (and black $DyCl_2$ reported by CORBETT and McCOLLUM, 1966) are also M(II) salts whereas it is doubtful whether PrI_2 is a metal containing Pr(III) or contains Pr(II) (cf. again Table 6.1). On the other hand, LaI_2 and CeI_2 are definitely metals having the 5d-electron delocalized to a conduction band (CORBETT et al. 1961; DWORKIN et al., 1962). The metallic alloy analyzing as $GdCl_{1.6}$ (MEE and CORBETT, 1965) is an interesting case. It has exactly the magnetic moment required for Gd(III), but it may contain metallic clusters Gd_3^{+5} bound by four 5d-like electrons per cluster. One of the difficulties by preparing lanthanide metallic elements by electrolysis of molten halides is a strong tendency for the reduced forms to disperse in the melt (recently, elemental lanthanides are normally made in crucibles by reduction with other metals in an aluminothermic type of reaction) which is also true for calcium and strontium. The chemical context of this solubility is not yet clarified.

KLEMM and BOMMER (1937) emphasized that the properties of europium metal are very different from those of the neighbour elements which interpolate from lanthanum to lutetium, with the exception of ytterbium. The molar volumes of Eu and Yb are considerably larger than of the other lanthanides, their boiling-points are lower and their chemical reactivity much larger; thus, they dissolve in liquid ammonia forming M(II) and the blue, solvated electrons. Said in other words, they resemble Ba much more than La. One would not normally speak about sodium (I), europium (II), gadolinium (III) or holmium (III) electronides in the case of metallic Na, Eu, Gd, Ho. This nomenclature would not by itself be unreasonable, but it runs into difficulties, as we see below, when we are going to classify metallic Cr, Mo or U. On the other hand, there is something so barium-like about metallic Eu and Yb that is tempting to ascribe bivalency in some way. Professor K. A. JENSEN, University of Copenhagen, suggested to write oxidation states determined from the preponderant configuration of the groundstate by magnetic measurements or comparable physical properties in brackets Eu[II] and Gd[III]. We call these, not directly spectroscopic, oxidation states *conditional oxidation states* since they are derived from the measurements related to the *partly filled shell with smallest average radius*. It would be appropriate to consider the configurations $3d^64s4p$ and $3d^64s\ 117d$ of the iron atom as Fe[II] whereas $3d^74s$ would be Fe[I]. It is remarked that this definition makes also the groundstate of the iron atom belonging to the preponderant configuration $3d^64s^2$ Fe[II] whereas the groundstate of the chromium atom $3d^54s$ would be Cr[I] and of the palladium atom $4d^{10}$ Pd[0]. Many parameters in atomic spectroscopy vary in a step-wise fashion with the number of electrons in the partly filled shell with smallest average radius, and it may be quite practical to introduce this new notation.

There is little doubt that "good" metals with approximately one conduction electron per atom have one-electron functions similar to those of free electrons in a constant potential, i.e. plane waves, just modified by the orthogonalization on the filled orbitals in the relatively small percentage of the volume occupied by the atomic cores. The writer's reluctance to accept the energy-band theory does not extend to such clear-cut cases as the alkali metals or the coinage metals which clearly are Cu[I], Ag[I] and Au[I]. But the majority of metals is in a much more complicated and confusing situation. Already the "bivalent" metals Be, Mg, Ca, Sr, Ba, Zn, Cd and Hg need strong mixing of ns and np functions in order to be metals at all; and actually, mercury is not far from behaving like a condensed, noble gas (and the other way round, liquid radon or ($Z = 118$) may be rather mercury-like). When high pressure is applied to metallic strontium or ytterbium, the electric conductance fluctuates several orders of magnitude, and some areas of the temperature-pressure diagram of state actually involve semi-conducting phases (cf. DRICKAMER, 1965). This is a very different situation from the sudden decrease of the molar volume of metallic caesium at the phase transition close to 41000 atm. The general opinion (first suggested by STERNHEIMER) is that this involves the transformation from alkali-metal behaviour built on 6s-conduction to transition-group behaviour having a conduction band constructed mainly of 5d-electrons.

It is not generally possible to ascribe conditional oxidation states to d-group metals. The number of d-electrons does not seem to be an integer because of the excessive delocalization. Magnetic measurements can yield some clues. Thus, nickel heated above its ferromagnetic Curie temperature has $S = 1$ which is a rather strong argument for Ni[II]. But since the five d-orbitals may have somewhat different energy in the earlier 3d-elements (this may be studied by X-ray spectroscopy) even $S = 1$ as the asymptotic magnetization value for iron can always be construed to involve differently localized magnetic moment for each of the two sub-shells. The weak Pauli-paramagnetism of metallic palladium disappears as a linear function by introduction of copper or hydrogen that the final alloy is diamagnetic Pd[0] but the starting material may contain, by delocalization, about half Pd[I] and half Pd[0]. Gaseous palladium atoms are, of course, a kind of a noble gas.

Going along the beginning of a transition group, such as the metals Ca, Sc, Ti, V, Cr; or Sr, Y, Zr, Nb, Mo; or Ra, Ac, Th, Pa, U; the molar volumes decrease rapidly, and the boiling points and heats of atomization increase rapidly. It is not very easy to argue that these materials contain exactly 2,3,4,5 and 6 conduction electrons per atom; as a matter of fact, most of them conduct much less than the alkali metals. One may make a scale of how d-group elements are expected to behave though it would be temerary to talk about Cr[VI] and Mo[VI]; and on such a scale, uranium has difficulty of "keeping up with the Joneses" though it has certainly a much smaller molar volume than a comparison with Nd and Ac would suggest if it was U[III]. The magnetic properties of U are also very d-like. The examples chosen are followed by a plateau of almost invariant molar volume, such as Mo, Tc, Ru, Rh or U, Np, Pu. On the other hand, the lower densities of americium and curium metal are compatible with Am[III] and Cm[III].

Whereas almost all metallic alloys of the 4f group which have been studied have magnetic moments clearly indicating either M[II] or M[III], metallic cerium has a highly unusual phase transition recently further studied by JAYARAMAN (1965). γ-Ce has the atomic volume expected for Ce[III], i.e. one 4f electron, whereas the modification stable at ordinary pressure below $-100°$ C (determined by TROMBE and FOËX 1944) α-Ce has 10% smaller atomic volume and behaves as one would expect by an extrapolation from Y, Zr and La that Ce[IV] would behave. The astonishing fact is that the two phases are isomorphous. One would have expected a gradual variation in the occupation number of 4f- and 5d (or conduction)-like electrons as external pressure or temperature vary; but the choice between the two states is apparently a collective effect. However, JAYARAMAN (1965) reports a critical point at 17500 atm. and 280° C where the two phases have identical volumes.

The anti-ferromagnetic or ferromagnetic coupling effects are much stronger in the 5f group than in the 4f group where the Néel or Curie temperatures always are situated below room temperature (with exception of metallic Gd and GdI_2) and usually close to liquid helium temperatures. Consequently, it is not always easy to recognize the preponderant electron configuration of undiluted 5f-group materials. Further on, the ground J-level is split much more by "ligand field" effects into sub-levels having their individual values of magnetic moment (specifically being only temperature-independent paramagnetic in the case of non-degenerate states for an even number of f electrons) which then are populated as a function of temperature in a Boltzmann distribution. However, the general tendencies are toward M[III] and M[IV], even in metallic alloys. In the search for compounds containing thorium in a lower oxidation state than IV, many metallic compounds have been found. Thus, SCAIFE and WYLIE (1964) described brass-golden ThI_2 probably containing Th[IV] and two conduction electrons as also ThS and ThSe(NaCl type) and ThTe(CsCl type). The non-metallic ThI_3 showing strong dichroic effects (olive-green to violet) in thin flakes on a polarization microscope may contain $Th(III)5d^1$ coupled in some way to explain the diamagnetism (gaseous Th^{+3} has 5f some 10 kK below 6d) or may contain clusters. The cubic structure represented by semi-conducting Th_3P_4 and Th_3As_4 has a strongly distorted $Th(IV)X_8$ coordination. FLAHAUT and collaborators (1965) have studied many 4f-group sulphides and selenides, among which many crystallize in the Th_3P_4 type. Also mixed sulphides such as CaM_2S_4, SrM_2S_4 and BaM_2S_4 (M = La, Ce, Pr, Nd) crystallize in this lattice and seem semi-conducting, whereas Ce_3S_4 is metallic. It is interesting to note that one of the modifications of Ce_2S_3 is a defect version of Ce_3S_4 lacking one out of nine cerium atoms. These cation vacancies are statistically distributed, and the material is not metallic.

Coming back to the 5f group, UN_2 crystallized in CaF_2 type like UO_2 but has a smaller lattice parameter, and seems to contain U(VI) and N(-III). Metallic US and USe (NaCl type), U_3Se_4(Th_3P_4 type) are ferromagnetic and probably contain U(IV), whereas UP, UAs and USb (NaCl type) are anti-ferromagnetic of less certain oxidation state. In metallic alloys, uranium seems to be most frequent as U[IV].

HULLIGER (1968a) recently published an excellent review on d-group (including Lu, Th, U) compounds of S, Se, Te, P, As, Sb and Bi. Many of these contain poly-anions or poly-cations having chemical bonds between atoms of the same element. Poly-anions in solution were early recognized, such as S_2^-, I_3^- or N_3^-. ZINTL dissolved metallic alloys of sodium with a variety of elements in liquid ammonia, where strongly coloured anions such as Sn_9^{-4}, Sb_7^{-3}, Pb_7^{-4}, Pb_9^{-4}, Bi_3^{-3} and Bi_5^{-3} are formed. It is striking that the negative charge of these anions are as if one atom had the most negative oxidation state conceivable, whereas the other atoms were neutral. However, it is not yet known whether this is an accident, and the possible distribution of oxidation states has not been established. We return in Chapter 10 to the question of cluster compounds. MOOSER

---

and PEARSON (1960) and HULLIGER and MOOSER (1965) previously discussed the necessary condition for non-metallicity that all electrons are forming definite bonds (two electrons for each bond) between differing or identical elements, or are ascribed to definite shells of the individual atoms. This is not a sufficient condition, though many metallic materials apparently fulfilling the conditions for being normal valence compounds may be "accidental" metals or slightly non-stoichiometric. The fundamental hypothesis of MOOSER and PEARSON is that either are two atoms of the same element forming a two-electron bond, or otherwise, that no electrons are employed at all. It is quite true that crystallography gives the strong impression that distances between two adjacent atoms are either sufficiently short to indicate a chemical bond or are considerably longer, corresponding to the so-called Van der Waals-radii. However, the difficulty is that in many cases, one does not have a reasonable standard of comparison for how long a bond should be. If MOOSER and PEARSON's rules are applied to $As(C_6H_5)_4^+ I_3^-$ known to contain a linear anion I_3^- with two identical $I-I$ distances, metallic bonding is expected, unless the anion is really considered to be the di-iodo complex of iodine (I). A physically more significant difficulty is that CdI_2 crystals have so short $I-I$ contacts that it is surprising that the material is not metallic though, actually, it is even colourless. It is not very attractive that one has to accompany the crystallographer with a specialist having to judge whether two adjacent atoms are bound or not. However, HULLIGER (1968a) obtained a most useful classification of a rich material of experimental facts. We are not here making a distinction between cubic pyrites, orthorhombic marcasites, and the arsenopyrites MXY and other types containing two different metalloid elements. These types have all M surrounded by an octahedron of six atoms, and the non-M atoms occur in pairs. If X is a chalcogen S, Se or Te; and Y a pnigogen P, As, Sb or Bi, HULLIGER calculates the number of d-like electrons left on M if the twin anions are X_2^{-2}, XY^{-3} and Y_2^{-4}. This number agrees with the value of S found from the magnetic properties of the semi-conducting materials, whereas materials with apparent high number of d electrons may be metallic and diamagnetic:

d^2 $S=1$, antiferromagnetic:
   $CrSb_2$ Cr[IV]
d^4 $S=0$, cation pairing?
   FeP_2, $FeAs_2$, $FeSb_2$ Fe[IV]; RuP_2, $RuAs_2$, $RuSb_2$ Ru[IV]; OsP_2, $OsAs_2$, $OsSb_2$ Os[IV]
d^5 $S=5/2$, antiferromagnetic:
   MnS_2, $MnSe_2$, $MnTe_2$ Mn[II]
   $S=0$, cation pairing:
   $FePS$, $FeAsS$, $FeSbS$, $FePSe$, $FeAsSe$, $FeSbSe$, $FeAsTe$, $FeSbTe$ Fe[III];

$CoAs_2$, $CoSb_2$ Co[IV]; RuPS, RuAsS, RuSbS, RuPSe, RuAsSe, RuSbSe, RuAsTe, RuSbTe Ru[III]; RhP_2, $RhAs_2$, $RhSb_2$ Rh[IV]; OsPS, OsAsS, OsSbS, OsPSe, OsAsSe, OsSbSe, OsBiSe, OsAsTe, OsSbTe, OsBiTe Os[III]; $IrAs_2$, $IrSb_2$, $IrBi_2$ Ir[IV]

d^6 $S = 0$:
FeS_2, $FeSe_2$, $FeTe_2$ Fe[II]; CoPS, CoAsS, CoAsSe Co[III]; $NiAs_2$ Ni[IV]; RuS_2, $RuSe_2$, $RuTe_2$ Ru[II]; RhPS, RhAsS, RhSbS, RhBiS, RhPSe, RhAsSe, RhSbSe, RhAsTe, RhSbTe, RhBiTe Rh[III]; OsS_2, $OsSe_2$, $OsTe_2$ Os[II]; IrPS, IrAsS, IrSbS, IrBiS, IrPSe, IrAsSe, IrSbSe, IrBiSe, IrAsTe, IrSbTe, IrBiTe Ir[III]; PtP_2, $PtAs_2$, $PtSb_2$ Pt[IV]
$S = 0$, metallic:
NiP_2, $NiSb_2$, $PdAs_2$, $PtSb_2$, $PtBi_2$

d^7 $S = 1/2$ or 0, metallic:
CoS_2, $CoSe_2$, $CoTe_2$, NiPS, NiAsS, NiSbS, NiAsSe, NiSbSe, NiBiSe, $RhTe_2$, PdAsS, PdSbS, PdAsSe, PdSbSe, PdBiSe, PdSbTe, PdBiTe, PtAsS, PtSbS, PtSbSe, PtBiSe, PtSbTe, PtBiTe, $AuSb_2$

d^8 $S = 1$:
NiS_2 Ni[II]
metallic:
$NiSe_2$, $NiTe_2$

d^9 $S = 0$, metallic:
$CuSe_2$.

The d^4 and d^5 systems show various signs of strong coupling between the M constituents though the diamagnetism of the d^4 compounds may also be due to lower energy of (xz, yz) than of (xy). The metallic character of the five d^6-systems NiP_2,... may be accidental. Quite generally, the σ-anti-bonding electron which would be present in d^7 and d^9 pyrites delocalizes to a conduction band producing metallic character. Thus, the evidence is rather for the lower sub-shell containing six electrons remaining fairly localized, though $CoTe_2$ may be an accidental metal. Obviously, $AuSb_2$ is a rather typical metallic alloy, and there is no reason to assume Au[IV].

The cubic skutterudites CoP_3, $CoAs_3$, $CoSb_3$, RhP_3, $RhAs_3$, $RhSb_3$, IrP_3, $IrAs_3$ and $IrSb_3$ are semi-conducting d^6-systems containing octahedral Co(III)As_6 ... chromophores and the pnigogen atoms forming quadratic Y_4^{-4}. NiP_3 and PdP_3 are metallic skutterudites having their seventh d-like electron delocalized. It is quite general that polyanions imitate the stereochemistry of the isoelectronic neutral elements. Thus, X_2^- may be compared with diatomic halogens, the cyclic Y_n^{-n} with cyclic sulphur modifications, and K_4Si_4 and K_4Ge_4 containing tetrahedral Si_4^{-4} and Ge_4^{-4} comparable with P_4. This is also true for $BaSi_2$ whereas in $CaSi_2$, the silicon atoms form a layer structure very similar to that of elemental As. It may even be argued that the thallium atoms in metallic NaTl form a diamond lattice as one might expect for Tl($-$I) isoelectronic with the neutral IV group atoms.

The d^0-systems TiS_2, $TiSe_2$, $TiTe_2$, ZrS_2, $ZrSe_2$, $ZrTe_2$, HfS_2 and $HfSe_2$ crystallize in the CdI_2 type obviously containing $M(IV)X_6$ chromophores. Of these materials, only $TiTe_2$ is metallic. Various d^1-systems tend to form weak bonds between pairs of M atoms, as already known from VO_2 being a distorted modification of rutile TiO_2. One example is patronite VS_4 having S_2^{--} groups, but it is diamagnetic and has short $V - V$ distances. $PbCl_2$ contains $Pb(II)Cl_9$ chromophores; TiP_2, ZrP_2, $ZrAs_2$, $MoGe_2$, HfP_2, ThS_2, $ThSe_2$, US_2 and USe_2 crystallize in this structure. Apparently, the pnigogen anions must form mutual bonds to some extent. Tetragonal PbFCl well known from lanthanide MOX occurs for $ThSb_2$, UP_2, UAs_2, ThPS, ThAsS, UPS, UAsS, ThOS, UOS and ThNI of which the seven first are metallic and genuinely electron-deficient, the average anionic charge available being only two, since the uranium compounds are known to contain U[IV]. Isotypic YSbSe and YSbTe are undoubtedly in the same situation.

Some of the thio-spinels M_3S_4 present very interesting cases of genuine electron-deficiency. The metallic materials $CuRh_2S_4$ and $CuRh_2Se_4$ might conceivably have contained Cu(II) and Rh(III). However, the diamagnetism and the fact that Cu(II) is known to produce deviations from cubic symmetry in other spinels makes it much more probable that the average number of electrons available to each sulphur atom is 0.25 lower than in ordinary, closed-shell sulphides (LOTGERING and VAN STAPELE, 1968). Hence, the conditional oxidation states are Cu[I] and Rh[III], and the sulphide has been collectively oxidized. It is worth stressing that this depletion of closed-shell anions sometimes *can* take place without formation of definite anion-anion bonds. $MnCr_2S_4$ and anti-ferromagnetic $CoRh_2S_4$ are non-metallic containing $Mn(II)S_4$, $Cr(III)S_6$, $Co(II)S_4$ and $Rh(III)S_6$ chromophores, whereas $NiRh_2S_4$ and $NiIr_2S_4$ behave in a rather unexpected way according to HULLIGER (1968a) having temperature-independent paramagnetism. Rh_2S_3 itself contains octahedral $Rh(III)S_6$ but pairs of octahedra share faces like in $W_2Cl_9^{-3}$.

In the case of NiAs and the closely related MnP type, it may be discussed whether cation-cation bonds occur or not. In NiAs, the nickel atom is surrounded by six arsenic atoms; but short Ni-Ni distances occur along one trigonal axis of the octahedron. It is quite typical that d^7-systems such as CoS, CoSe, CoTe, RhSe, RhTe, NiAs, NiSb, NiBi, PdSb, PtSb and PtBi are common in this structure. However, $3d^1$ ScTe and TiSb; $3d^3$ VS, VSe, VTe, CrSb; $3d^4$ CrTe, MnSb, $3d^5$ MnTe and $3d^6$ FeSe and CoSb are other examples. The conditional oxidation state is not easy to define in this case, partly because several compounds are semi-conductors with so small an energy-gap that it is doubtful whether they are actually metallic.

Diamagnetic, semi-conducting PdS_2, PdSSe and $PdSe_2$ are distorted pyrites in such a way that quadratic $Pd(II)X_4$ are formed. Quadratic coordination is also found in semi-conducting NiP_2, PdP_2 and $PdPAs$, and in PdS and PtS. Metallic covellite CuS has an extraordinary complicated structure with the constituents in the ratio two Cu(I), one Cu(II), S_2^{--} and S(−II). CuSe is isotypic. We are not going here to discuss the more metal-rich compounds which have been reviewed thoroughly by HULLIGER (1968a).

In many cases, the weak distortions from higher crystal symmetry demonstrates some intermetallic interaction though it is difficult to maintain that polycations are formed. A typical case is yellow indium (I) chloride studied by VAN DEN BERG (1966) as a cubic superstructure of NaCl with the $5s^2$ lone-pairs probably mixed with some 5p character and sticking away from each group of four relatively close indium atoms. The typical behaviour of post-transition group elements is to form pairs, such as Hg_2^{++} which is even known in aqueous solution, or colourless $Cd_2^{++}(AlCl_4^-)_2$, or the yellow GaS, orange GaSe and dark red GaTe having each gallium surrounded by three chalkogen and one gallium atom. Even Zn in black ZnP_2 (MOOSER and PEARSON, 1960) show this tendency, whereas the red modification of ZnP_2 contains $[P^-]_n$ spiral chains isoelectronic with neutral sulphur.

Colorless compounds containing N(−III) are known such as AlN (wurtzite), Mg_3N_2 (anti-C-M_2O_3-type) and Ca_3N_2. Obviously, nitride anions have the same difficulties as oxide to a much more pronounced extent, and they need an extremely strong Madelung potential for being stabilized. It is quite characteristic that Li_3N is reported to be red, i.e. has low-lying 2p → 3s transitions. This remark should not be taken to imply that the fractional charge δ on nitrogen is by far close to −3; it is probably closer to −1. It is very interesting that colourless carbides also exist such as Be_2C and Al_4C_3 which must be ascribed the oxidation state C(−IV). Some carbides such as CaC_2 contain the diatomic C_2^{--} derived from acetylene and isoelectronic with N_2. BUTHERUS and EICK (1968) recently described cubic Nd_4O_3C thought to contain C(−IV) but having metallic properties because of the four Nd(III) allow two electrons to form a conduction band.

The closed-shell systems can sometimes be recognized in semi-conductors having very low energy-gap. Thus, Mg_2Sn crystallizing in CaF_2 type can be thought of as Sn(−IV)Mg_8 and Mg(II)Sn_4 in close analogy to Be_2C. It is an interesting question to what extent transition group elements can be ascribed negative oxidation states of the same reason. When SPICER, SOMMER, and WHITE (1959) find that CsAu is a red semi-conductor with relatively high energy-gap crystallizing in the CsCl type, it is tempting to describe the situation as Cs(I)Au(−I) isoelectronic with the mercury atom. Metallic alloys such as TlBi crystallizing in CsCl type may conceivably involve comparable distributions, but it is obviously much less certain than in the case of semi-conductors. There is a persistent argument that certain platinum alloys such as $AlPt_3$ might contain Pt in negative oxidation state. (cf. eq. 2.5).

Many oxides, nitrides and carbides are frequently described as interstitial compounds of the d-group elements. These materials may have a wide range of varying stoichiometry and may show sharp metal / semiconductor transitions by cooling such as VO discussed by MOTT (1964) (cf. also ADLER et al., 1967). The compounds TiN, TiC, VC, ZrN and ZrC crystallize in NaCl type but are not typical nitrides and carbides . The question of oxidation state may not be readily answered as yet.

Mixed oxidation states frequently produce spectacular colours which have been reviewed by ROBIN and DAY (1967), HUSH (1967) and ALLEN and HUSH (1967). In certain cases, the materials are *electronically ordered*, definite sites having definite oxidation states. Thus, ROBIN (1962) discussed the absorption spectrum of prussian blue $K^+[Fe^{II}(CN)_6Fe^{III}]^-$ where each $Fe(II)C_6$ has $S=0$ and each $Fe(III)N_6$ has $S=^5/_2$. In such cases, the element in one of the oxidation states can be replaced by other elements, such as purple $[Ru^{II}(CN)_6Fe^{III}]^-$ and the electron transfer band in the visible is caused by $Fe^{II}Fe^{III} \rightarrow Fe^{III}Fe^{II}$ and $Ru^{II}Fe^{III} \rightarrow Ru^{III}Fe^{II}$ in the two cases. This is quite comparable to the deep colours of silver (I) and thallium (I) salts (JØRGENSEN, 1961 and 1963)

CrO_4^{--}	yellow	Ag_2CrO_4	red		
MnO_4^-	purple	$AgMnO_4$	blue		
$ReCl_6^{--}$	pale green	Ag_2ReCl_6	orange		
$OsCl_6^{--}$	yellow	Ag_2OsCl_6	brown	Tl_2OsCl_6	olive-green
$OsBr_6^{--}$	tomato-red			Tl_2OsBr_6	black
$IrCl_6^{--}$	orange	Ag_2IrCl_6	blue	Tl_2IrCl_6	dark green

caused by the transfer of one of the ten 4d electrons of Ag(I) or one of the two 6s electrons of Tl(I) to the low-lying orbitals of the oxidizing anion. Thus, the electron transfer is between two metallic elements, e. g. $Ag^IIr^{IV} \rightarrow Ag^{II}Ir^{III}$, rather than between a non-metallic and a metallic element, such as in the isolated complex anions. In solution or in solids, the simultaneous presence of Fe(II) and Ti(IV) produces similar brown colours (blue in crystals) corresponding to the excited state $Fe^{III}Ti^{III}$ (REYNOLDS, 1965) and BRATERMAN (1966) recently studied the electron transfer from $Fe(CN)_6^{-4}$ to a variety of oxidizing entities such as Co(II) and Cu(II). The dark red colour of copper (II) and uranyl (VI) ferrocyanides is known from qualitative analysis and has a similar origin. LARSSON (1967) detected a new absorption band at 23 kK of the ion-pair formed between Co en_3^{+3} and $Fe(CN)_6^{-4}$.

It is worth remembering the existence of electron transfer from atoms of one metallic element to another when discussing mixed oxidation states. It may at first seem paradoxical that electron transfer $Fe^{II}Fe^{III} \rightarrow Fe^{III}Fe^{II}$ may at all correspond to absorption bands in the visible rather than at zero wave-number. What ROBIN and DAY (1967) call trapping of valencies is caused by the internuclear distances $Fe(II)X_6$ being longer than in $Fe(III)X_6$. At room temperature, the activation energy for re-arranging the internuclear distances $Fe-X$ to an intermediate value, favouring the electron transfer, is so high that the reaction between $Fe(H_2O)_6^{+2}$ and $Fe(H_2O)_6^{+3}$ can be shown to take some seconds and between the Eu(II) and Eu(III) aqua ions even slower (cf. the review by SYKES, 1967). This is so much more remarkable since the exchange of an individual water molecule in the europium (III) aqua ion takes less than a microsecond, and Eu(II) is so similar to Ba(II) that one may even discuss whether a definite number of aqua ligands is fixed at all. One should not confuse the two aspects of the problem; $Eu(H_2O)_9^{+3}$ is perhaps extremely readily rearranged to other forms having eight or ten water molecules coordinated, whereas the simultaneous loosening of all nine water molecules imitating the higher ionic radius of Eu(II) needs a high activation energy. VLČEK (in press) has written a review on the kinetics of such electron transfer, which is of great importance in polarography and in general electrochemical problems.

Most common minerals and rocks are dark coloured by the simul-taneous presence of Fe(II) and Fe(III). It is also known that pale green $Fe(OH)_2$ oxidizes to a dark green intermediate before finally forming orange rusty $Fe(OH)_3$. We have reasons to believe that most of these cases involve individual Fe(II) and Fe(III) though the relative high intensity of the absorption bands indicate a certain electronic delocaliza-tion. DAY (1963) also studied the blue colours of Sb(III) and Sb(V) in A_2SnCl_6 and dark red colours in A_2SnBr_6 and related materials. We know today (ROBIN and DAY, 1967) that these crystals contain individual groups $SbCl_6^{-3}$ and $SbCl_6^-$ with shorter Sb^VCl than $Sb^{III}Cl$ distances, partly from far infra-red spectra of the diluted samples and partly from the crystal structure of undiluted $(NH_4)_2SbBr_6$. Again, the difference in internuclear distances produces the unexpectedly high wavenumbers of the electron transfer bands $Sb^{III}Sb^V \rightarrow Sb^{IV}Sb^{IV}$ though it is not yet understood why the intensities are so large when mixed oxides such as SbO_2 are colourless. We call the situation of definite oxidation states *electronic ordering* in analogy with atomic ordering of superstructures such as the cubic pyrochlore $Sm_2^{III}Zr_2^{IV}O_7$ containing $Sm(III)O_8$ and $Zr(IV)O_6$ which is also known as a statistically disordered fluorite $Sm_{0.5}Zr_{0.5}O_{1.75}$. In mixed oxides of lanthanides with quadrivalent me-tals, and in the numerous compounds CeO_x, PrO_x and TbO_x ($1.5 < x < 2$)

a certain confusion exists between the black colours which seem to be due to genuine mixed oxidation state behaviour, and the deep colour (purple in ThO_2, chamois in CeO_2, orange in Y_2O_3) due to oxide $\rightarrow 4f$ electron transfer of traces of the monomeric, cubic chromophores $Pr(IV)O_8$ and $Tb(IV)O_8$ (JØRGENSEN and RITTERSHAUS, 1967). Already in 1915, HOFMANN and HÖSCHELE noted the dark blue colour of the disordered fluorites $Ce_xU_{1-x}O_2$ which most probably can be ascribed to electron transfer $U^{IV}Ce^{IV} \rightarrow U^{V}Ce^{III}$. VERWEY et al. (1947) pointed out that the spinel magnetite Fe_3O_4 has tetrahedral $Fe(III)O_4$ and mobile electrons between the octahedral sites $Fe(II)$, $Fe(III)O_6$ at room temperature, but that at liquid air temperature, the octahedral sites also show electronic ordering between $Fe(II)O_6$ and $Fe(III)O_6$. DRIESSENS (1967) gives convincing arguments that the tetragonally distorted spinel hausmannite Mn_3O_4 contains $Mn(III)O_6$ and $Mn(II)O_4$ at room temperature, but above $1120°\,C$ transforms to a cubic spinel with electronic ordering $Mn(IV)O_6$, $Mn(II)O_6$ and $Mn(II)O_4$. It is rare to see increasing temperature producing higher ordering.

Certain oxides with high dielectric constants, such as TiO_2 and WO_3, turn purple or blue, respectively, by very slight reduction to non-stoichiometric forms TiO_{2-x} and WO_{3-x} (CHOPOORIAN et al., 1966). It is conceivable that these materials contain electrons trapped collectively by a large number of atoms. This type of explanation is always somewhat doubtful because of the necessity for the additional electrons to occupy orbitals orthogonal on all the previously filled orbitals. However, it may be true in these cases as well as in the highly unusual perovskite $SrTiO_3$ which is a perfectly normal, colourless $Sr(II)O_{12}$ and $Ti(IV)O_6$ system, but which can be reduced in the presence of small amounts of lanthanum $Sr^{II}_{1-x}La^{III}_x Ti^{IV}_{1-x}Ti^{III}_x O_6$ forming an accidental metal which is even superconducting at sufficiently low temperature. The perovskite $La^{III}Co^{III}O_3$ also has highly unusual electric properties; the general question of localized, electronically ordered atoms with a definite number of d-electrons [and hence well-defined oxidation state] vs. metallic, delocalized situations have been treated by GOODENOUGH (1963 and 1968) who considers it mainly to be a question of the average radius of the partly filled d-shell compared with the internuclear distances of the metallic elements. The tungsten bronzes $(M^{+z})_xWO_3$ crystallize in most cases in the disordered cubic perovskite lattices having M on the twelve-coordinated positions and W in octahedral groups of six oxygen atoms. They have recently been reviewed by DICKENS and WHITTINGHAM (1968). M is most frequently an alkali metal but can also be multivalent ions provided the ionic radii are sufficiently large, e. g. lanthanides. The tungsten bronzes are good metals; there has been much discussion about the constitution of the conduction band which does not seem to involve significantly s-electrons of the M-atoms. FERRETTI, ROGERS and GOODENOUGH (1965) studied the related problem of ReO_3. This cubic crystal is a perovskite with all the twelve-coordinated positions vacant. This diamagnetic, metallic material would contain one 5d electron and $Re(VI)O_6$ in a conventional description. It was concluded that the 5d electrons are delocalized, not via direct Re — Re bonds which would be excessively long, but via the M. O. formation with the oxide ligands. The compound was compared with the perovskite Sr_2MgReO_6 which turned out to be a semiconductor with localized $Re(VI)O_6$ and $Mg(II)O_6$ groups.

The various reduced forms of niobium, molybdenum and tungsten blue having average oxidation numbers slightly below Nb(V), Mo(VI) and W(VI) are undoubtedly related to the crystalline compounds above. However, from a purely chemical point of view, one has the feeling that more distinct species sometimes may be involved. Thus, the colourless hetero-dodecatungstates $SiW_{12}O_{40}^{-4}$ and $PW_{12}O_{40}^{-3}$ can be reduced by one, two, ... electrons to dark blue species having strong absorption bands in the near infra-red and the red (POPE and VARGA, 1966). Though the aspect of these bands is not very different from the broad absorption due to so-called "free charge carriers" said to produce the brown colour of Ag_2O and CdO and black colour of Tl_2O_3 which might otherwise have been expected to be colourless closed-shell systems, there is little doubt that they are localized to a small number of tungsten atoms, perhaps even smaller than twelve. In this sense, they are comparable with cluster complexes such as $Nb_6Cl_{12}^{+z}$ to be mentioned in Chapter 10. On the other hand, the absorption bands observed are far more intense than of the pale green, monomeric NORDENSKJÖLD's anions $MoOX_5^{--}$ and WOX_5^{--}. Like in several other cases of electron transfer between rather distant atoms, it is not easy to explain how the relatively large oscillator strength originates.

Though mixed oxidation states, and in particular the black colour due to general absorption of the whole visible region in the case of electronically disordered materials, are among the main sources of strongly coloured solids, it must be recognized that stoichiometric compounds under special circumstances may show comparable absorption bands. Thus, CLARK (1964) studied the deep violet colour of one of the modifications of anhydrous $TiCl_3$. Though the brown colour of other modifications of $TiCl_3$ are suspected to be due to deviations from stoichiometry, there is little doubt that the transitions in violet $TiCl_3$ are due to transitions of the 3d electron of each Ti(III) to a delocalized conduction band which may partly consist of 4s orbitals, but certainly has other constituents in view of the high intensity. The analogous light absorption occurs in the visible of low oxidation state compounds such as $TiBr_3$, $ZrCl_3$ and $ZrBr_3$, and probably in the near ultra-violet of anhydrous $CoCl_2$ containing octahedral chromophores $Co(II)Cl_6$ in spite of the pale blue colour. There is a close analogy between the dark violet colour of $TiCl_3$ and the inverted electron transfer bands of titanium (III) acetylacetonate though, of course, the necessity of d-electrons being present makes a certain relation with cluster complexes such as $Nb_6Cl_{12}^{+z}$ conceivable.

We are not going here to discuss at length the strong colours observed of certain quadratic d^8-systems in the solid state, since they influence the assignment of oxidation states only to an insignificant extent. A typical case is the red colour of nickel (II) dimethylglyoximate (BASU et al., 1964) which is yellow in organic solvents; or the characteristic yellow or red colours of salts of $Pt(CN)_4^{--}$ which frequently correspond to bright fluorescence, the colour of which may depend on the number of molecules of water of crystallization and other factors influencing the perpendicular $Pt-Pt$ distance between two anions (BERGSØE, 1962). Though a one-dimensional conduction band conceivably might become as prominent as to take over the $(z^2 - \frac{r^2}{3})$ 5d-like electrons with remaining Pt[IV] (this would be happening because of the mixing with 6pz-orbitals) this seems to be far removed from the comparatively weak interactions observed; but it is worth remembering that these strong colours of certain quadratic Ni(II) and Pt(II) complexes are not due to deviations from regular stoichiometry. Actually, DAY et al. (1965) give convincing evidence that MAGNUS' green salt $[Pt(NH_3)_4][PtCl_4]$ formed from colourless cations and pink anions $PtCl_4^{--}$, and related substances, actually have their visible spectra determined by a shift of the internal $5d^8$-transitions of the anion. However, ANEX et al. (1967) found a very intense band at 34.5 kK corresponding to a genuine cooperative effect.

The deviations from the principle of additivity of colours of ionic constituents are very interesting to study. The two main classes recognized today are the electron transfer from one atom to another of the same or of two different metallic elements; and the excitation of a localized electron to a delocalized conduction band. If one took the energy-band description very seriously, one would expect these transitions to have lower wavenumbers than actually observed. One reason for this discrepancy is the re-adaption of internuclear distances producing vertical optical transitions in the visible of the types $Fe^{II}Fe^{III} \rightarrow Fe^{III}Fe^{II}$ and $Sb^{III}Sb^V \rightarrow Sb^{IV}Sb^{IV}$ discussed above; another reason is the much smaller electron affinity than ionization energy of partly filled shells.

The chemist has many reasons to believe that two materials may have nearly the same type of chemical bonding though one is metallic and another is not. A striking example are the semi-conducting alloys Sb_xBi_{1-x} isomorphous with the metallic Sb and Bi. By the same token as complicated absorption spectra can be recognized of definite chromophores MX_N in brightly coloured, vitreous materials, there is much evidence, as also emphasized by Mooser and Pearson (1960) that local order is much more important for the choice between metallic and semi-conducting behaviour than the extended crystal structure. Actually, some physicists seem to forget that liquid metals exist. The behaviour of sodium vapour under high pressure is not essentially different from that of mercury under normal conditions.

Rather unusual metallic materials are known. Thus, McDonald and Thompson (1966) describe golden-red metallic $Li(NH_3)_4$ having very high mobility though the density is only 0.5 g/cm³. Littlehailes and Woodhall (1967) described amalgams of tetra-alkylammonium $N(C_nH_{2n+1})_4$ with $n = 1,2,3$ and 4 and about 12 Hg per N. There is little doubt that these amalgams can be considered as the solutions of the cations NR_4^+ in the delocalized electronic cloud. However, it is worth remembering that Wan (1968) has discussed the possible existence of NH_4 as a gaseous species. Though this would involve one electron in a 3s-like Rydberg-orbital, it would not be much more surprising than CH_5^+ which is known both in gaseous state and in solution, as discussed in Chapter 9. In this connection, it may be mentioned that CCl_4 has a considerable electron affinity in gaseous state forming CCl_4^- (cf. Jørgensen, 1967).

Bibliography

Adler, D., J. Feinleib, H. Brooks, and W. Paul: Phys. Rev. **155**, 851 (1967).

Allen, G. C., and N. S. Hush: Progr. Inorg. Chem. **8**, 357 (1967).

Anex, B. G., M. E. Ross, and M. W. Hedgcock: J. Chem. Phys. **46**, 1090 (1967).

Basu, G., G. M. Cook, and R. Linn Belford: Inorg. Chem. **3**, 1361 (1964).

Bergsøe, P.: Acta Chem. Scand. **16**, 2061 (1962).

Braterman, P. S.: J. Chem. Soc. (A) 1471 (1966).

Butherus, A. D., and H. A. Eick: J. Am. Chem. Soc. **90**, 1715 (1968).

Chopoorian, J. A., G. H. Dorion, and F. S. Model: J. Inorg. Nucl. Chem. **28**, 83 (1966).

Clark, R. J. H.: J. Chem. Soc. 417 (1964).

Corbett, J. D., L. F. Druding, and C. B. Lindahl: J. Inorg. Nucl. Chem. **17**, 176 (1961).

—, and B. C. McCollum: Inorg. Chem. **5**, 938 (1966).

Day, P.: Inorg. Chem. **2**, 452 (1963).

—, A. F. Orchard, A. J. Thomson, and R. J. P. Williams: J. Chem. Phys. **42**, 1973 and **43**, 3763 (1965).

Dickens, P. G., and M. S. Whittingham: Quart. Rev. (London) **22**, 30 (1968).

Didchenko, R., and F. P. Gortsema: J. Phys. Chem. Solids **24**, 863 (1963).

Drickamer, H. G.: Solid State Phys. **17**, 1 (1965).

Driessens, F. C. M.: Inorg. Chim. Acta **1**, 193 (1967).

Druding, L. F., and J. D. Corbett: J. Am. Chem. Soc. **81**, 5512 (1959).

Dworkin, A. S., H. R. Bronstein, and M. A. Bredig: J. Phys. Chem. **66**, 1201 (1962).

Ferretti, A., D. B. Rogers, and J. B. Goodenough: J. Phys. Chem. Solids **26**, 2007 (1965).

FLAHAUT, J., M. GUITTARD, M. PATRIE, M. P. PARDO, S. M. GOLABI, and L. DOMANGE: Acta Cryst. **19**, 14 (1965).

GOODENOUGH, J. B.: Magnetism and the Chemical Bond. New York: Interscience 1963.

— J. Appl. Phys. **39**, 403 (1968).

HULLIGER, F., and E. MOOSER: Progr. Solid State Chem. **2**, 330 (1965).

— Struct. Bonding **4**, 83 (1968a).

— Helv. Phys. Acta **41**, 945 (1968b).

HUSH, N. S.: Progr. Inorg. Chem. **8**, 391 (1967).

JAYARAMAN, A.: Phys. Rev. **137**, A 179 (1965).

JØRGENSEN, C. K.: Mat. fys. Medd. Dan. Vid. Selskab, **30**, no. 22 (1956).

— Mol. Phys. **4**, 235 (1961).

— Acta Chem. Scand. **17**, 1034 (1963).

— Mol. Phys. **7**, 417 (1964).

— Halogen Chemistry **1**, 265 (1967). London: Academic Press.

—, and E. RITTERSHAUS: Mat. fys. Medd. Dan. Vid. Selskab, **35**, no. 15 (1967).

KLEMM, W., u. H. BOMMER: Z. Anorg. Allgem. Chem. **231**, 138 (1937).

LARSSON, R.: Acta Chem. Scand. **21**, 257 (1967).

LITTLEHAILES, J. D., and B. J. WOODHALL: Chem. Commun. 665 (1967).

LOTGERING, F. K., and R. P. VAN STAPELE: J. Appl. Phys. **39**, 417 (1968).

McCLURE, J. W.: J. Phys. Chem. Solids **24**, 871 (1963).

McDONALD, W. J., and J. C. THOMPSON: Phys. Rev. **150**, 602 (1966).

MEE, J. E., and J. D. CORBETT: Inorg. Chem. **4**, 88 (1965).

MOOSER, E., and W. B. PEARSON: Progr. Semiconductors **5**, 103 (1960).

MOTT, N. F.: Advan. Phys. **13**, 325 (1964).

POPE, M. T., and G. M. VARGA: Inorg. Chem. **5**, 1249 (1966).

REID, F. J., L. K. MATSON, J. F. MILLER, and R. C. HIMES: J. Phys. Chem. Solids **25**, 964 (1964).

REYNOLDS, M. L.: J. Chem. Soc. 2991 and 2993 (1965).

ROBIN, M. B.: Inorg. Chem. **1**, 337 (1962).

—, and P. DAY: Advan. Inorg. Chem. Radiochem. **10**, 248 (1967).

SCAIFE, D. E., and A. W. WYLIE: J. Chem. Soc. 5450 (1964).

SCHUMACHER, D. P., and W. E. WALLACE: Inorg. Chem. **5**, 1563 (1966).

SCLAR, N.: J. Appl. Phys. **35**, 1534 (1964).

SPICER, W. E., A. H. SOMMER, and J. G. WHITE: Phys. Rev. **115**, 57 (1959).

SYKES, A. G.: Advan. Inorg. Chem. Radiochem. **10**, 153 (1967).

TROMBE, F., et M. FOEX: Ann. Chim. (Paris) **19**, 417 (1944).

VAN DEN BERG, J. M.: Acta Cryst. **20**, 905 (1966).

VERWEY, E. J., P. W. HAAGMAN, and F. C. ROMEIJN: J. Chem. Phys. **15**, 181 (1947).

VLČEK, A. A.: Revue chimie min. **5**, 299 (1968).

WAN, J. K. S.: J. Chem. Educ. **45**, 40 (1968).

9. Closed-Shell Systems, Hydrides and Back-Bonding

Once the spectroscopic oxidation states have been firmly established for nearly all systems containing partly filled f shells and most systems containing partly filled d shells, we are in the rather paradoxical situation that the classification of closed-shell systems is more uncertain and to some extent derived by analogy. Thus, there is no serious doubt that the colourless copper (I), zinc (II), gallium (III) and germanium (IV) compounds are d^{10}-systems with the completely filled M.O. well separate energy-wise from the empty M.O. However, one already runs into a problem when considering ligands specializing in forming low oxidation states of the central atom, such as CO or PF_3 when entering colourless iron ($-$II), cobalt ($-$I) and nickel (0) complexes. Professor K. A. JENSEN once, half seriously, suggested that $Ni(CO)_4$ is a nickel (VIII) compound of the ligand CO^{--} formed by complete deprotonation of monomeric formaldehyde H_2CO. This rather extreme example illustrates a very important point about complexes having no low-lying excited levels: if there is any doubt about the oxidation state of the ligands, the acceptable values usually jump by two units with the result that the oxidation state of the central atom jumps 2N units if N ligands occur. Thus, $V(CO)_6^-$ and $Cr(CO)_6$ would involve V(XI) and Cr (XII) if the description as CO^{--} complexes was assumed, which is entirely unacceptable. In this connection, it may be mentioned that the highest oxidation state known is $+8$ in the tetroxides and a few other oxo-complexes of Ru(VIII), Xe(VIII) and Os(VIII). The element having the greatest chance of being further oxidized would be iridium (IX) conceivably forming IrO_3F_3 or salts of IrF_6^{+3} of non-oxidizable anions providing a large Madelung potential such as SiF_6^{--} (cf. JØRGENSEN, 1963).

If we apply the definition of spectroscopic oxidation states given on p. 145 to molecules such as CF_4 or the isoelectronic BF_4^- and NF_4^+, it becomes a very important argument that they are *colourless* in the generalized sense of not having any excited electronic levels within 25 kK from the groundstate. This statement is slightly more restrictive than it needs to be; all our following arguments might as well apply if the compounds were lemon-yellow with an absorption band at 24 kK; but if they were dark green and suspected of having further absorption bands in the near infra-red, our arguments would not apply. Obviously, the filled M.O. are well separated in energy from the empty ones in such a way that the excitation energies (which usually are smaller than the differences between the ionization energies as expressed by eq. 7.4) are above 25 kK. Since CF_4 is known from a variety of techniques (electron diffraction; infra-red and Raman spectroscopy, etc.) to possess regular tetrahedral symmetry, group-theory poses very stringent conditions on the sets of degenerate M.O. which are filled. The obvious suggestion is a fluoride $(-I)$ of carbon (IV); this is in agreement with everything we know about the order and symmetry types of the filled M.O. in CF_4 and related molecules (cf. JØRGENSEN, 1967 b). A proposal which is close to science-fiction would be to have a carbon (II) complex of collectively oxidized F_4^{-2}. In M.O. formulation, this would correspond to a totally symmetric orbital a_1 being essentially a carbon 2s-orbital, whereas the corresponding linear combination of fluoride σ-orbitals is absent. This proposal looks unreasonable of many reasons, one being the stereochemical fact that nitrogen (III), phosphorus (III) and sulphur (IV) compounds all are known to be highly distorted, leaving space for the s-lone-pair to get mixed with higher l-values. On the other hand, a carbon $(-IV)$ compound of fluorine (I) would show the required T_d symmetry, and the two sets of three degenerate orbitals having the symmetry type t_2 and the solitary a_1 orbital would be ascribed to the carbon atom. Said in other words, the four ligands had been deprived of the σ-lone-pairs and contain each two π-orbitals as the loosest bound electrons. However, this proposal is chemically unacceptable, because we know that the σ-orbitals t_2 and a_1 are concentrated much more on the fluorine atoms than on the carbon atom. It is worth remembering that $C(-IV)$ F(I) might be equally colourless as the observed molecule CF_4 but that arguments on the relative electronegativity makes $C(IV)F(-I)$ far more preferable.

The well-defined oxidation states in CF_4 and CCl_4 are closely related to the simultaneous presence of two π-orbitals (loosest bound) and one σ-orbital in each $F(-I)$ and $Cl(-I)$. As discussed by JØRGENSEN (1964 a) the situation is much less clear-cut in ligands have only one σ-bonding lone-pair and no π-orbitals. The order of M.O. energies in CH_4

$$1a_1 \ll 2a_1 < 1t_2 \ll 3a_1 < ... \qquad (9.1)$$

can be described in two different ways. There is no doubt that $1a_1$ corresponds to an inner 1s-orbital of the carbon atom. It might then be argued either that $2a_1$ is predominantly a 2s-orbital of $C(-IV)$ and the three $1t_2$-orbitals essentially the 2p-shell of $C(-IV)$ perturbed by the positive ligands H(I); or otherwise, the central atom is $C(IV)$ and $2a_1$ and $1t_2$ are linear combinations of the four lone-pairs of the $H(-I)$ ligands. In the latter case, the origin of the lower energy of $2a_1$ than of $1t_2$ is partly the angular node-plane producing hydrogen-hydrogen anti-bonding in the three latter orbitals, and partly the content of 2s with lower diagonal element of energy in the $2a_1$-orbital. As we are discussing in Chapter 11 on quanticule oxidation states, it is not necessary to restrict L.C.A.O. analysis of M.O. to the basis set of strongly bound atomic orbitals; one might also include a third component having positive sign in the bond region of bonding M.O. In this sense, the $2a_1$ and $1t_2$ orbitals might partly consist of such bond-region orbitals. From a purely chemical point of view, it is more satisfactory to consider methane as a carbon (IV) hydride because the halogen-substituted molecules are so obviously more similar than are alkali metal alkyls. Whereas the electronegativity x of the 2p-orbitals of carbon is lower than that of the 1s-orbital of hydrogen, MULLIKEN pointed out that the valence state electronegativity corresponding to the mean value $\frac{1}{4}x_{2s}+\frac{3}{4}x_{2p}$ is higher than of hydrogen. KLOPMAN (1964) has recently thoroughly discussed this problem. It is certainly true that acetylene HCCH has a much higher x of the carbon atoms (connected with a triple bond) than usually, and HCC^- and C_2^{--} are perfectly well known in complexes such as quadratic $Ni(CCH)_4^{--}$ (cf. a recent review by BOWDEN and LEVER, 1968) and in solids such as (in principle) colourless CaC_2 and the red, explosive Cu_2C_2.

In diatomic hydrides, the ambiguity of the spectroscopic oxidation states being $M(I)H(-I)$ or $M(-I)H(I)$ is even more striking (JØRGENSEN 1964). In the 2p group hydrides BH, CH, NH, OH and FH, the order of M.O. energies analogous to (9.1) is

$$1\sigma \ll 2\sigma < 3\sigma < 1\pi \ll 4\sigma < ... \tag{9.2}$$

BH is already rather unusual by having the three first σ-orbitals filled (1σ corresponding closely to the 1s orbital of boron) and being isoelectronic with Li_2 (where $1\sigma_g$ and $1\sigma_u$ replace 1σ and 2σ, both being inner 1s shells) and with BeHe which presumably dissociate spontaneously to atoms. The reason why BH is fairly stable (though, of course, it condenses to boron and to various boranes) is that though 3σ formally is σ-anti-bonding (and it is usually fatal for a molecule, such as He_2, to contain as many anti-bonding as bonding electrons) it consists of a mixture of 2s and 2p character that is nearly exclusively concentrated on the rear side of the boron atom and behaves like the non-bonding σ electron in the diatomic molecules in Table 6.4. However, there is no doubt that BH should be classified as boron (I) hydride, 3σ representing the lone-pair. The lowest configuration of NH is $1\sigma^2 2\sigma^2 3\sigma^2 1\pi^2$ having two electrons available for two equivalent π orbitals like in O_2 and consequently, the groundstate $^3\sum^-$ has $S=1$. It is seen from (9.2) that one may think of the 2p shell as being split by the hydride ligand in a strongly σ-anti-bonding orbital 4σ (which would be half filled in the unstable molecule NeH) and in two non-bonding orbitals 1π This is exactly analogous to the behaviour of the partly filled d-shell in complexes of ligands having one σ-bonding lone-pair such as ammonia. However, one may also consider (9.2) as the 2p orbital 3σ being stabilized more by the positive charge of the proton relative to the non-bonding 2p orbitals 1π. According to the photo-electron spectroscopic results for HF discussed in eq. (7.3), there is no doubt that the latter explanation is the more suitable for hydrogen (I) fluoride ($-$ I). The question whether CH and NH are hydrides or proton adducts would be decided if it was known with certainly whether 4σ or 3σ contains the largest contribution of 2p-character. In the former case, 3σ is closest to represent the σ-bonding electron pair of hydride.

From the practical point of view of nomenclature, it may be the most reasonable to decide *ex officio* that only $F, Cl, Br, I, O, S, Se, Te$ and N form protonic adducts and all other elements hydrides. This would be in agreement with the spectroscopic oxidation states when they are known. The reason why N cannot readily be described as N^{III} in NH_3 is that it would be N^V in NH_4^+ (we are here writing the somewhat more conventional oxidation numbers as superscripts). It is less offending that PH_3 under somewhat unusual circumstances can form PH_4^+ according to eq. (2.9). Actually, the stereochemistry of $P^{III}X_3$ having $X = H, F, Cl, Br, ...$ is so similar that there is no strong reason not to accept P(III) hydride. If it were not for the solid tellurides important in mineralogy and solid-state physics, it might be argued that H_2Te contains Te^{II} rather than Te^{-II}. In this connection, it must be remembered that H_2O, OF_2 and Cl_2O have very similar stereochemistry. The oxidation states hydrogen (I) oxide ($-$ II); oxygen (II) fluoride ($-$ I); and chlorine (I) oxide ($-$ II) have only been determined from the knowledge of the predominant constituents of the bonding M.O. From a spectroscopic point of view, the similarity of stereochemistry and relative order of M.O. having definite symmetry types sometimes suggest much more far-reaching analogies than acceptable to the chemist. Thus, WALSH (1953) pointed out that spectroscopically, there is a close similarity between NO_2^-, O_3 and SO_2. Though nitrite undoubtedly is a di-oxo complex of nitrogen (III), it is not easy to accept ozone as oxygen (IV) dioxide. Such examples can be made into caricatures; from the legitimate ideas that CO_2 is carbon (IV) dioxide and the di-anion of cyanamide NCN^{--} a di-nitrido complex of carbon (IV), one may extrapolate that the isosteric (linear) and isoelectronic azide N_3^- is a di-nitrido complex of nitrogen (V). It is not making life easier that N_3^- may function as a pseudo-halide ligand also for central atoms such as C(IV) and P(V) forming $C(N_3)_3^+$ and $P(N_3)_6^-$. The only conclusion possible for a chemist is that spectroscopic similarities sometimes may be compatible with *different* distributions of oxidation numbers if the electronegativities of the constituent atoms are allowed to change the sign of their differences. This is particularly true if two of the connected atoms at two inequivalent sites are of the same element. Thus, whereas thiosulphate $S_2O_3^{--}$ really seems to involve sulphur $(-II)$ bound the central atom sulphur (VI), one would not expect an analogous tetroxo complex of oxygen (VI) O_5^{--} to exist; and the known S_5^{--} is a chain to be discussed in Chapter 10. An analogous problem is I_3^- which is best considered as a linear di-iodo complex of iodine (I) but which might also be considered as an adduct of I_2 and I^- (cf. JØRGENSEN, 1967b).

The only monomeric transition complexes containing only hydride ligands are TcH_9^{--} and ReH_9^{--} (ABRAHAMS, GINSBERG and KNOX, 1964). These colourless species are d^0-systems Tc(VII) and Re(VII) with surprisingly high wavenumbers of the electron transfer bands from which a somewhat uncertain value of the optical electronegativity $x_{opt} \sim 3.2$ for $H(-I)$ can be derived (cf. Table 7.1 and the review by JØRGENSEN, 1968a). Many complexes contain several ligands plus one hydride ligand. When it was established that CHATT's colourless complex Pt $HCl[P(C_6H_5)_3]_2$ is quadratic, this could be taken as stereochemical evidence for low-spin $5d^8$-behaviour and consequently the chromophore Pt(II)HP_2Cl. If this complex had been a protonic adduct of $5d^{10}$, one would have expected a tetrahedral chromophore. HIEBER had previously introduced the idea of "pseudo-atoms" when remarking that mixed carbonyl-hydrides $HMn(CO)_5$ and $HCo(CO)_4$ exist of atoms with odd atomic number Z which can be compared with the compounds of the central atom $(Z+1)$ containing the same number of carbon monoxide ligands $Fe(CO)_5$ and $Ni(CO)_4$. From a quantum-mechanical point of view, it is not easy to know whether protons might not penetrate the central atomic core to a relatively short distance from the nucleus. However, quite recent progress in the diffraction technique has permitted to resolve this old problem. First of all, $HMn(CO)_5$ is octahedral and $HCo(CO)_4$ trigonally bipyramidal. This shows that the hydrogen atom takes up as much space as any other ligand, and the stereochemistry strongly suggests $3d^6$ Mn(I) and $3d^8$ Co(I) hydrides. Secondly, the distance Mn $-$ H is known to have the moderate value 1.43 Å (ROBIETTE et al., 1968). This is definitely shorter than the distance Mn $-$ Cl in comparable complexes, but longer than compatible with the penetration of the proton in the atomic core. The only aspect left of the pseudo-atom concept is the fact familiar from organic chemistry that say one CH group of benzene is replaced by one N atom in pyridine.

For the chemist, there is one unfamiliar aspect of the transition group hydride complexes. Though containing hydrogen $(-I)$, they can be deprotonated by bases which consequently decrease the oxidation state of the central atom by two units for each proton removed. Quite generally, there is no universal relation between the acidity, and specifically the pK value in aqueous solution, and the fractional charge on the hydrogen atom in the acid. Thus, HCl, HBr and HI are far stronger acids in aqueous solution than HF though gaseous HF has a much larger dipole moment and other evidence for relatively positive charge on the hydrogen atom relative to the nearly electroneutral HI. By the same token, the acidity in aqueous solution increases in the order $H_2O < H_2S < H_2Se < H_2Te$ though, as we have seen, the latter compound might as well be described as tellurium (II) hydride or as hydrogen (I) telluride $(-II)$. This behaviour is connected with the strong solvation effects on ions in water (cf. also KLOPMAN, 1968). Returning to the reactions

$$\begin{aligned}
HCo^I(CO)_4 &= [Co^{-I}(CO)_4]^- + H^+ \\
H_2Fe^{II}(CO)_4 &= [HFe^0(CO)_4]^- + H^+ \\
[HFe^0(CO)_4]^- &= [Fe^{-II}(CO)_4]^{--} + H^+
\end{aligned} \tag{9.3}$$

having the pK values in aqueous solution below 2, 4.4 and 14, it must be realized that they are related to the general reactions

$$\begin{aligned}
H^- + H^+ &= H_2 \\
2e^- + H^+ &= H^-
\end{aligned} \tag{9.4}$$

which are at the same time acid-base neutralizations and redox reactions as pointed out by BRØNSTED. Actually, the carbonyl-complexes may get involved in other redox reactions affecting the CO ligands such as

$$Fe(CO)_5 + 4\, OH^- = Fe(CO)_4^{--} + CO_3^{--} + 2\, H_2O . \tag{9.5}$$

L'EPLATTENIER and CALDERAZZO (1967) showed that colourless $Ru(CO)_5$ and $Os(CO)_5$ indeed do exist, though they readily decompose to the trimeric cluster compounds red $Ru_3(CO)_{12}$ and orange $Os_3(CO)_{12}$ and that they react with halogens X_2 to form $Ru(CO)_4X_2$ and $Os(CO)_4X_2$ which, like the corresponding iron (II) complexes, always are *cis*. It was found that $H_2Os(CO)_4$ is a colourless, mobile liquid also with the two hydride ligands in *cis*-position. It might be argued that this is in a sense an adduct of H_2 with the group $Os(CO)_4$ known from its trimer. However, in the writer's opinion, there is no doubt that the compound is a low-spin $5d^6$ Os(II) complex of hydride. Bases can deprotonate it to the osmium (0) hydride complex $HOs(CO)_4^-$. There are cases known of colourless monomeric iridium (III) hydrides such as *mer*- and *fac*-$IrH_3[P(C_6H_5)_3]_3$ (MALATESTA, 1967) evolving H_2 with $HClO_4$ and forming five-coordinate $IrH_2L_3^+$. Also IrH_3L_2 is known. Recently, there has been a considerable interest in Ru, Rh and Ir hydride complexes because of their catalytic activity, e.g. in the hydrogenation of olefins. In many cases, the reactions can be considered as quadratic Rh(I) or Ir(I) complexes reducing H_2 to octahedral Rh(III) and Ir(III) hydride complexes (cf. STROHMEIER 1968). WILKINSON and collaborators (OSBORN et al., 1966; BAIRD et al., 1967) have also reported cases of a quadratic Rh(I) complex having a tendency of dissociating to a tricoordinate, highly reactive intermediate (adding olefins remaining Rh(I) or H_2 forming Rh(III))

$$RhCl\,[P(C_6H_5)_3]_3 = RhCl\,[P(C_6H_5)_3]_2 + P(C_6H_5)_3 \qquad (9.6)$$

but we are not going to discuss these catalytic problems further.

On the other hand, the questions of extreme conditions of acidity have interesting aspects. In gaseous state, methane is known to add protons to form CH_5^+ much in the same way as H_3^+, KrH^+ and XeH^+ which are stable toward dissociation of protons in the gaseous state. Quite recently, OLAH and SCHLOSBERG (1968) demonstrated that mixtures of fluorosulphonic acid $FSO_2(OH)$ and SbF_5 are able to stabilize CH_5^+ in solution though the ion has a certain tendency to dissociate to H_2 and CH_3^+. Quite generally, alkanes form *carbonium ions* R_3C^+ in this solvent as previously known from the solutions of $(C_6H_5)_3COH$ in strong acids forming $(C_6H_5)_3C^+$. If such species had been the only known hydrocarbons, we would have the embarrassing choice between CH_5^+ being a protonic adduct of carbon $(-IV)$ and CH_3^+ being a tri-hydrido complex of carbon (IV). This is related to the difficulties for NH_4^+ and PH_4^+ already mentioned. The *carbanion* CH_3^- would be a hydride of carbon (II) or a protonic adduct of $C(-IV)$.

Returning to H_2 it is frequently thought that the first excited states occur above 90 kK corresponding to higher excitation energy than 82 kK for $1s \rightarrow (2s, 2p)$ in the hydrogen atom. These excited Rydberg states are indeed the first corresponding to minima of the potential curves as a function of the internuclear distance. However, H_2 has also dissociating anti-bonding states dissociating to the same atomic configurations, $1s + 1s$, of the isolated hydrogen atoms as the groundstate. The vertical excitation to the lowest excited state $^3\sum_u^+$ having $S = 1$ and belonging to the M.O. configuration $(1\sigma_g)^1(1\sigma_u)^1$ produced by one-electron excitation of the ground configuration $(1\sigma_g)^2$ occurs well above 60 kK. The emission to the ground state corresponds to a broad continuum in the ultraviolet 25—50 kK which is used in spectral lamps filled with H_2 or D_2 as a light source without spectral lines after the limit of the Balmer series of H atoms at 27.4 kK.

If we apply the definition p. 145 of spectroscopic oxidation states to H_2 we see that it is reduced *ad absurdam*; the isolated groundstate $^1\sum_g$ with the first excited levels at much higher energy would only be compatible with *one* of the "resonance structures" hydrogen (I) hydride $(-I)$. This is a very unsatisfactory situation; hydrogen (0) twice would correspond to $^3\sum_u$ and $^1\sum_g$ of comparable energy, as is the case for H_2 stretched to a long internuclear distance. The same disastrous extrapolation holds for the diatomic halogens; Cl_2 has exactly the low-lying isolated $^1\sum_g$ one would expect for *one* of the structures chlorine (I) chloride $(-I)$. If heteroatomic interhalogens are considered, the spectroscopic oxidation states are arranged according to the relative electronegativity, for instance ClBr being bromine (I) chloride $(-I)$. One might propose the unreasonable argument that since the ionization energy of deuterium atoms is 270 parts per million higher than of usual hydrogen atoms, HD would be protium (I) deuteride $(-I)$. However, before being too discouraged by this failure of the concept of spectroscopic oxidation states in the case of homonuclear molecules, it must be realized that the conventional valence-bond description meets far more serious difficulties (cf. JØRGENSEN, 1962). Thus, carbon atoms in the electron configuration $1s^2 2s^2 2p^2$ correspond to 15 states, $1s^2 2s 2p^3$ to 40 states, $C^+ 1s^2 2s^2 2p$ to 6 states, $C^- 1s^2 2s^2 2p^3$ to 20 states, etc. It is not very easy to see how diamond succeeds in being colourless, the first excited level of the pure material occurring above 45 kK. Again, it must be admitted that diamond has no sensible spectroscopic oxidation state because half of the atoms being carbon (IV) and half carbon $(-IV)$ is an unacceptable proposal. However, at least in inorganic chemistry, it is a minority of compounds having strictly homonuclear bonds between two identical atoms, or heteronuclear bonds with hardly any difference between the electronegativities. We are reserving Chapters 10 and 11 for the treatment of the problems related to these almost non-polar bonds.

Hydride anions may form bridges between two or more atoms of other elements. Thus, solid LiH, NaH and KH crystallizing in the NaCl type has octahedral $H(-I) M_6$. The perovskites $LiBaH_3$ and $LiEu^{II}H_3$ contain $Li(I)H_6$, $M(II)H_{12}$ and six-coordinate hydride. Many heavy elements form volatile borohydrides such as $U(BH_4)_4$ and $Np(BH_4)_4$. $Zr(BH_4)_4$ seems to contain $Zr(IV)H_{12}$ bridges and can be formulated $Zr(H_3BH)_4$ (BIRD and CHURCHILL, 1967). The gaseous aluminium (III) compound has been shown by electron diffraction to have perfect or almost trigonal prismatic symmetry $Al(H_2BH_2)_3$ (ALMENNINGEN et al., 1968). ROSSMANITH (1964) distilled the yellow oil $Ho(BH_4)_3$ and the pink oil $Er(BH_4)_3$ in vacuo. Preliminary measurements of R. PAPPALARDO and S. LOSI show a moderate nephelauxetic effect in such compounds. The symmetry is not O_h.

We are not going here to discuss the fascinating question of hydrogen bonding with positive fractional charges on the hydrogen. There seems to be no cases involved where the oxidation state of any element is modified by hydrogen bonding. The metallic alloys of vanadium, chromium, palladium, etc. with hydrogen have been subject for complicated discussions. One might hope by physical measurements to get an idea about the electronic density close to the proton, and hence, to get a measure of how much of the conduction electron density has been concentrated by the penetrating proton; but this question is far from resolved. This is even less so for the compound UH_3 known to have twelve hydrogen atoms adjacent to each uranium atom. On the other hand, JONES and SATTEN (1966) measured absorption spectra of Pr(III) and Nd(III) in PrF_3 and NdF_3 isotypic with LaF_3 where some F^- have been replaced by H^-. The crystal structure of LaF_3 is now known (MANSMANN, 1965) it may be discussed whether nine or eleven fluoride anions (including two at longer distances) are coordinated to La(III). The nephelauxetic effect (cf. Tables 5.1 and 5.2) of replacing just one F^- by H^- is perceptible and suggests relatively strong tendency toward covalent bonding of $H(-I)$ even to the 4f group central atoms.

Inorganic complexes, and in particular those of which the absorption spectra have been most intensively studied in recent years, have the ligands bound with strong σ-bonds, and any effect of π-bonding or π-anti-bonding are superposed as a minor correction. However, there exist a group of ligands hardly forming any σ-bonds at all, and having very strong π-bonding. This is a rather distinct group among the *organometallic complexes* (having $M - C$ bonds; certain authors restrict the concept of organometallic compounds to those having normal σ-bonds) very different from the σ-bonded complexes. The problems regarding the bonding of CH_3 and other alkyl groups to transition group complexes are essentially the same as those of hydrides. Recently, it has been possible to react chromium (II) aqua ions with halogen-substituted methanes to form species such as $Cr(CH_2Cl)(H_2O)_5^{+}$, $Cr(CHCl_2)(H_2O)_5^{++}$ and the analogous bromine and iodine derivatives (DODD and JOHNSON, 1968). The visible spectra are strictly comparable to $Cr(CN)(H_2O)_5^{++}$ (KRISHNA-MURTHY, SCHAAP and PERUMAREDDI, 1967) and can be extrapolated to $Cr(CH_3)_6^{-3}$ discussed in Chapter 5. In all cases, we have a perfectly normal $3d^3$ Cr(III) complex of a carbanion ligand such as CH_2Cl^-, CH_3^- (isoelectronic with NH_3) etc. having more pronounced covalent bonding because of the lower electronegativity of the lone-pair of the carbanion. The same arguments apply to complexes of CF_3^- and other fluorinated alkyl groups, where the electronegativity of the ligand is higher and comparable to Cl^-. There is no reason to assume that aryl groups such as $C_6H_5^-$ behave essentially differently; the non-transition group phenyl complexes are colourless as expected of systems where conjugation through the central atom is not particularly important; and the orange colour of $Cr(C_6H_5)_6^{-3}$ and red colour of $Cr(C_6H_5)_3$ (tetrahydrofuran)$_3$ is compatible with Cr(III)-carbon σ-bonds.

14*

The situation is entirely different in *olefinic complexes*. ZEISE discovered 1830 the yellow $PtCl_3(C_2H_4)^-$ having the absorption bands of $PtCl_4^-$ shifted to slightly higher wavenumbers (DENNING, HARTLEY and VENANZI, 1967). This complex is normally described as a quadratic $5d^8$ platinum (II) chromophore. This is perhaps slightly abusive propaganda; though the three chloride ligands are situated in the plane containing the central atom, the fourth coordination positive is occupied by the *middle*, the centre of inversion, of the ethylene molecule. It is perhaps not too surprising that if the π-bonding to empty orbitals of the ethylene molecule (specifically of 5d (xz) to the π-orbital having a C — C node-plane coinciding with the $PtCl_3$ plane) keeps the complex together, the stereochemistry is highly different from that of usual σ-bonds. It may be worth analyzing the concept of the *coordination number* N. For a crystallographer, this is the number of adjacent atoms either all at the shortest distance from a given central atom, or at roughly the same distance. The crystallographer does not discuss the nature of the chemical bonding at this point; he is only interested in the nuclear positions (or strictly speaking the baricenters of the almost spherical electronic distributions). Thus, N = 6 without discussion both for sodium and chlorine in NaCl though the chemical bonding may be non-directional electrostatic. The isotypic calcium metal and solid argon both have N = 12. There is the difficulty that the crystallographer observes a time-average of atomic positions; it is conceivable that the instantaneous picture of PbS would not most probably be close to octahedral $Pb(II)S_6$ though the solid crystallizes in the NaCl type (cf. JØRGENSEN, 1967a). It is obviously a matter of judgment how much longer the next-shortest distances M—X can be than the shortest before one would no longer include them in N. Thus, one would normally say N = 8 for CsCl but the iron modification magnified in Brussels, or tungsten, have the same atomic positions though one tend to include the six slightly longer distances and say N = 14. The reason is partly that all fourteen neighbour atoms are identical in Fe and W whereas there are eight Cl and six Cs around each caesium in CsCl. It is evident that crystals with relatively low symmetry, or vitreous materials or liquids without any long-range order at all, may not show any definite N at all. Thus, the ten first neighbours may have their distances distributed over a broad interval. Actually, it is more surprising that d- and f-group absorption spectra so frequently indicate the presence of chromophores MX_N with a definite symmetry in such vitreous or liquid phases.

For the crystallographer, N is indiscutably 5 for platinum in $PtCl_3$ $(C_2H_4)^-$. We are discussing below why it is not highly recommended to ascribe coordination positions, not to definite atoms of a polyatomic ligand, but rather to intermediate points of the ligand, though it can sometimes be a strong temptation. Now coming to the question of the oxidation states in this complex, if any, one may argue that the olefin has retained its individuality, as suggested by the rapid exchange with other olefins, and a variety of other arguments, partly based on infra-red spectra. It would be interesting to know how close the four hydrogen atoms of C_2H_4 are to be in the same plane as the two carbon atoms. In the unprobable case that σ-back-bonding occurred involving a strong chelate of Pt(IV) with the di-anion $C_2H_4^{--}$ derived by deprotonation of ethane, one would expect the four hydrogen atoms not to be co-planar with the carbon atoms. However, CRAMER and PARSHALL (1965) added one mole of tetrafluoroethylene C_2F_4 to VASKA's reactive iridium (I) complex quadratic $Ir(CO)Cl[P(C_6H_5)_3]_2$ and considered the product as intermediate between a Zeise-type adduct of Ir(I) and an Ir(III) complex of $C_2F_4^{--}$ forming two σ-bonds in the chromophore Ir(III)C_3P_2Cl. GREEN et al. (1966) treated C_2F_4 with the platinum (0) complex $Pt[P(C_6H_5)_3]_4$ (in the following, triphenylphosphine is called L) and obtained $L_2PtC_2F_4$ having two σ-bonds to $C_2F_4^{--}$. ROUNDHILL and WILKINSON (1968) described related species. Surprisingly enough, hexafluoroacetone $(CF_3)_2CO$ forms an adduct with L_2Pt (see GREEN et al., 1966) having σ-bonds to both O and C forming a triatomic cycle PtOC. According to NYMAN, WYMORE and WILKINSON (1968) PdL_4 and PtL_3 (the latter species is known to contain a roughly planar, triangular chromophore Pt(O)P_3, ALBANO et al., 1966) react with O_2 to form green L_2PdO_2 and pale orange L_2PtO_2. These complexes are most readily described as M(II) complexes of bidentate O_2^{--}. They react with CO_2 to form L_2MCO_3 and $(C_6H_5)_3PO$. BAIRD, HARTWELL and WILKINSON (1967) described the unusual reactions of d-group complexes with CS_2, which may coordinate with one S; or with one C and one S; or decompose to CS complexes. Both Rh^I Cl(CS)L_2 and Rh^{III} X_3(CS)L_3 were detected.

Tetracyanoethylene $(NC)_2CC(CN)_2$ has a low-lying, empty M.O. (for a review of the chemistry, see DHAR, 1967) and forms strongly coloured adducts with organic molecules having relatively low ionization energy. However, the complexes $L_2M(CO)ClC_2(CN)_4$ (M = Rh and Ir) prepared by BADDLEY (1966) are approximately octahedral Rh(III) and Ir(III) complexes of $C_2(CN)_4^{--}$ bound by two σ-bonds. PANATTONI et al. (1968) studied the crystal structure of $L_2PtC_2(CN)_4$ where the ligand is sufficiently modified, relative to free $C_2(CN)_4$, in such a way that the complex is an intermediate case between π-back-bonding and the platinacyclopropane structure (the angle CPtC is only 42°). McGIN-NETY and IBERS (1968) discussed the related problem of $L_2Ir(CO)Br$ $C_2(CN)_4$. The two planes containing each a half $[-C(CN)_2]$ of the ligand form the dihedral angle 110° and the central $C-C$ bond is as long as 1.51 Å whereas it is 1.34 Å in free tetracyanoethylene. McGINNETY and

IBERS also point out that VASKA's reversible oxygen adduct $L_2Ir(CO)ClO_2$ has the relatively short distance $O - O$ 1.30 Å (it is 1.207 Å in gaseous O_2) whereas the irreversible adduct $L_2Ir(CO)IO_2$ has $O - O$ 1.51 Å. It is tempting to consider the iodide as a six-coordinate Ir(III) peroxo-complex whereas VASKA's chloride may be an olefin-like O_2-adduct of Ir(I). The diphosphine $(C_6H_5)_2PCH_2CH_2P(C_6H_5)_2$ forms a cation Ir (diphos)$_2O_2^+$ with the extremely long $O - O$ distance 1.66 Å. SCOTT, SHRIVER and VASKA (1968) related the increasing stretching frequency of $C - O$ (and presumably its increasing triple-bond character and decreasing tendency toward π-back-bonding from the iridium central atom) with the electron affinity of O_2, SO_2, $C_2(CN)_4$ and BF_3. The latter adduct, $L_2Ir(CO)Cl$ BF_3 is thought to have σ-bonding in direction from Ir to B; we are returning below to the question of BF_3^-.

It may be remarked that $N = 5$ in $PtCl_3(C_2H_4)^-$ is nearly as unusual for Pt(II) as for Pt(IV) though the π-back-bonding from Pt(II) is the most plausible hypothesis. This influence of empty orbitals in the ligands is also invoked by CHATT, DUNCANSON and VENANZI (1955) and ORGEL (1956) in order to explain the *trans*-effect (for reviews, see BASOLO and PEARSON, 1961 and 1962) which both has kinetic and crystallographic aspects. Hydride and carbanion ligands have a loosening effect on the bond *trans* to the strong σ-bond. Thus, the fifth ammonia molecule is readily lost from $Rh^{III}H(NH_3)_5^{++}$ (OSBORN et al., 1966) and the opposite ligand may have unusually long bond-length. The first square-pyramidal Rh(III) complex $RhL_2I_2CH_3$ having CH_3 on the perpendicular axis $Rh - C$ 2.08 Å was reported by THROUGHTON and SKAPSKI (1968). These examples show that *sometimes* the *trans*-effect is due to σ-bonding.

Both in the case of π-back-bonding and of supplying an unsaturated hydrocarbon ligand with two electrons so that two of the carbon atoms are able to form σ-bonds to the central atom, the metallic element acts as a Lewis base, and the ligand as a Lewis acid, whereas we are normally accustomed to the coordination having the opposite direction. A most clear-cut case is the reaction between diborane B_2H_6 (known to be in equilibrium with a minute concentration of BH_3) and $Mn(CO)_5^-$ studied by PARSHALL (1964) where it is nearly certain that $Mn(-I)$ is oxidized to $Mn(I)$ and that the ligand in the final complex is BH_3^- isoelectronic with CH_3^- and NH_3:

$$Mn(CO)_5^- + BH_3 = [Mn^I(BH_3^{-2}) (CO)_5]^- \qquad (9.7)$$

where the Lewis acid (or as J. BJERRUM proposes, *anti-base*) BH_3 oxidizes $Mn(-I)$ to $Mn(I)$ much in the same way as the Brønsted acid H_3O^+ reacts with carbonyl anions of low oxidation states in eq. (9.3) increasing the oxidation states of the central atom by two units. PARSHALL (1964) also reacted $Re(CO)_5^-$ with B_2H_6 obtaining $Re(CO)_5(BH_3)_2^-$ which most probably is a seven-coordinate Re(III) complex of BH_3^-.

When acting as ligands, BH_3^- and CH_3^- may either be considered hydrides of boron(I) and carbon (II), the lone-pair being the mixture of 2s and 2p orbitals comparable to the loosest bound orbital of ammonia; or protonic adducts of $B(-V)$ and $C(-IV)$. However, one may also consider some of the complexes containing central atoms acting as Lewis bases as catenated compounds to be discussed in Chapter 10. Thus, NÖTH and SCHMID (1963) reacted $[(CH_3)_2N]_2BCl$ with $Mn(CO)_5^-$ obtaining orange $[(CH_3)_2N]_2BMn(CO)_5$. Under normal circumstances, one would regard this ligand as a cation R_2B^+ formed by removing one chloride from the neutral boron (III) compound. It might then be argued that $Mn(CO)_5^-$ is an anion re-arranging from trigonal-bipyramidal configuration to square-pyramidal offering the lone-pair on the sixth position of an octahedron. This impression becomes stronger when it is realized that NÖTH and SCHMID (1963) also prepared orange $(CH_3)_2NB[Mn(CO)_5]_2$. Red crystalline $ClSn[Mn(CO)_5]_3$ (TSAI et al., 1967) and very pale yellow $H_2Ge[Mn(CO)_5]_2$ (MASSEY, PARK and STONE, 1963) are other examples; and CLARK et al. (1966) performed the reaction

$$(CH_3)_3GeMn(CO)_5 + C_2F_4 = (CH_3)_3GeCF_2CF_2Mn(CO)_5. \qquad (9.8)$$

Rather than going to the extreme of considering these species as $Mn(-I)$, it may be more sensible to assign them the conditional oxidation states Ge[IV] and Mn[I] joined with a definite σ-bonding lone-pair, as discussed in Chapter 10.

One may ask the question why higher oxidation states do not form carbonyls. Actually, both $Mn(CO)_6^+$ and $Fe(CO)_6^{++}$ can be prepared; but the reaction with water is rather violent (cf. KRUCK and NOACK, 1964):

$$Mn(CO)_6^+ + H_2O \rightarrow HMn(CO)_5 + CO_2 + H^+. \qquad (9.9)$$

There are not many cases known outside of organometallic chemistry of d-group central atoms acting as Lewis bases. One would imagine that the $d(z^2 - \frac{r^2}{3})$ orbital perpendicular on the plane of quadratic low-spin d^8-complexes would have a good chance of showing Lewis basicity, but when e. g. BF_3 is reacted with nickel (II) dimethylglyoximate, the molecule is added to the oxygen atoms of the ligand and not to Ni(II). SHRIVER (1966) reviewed the attempts to produce $[Ni^{IV}(BF_3^-)(CN^-)_4]^{--}$ by adding BF_3 to $Ni(CN)_4^{--}$ but actually, the terminal nitrogen atoms act as Lewis bases forming $Ni^{II}(CNBF_3)_4^{--}$ much in same way as NH_3BF_3 and NH_3BH_3 (isoelectronic with ethane by the way).

Whereas the σ-back-bonding is only known in a very few cases such as (9.7), pure π-bonding and possibly simultaneous π-back-bonding is definitely very important in complexes of cyclopentadienide $cp^- = C_5H_5^-$; benzene C_6H_6; and the tropylium cation $C_7H_7^+$. The stereochemistry of cyclopentadienide complexes is far more complicated than first assumed (cf. the review by WILKINSON and COTTON, 1959 and the excellent textbook by COTTON and WILKINSON, 1966). Whereas the sandwich structure of the first complex prepared, ferrocene $Fe(C_5H_5)_2$ undoubtedly is represented by many M(II) and M(III) cyclopentadienides, many other structures are also conceivable. It is not surprising that Tlcp is a one-sided Danish-style sandwich; it is already more unexpected that $Becp_2$ has Be much closer to one five-membered ring than to the other much in the same way as a unsymmetric hydrogen bond (ALMENNINGEN et al., 1964) and recently, PREUSS has attempted M.O. calculations to explain this fact. The complexes formed by the earlier 3d-group elements, such as Ti cp_2, and certain 4d-group complexes, such as Rh cp_2, seem to be dimers possibly involving direct $M-M$ bonds (cf. SALZMANN and MOSIMANN, 1967).

In the regular sandwich structure, the sub-shells $d\sigma$ and $(d\delta c, d\delta s)$ have much lower energy than $(d\pi c, d\pi s)$. This can be readily explained by the filled set of two "π" orbitals having one node-plane on each ligand interacting with $d\pi$ which becomes the anti-bonding sub-shell. This effect is π-bonding in the direction from cp^- to M much in the same way as the strong π-bonding of one or two oxide ligands to d^0, d^1 and low-spin d^2-systems. Additional effects in the cyclopentadienides may be a mixing of the empty s orbital with $d\sigma$, like in $PtCl_4^-$, and π-backbonding in direction from M to cp^- via $d\delta$ interacting with the empty "π" orbitals on the ligands having two node-planes. The latter effect may be more important in complexes of the neutral benzene molecule with M in low oxidation states.

As first pointed out by ROBERTSON and McCONNELL (1960) and later elaborated by SCOTT and BECKER (1965) the separation of the three d-sub-shells conserves the quantum number λ as if the symmetry was linear $D_{\infty h}$. Actually, there is hardly any influence on the d-shell by the five-fold axis of the ligand because 2 and 5 have no common denominator higher than 1, and one cannot distinguish the eclipsed structure of the two ligands having the symmetry D_{5h} from the staggered structure turned 36° D_{5d}. Actually, the rotation of the two rings has an extremely small activation energy which has recently been determined (HAALAND and NILSSON, 1968). There is a close analogy between the three d-like orbitals of low energy and two strongly anti-bonding orbitals of Mcp_2 and Mcp_2^+ and the two sub-shells of an octahedral d-group chromophore. This analogy becomes even clearer if the octahedron is considered as a trigonal anti-prism with the z-axis directed along the trigonal axis in which case the lower sub-shell consists of $d\sigma$ and of a mixture of $d\pi$ and

$d\delta$ with definite coefficients. The anti-bonding character of the higher sub-shell is not larger than $\sim 4D$ in the case of Mn cp_2 since this compound has $S = 5/2$ though it has strong effects of antiferromagnetic coupling in the solid state. Vcp_2 has $S = 3/2$ and is a sandwich like $Mgcp_2$ whereas the titanium cyclopentadienides behave unexpectedly; even Ti cp_2^+ is not with certainty a sandwich compound. The most common preponderant configurations of this class are d^5 ($S = 1/2$) known in the blue ferricinium cation Fe cp_2^+ and analogous Ru cp_2^+ and d^6 ($S = 0$) Fe cp_2, Co cp_2^+, Ru cp_2, Rh cp_2^+, Os cp_2 and Ir cp_3^+. One anti-bonding electron occurs in the unstable Co cp_2 and two in the green Ni cp_2 ($S = 1$) which is readily oxidized to yellow Ni cp_2^+ ($S = 1/2$).

From the point of view of crystallography, the sandwich compounds have $N = 10$ and all M—C distances identical. We are discussing below whether O_2NO^- or NO when the central atom at long distance from both coordinated atoms can be said to be unidentate rather than bidentate. In this sense, one might argue that Mcp_2 and Mcp_2^+ have $N = 2$, the center of the regular pentagon being the effectively ligating point. By the same token, it might be argued that the titanium (IV) complexes $TiX_2 cp_2$ ($X = Cl, Br, I, \ldots$) have $N = 4$ rather than $N - 12$ in the crystallographic sense. The uranium (IV) complex Ucp_3Cl and the ytterbium (III) adducts $Yb cp_3L$ (where L is a variety of neutral ligands) presumably have $N = 16$. This is the highest value known for any monomeric complex (it may be remembered that if the ligands are spheres of the same radii as the central atom, $N = 12$ because this is the integer just below 4π) though metallic alloys are known with higher N. Thus, a samarium-beryllium alloy has 24 beryllium atoms surrounding each Sm. This is analogous to cubic $NaZn_{13}$, $ZrBe_{13}$, $CeBe_{13}$, $ThBe_{13}$ and UBe_{13} (BAENZINGER and RUNDLE, 1949). Many boranes and solid boron compounds contain fragments of a icosahedral structure B_{12}, each boron atom being connected with five other boron atoms in a pentagonal pyramid. If a few boron atoms in anions are replaced by carbon, sandwich-complexes of d-group atoms can be formed. Thus, $dcb^{--} = B_9C_2H_{11}^{--}$ forms not only diamagnetic $Fe^{II} dcb_2^{--}$ and $Co^{III} dcb_2^-$ but also brown ($S = 1/2$) $Ni^{III} dcb_2^-$ and yellow ($S = 0$) $Ni^{IV} dcb_2$ containing the chromophore $Ni(IV)B_6C_4$ (SCOTT, 1966; WARREN and HAWTHORNE, 1967)

It has been argued by some authors that there is a metaphoric sense in which Fe cp_2 has "N" $= 6$. This argument originated in the similarity between S-values for low-spin octahedral and for sandwich complexes. In the time of the hybridization theory, it was assumed that six orbitals such as $3d^2 4s 4p^3$ (the exponents indicating the number of orbitals and not the number of electrons) were responsable for forming six bonds in low-spin complexes such as $Co(NH_3)_6^{+3}$ or NiF_6^{-2}. The fact that half of the aromatic ligands in $Cr(C_6H_6)_2$ or in Fe cp_2 can be replaced by three

CO ligands to form $Cr(CO)_3(C_6H_6)$ or $Fe(CO)_3cp^+$ further contributed to this concept. The arene complexes and the more numerous mixed arene-carbonyl complexes have been reviewed by ZEISS, WHEATLEY and WINKLER (1966). For a long time, it was an intensely discussed question whether $Cr(C_6H_6)_2$ had hexagonal or trigonal symmetry, i.e. whether one of the two Kekulé structures predominates in the ligand. At least the time-average symmetry seems now to be fully hexagonal. The arguments about "N" $= 6$ in this compound and in Fe cp_2 are fundamentally unsound. However, it must be admitted that it is striking that d^6-systems form hexa-carbonyls, d^8 trigonal-bipyramidal penta-carbonyls and d^{10} tetrahedral $M(CO)_4$. If each ligand supplies the lone-pair to the central atom, it is true that these systems all contain 18 electrons which might occupy the shells $3d^{10}4s^24p^6$ between argon and krypton in agreement with SIDGWICK's old idea of effective electron numbers equal to those of the noble gases. However, there are many aspects by which $Co(NH_3)_6^{+3}$ or $Co(CN)_6^{-3}$ differ strongly from a krypton atom (JØRGENSEN, 1962) and one of the more striking is that Laporte-forbidden transitions occur at much lower wavenumbers than the Laporte-allowed absorption bands which is a definite proof that the 4p orbitals are far from being degenerate with 3d orbitals. Actually, there is no strong evidence available that the empty 4p orbitals are of any importance at all for the chemical bonding in 3d-group complexes, and stable complexes can be exceptions to Sidgwick's rule, such as $Cr(NH_3)_6^{+3}$ containing three electrons too few and $Ni(NH_3)_6^{++}$ two too many. The relatively stable $Cr(C_6H_6)_2^+$ has one electron lacking. The absorption spectrum of $Cr(CO)_6$ would be more compatible with the d^2sp^3 hybridization than nearly all other 3d-group complexes, but still, the effective electronic number seems to be an accidental consequence of other features of M.O. theory. Along this line of thought, the $3d^6$-system Fe cp_2 needs twelve electrons from the ligands, six from each. It is true that the number x of carbon monoxide ligands vary in a conspicuous way with the preponderant configuration of the central atom. Thus, the d^4-systems vanadium (I) have $x = 4$ in $V(C_6H_6)(CO)_4^+$ and $Vcp(CO)_4$ in apparent agreement with the proposition that these systems have "N" $= 7$ and need fourteen additional electrons to obtain the pseudo-krypton configuration proposed by SIDGWICK. In this connection, it may be remarked that the d^6-anion $V(CO)_6^-$ forms a salt of gold (I) which according to a kind communication from Sir RONALD NYHOLM is seven-coordinate $(C_6H_5)_3PAuV(CO)_6$ having the gold atom on one of the three-fold axes of the octahedral anion. This genuine example of N$=7$ could be treated in Chapter 10 on catenated compounds of the same kind as the linear mercury (II) complex $(OC)_4CoHgCo(CO)_4$. However, it should not be construed as an argument for "N" $= 3$ of each π-bonded aromatic ligand.

WILKINSON and collaborators also prepared "open oyster" complexes having two cp$^-$ and additional ligands of relatively small spatial requirement. They developed the proton nuclear magnetic resonance technique to detect $M - H$ bonds in hydride complexes both of this kind and of the platinum metals in general. The presence of such hydride ligands had frequently escaped notice and poses difficult problems for traditional quantitative analysis. A typical "open oyster" compound is the tantalum (V) complex Ta cp$_2$H$_3$ having the same constitution both in acid and alkaline solution, whereas

$$W^{IV}cp_2H_2 + H^+_{aq} = W^{VI}cp_2H^+_3$$
$$Re^{III}cp_2H + H^+_{aq} = Re^{V}cp_2H^+_2$$
(9.10)

are oxidized by aqueous acidic solutions in close analogy to (9.3). SHRIVER (1963) described the interesting compound Wcp$_2$H$_2$(BF$_3$) which may either be a tungsten (IV) complex of BF$_3$ or a tungsten (VI) complex of the hypothetical ligand BF$_3^{--}$. JOHNSON and SHRIVER (1966) prepared the analogous compounds Mo cp$_2$H$_2$(BF$_3$), Wcp$_2$H$_2$(BCl$_3$), Re cp$_2$H(BF$_3$) and Re cp$_2$H(BCl$_3$). Sn cp$_2$ and Pbcp$_2$ are also "open oysters".

The π-back-bonding in carbonyls has recently been treated by extensive M.O. calculations by NIEUWPOORT (1965) on the isoelectronic series Fe(CO)$_4^{--}$, Co(CO)$_4^-$ and Ni(CO)$_4$. This work has confirmed previous conclusions reached from infra-red and Raman spectra (STAMMREICH et al., 1960) that π-back-bonding is more important in the Fe($-$II) complex and decreases in importance when the oxidation state is increased. The π-back-bonding is less important in isonitrile complexes of RNC (MALATESTA, 1959) and is only perceptible in cyanide complexes of unusually low oxidation states such as the colourless Mn(CN)$_6^{-5}$ which must have the sub-shell energy difference Δ comparable to those for the isoelectronic Fe(CN)$_6^{-4}$ and Co(CN)$_6^{-3}$ (cf. Table 5.8). On the other hand, NIEUWPOORT's results do not suggest that a collectively reduced set of ligands occur in the carbonyls. The smallest number of electrons redistributed which would be compatible with the observed symmetry and lack of colour would make Cr(CO)$_6$ the Cr(VI) complex of (CO)$_6^{-6}$ and Ni(CO)$_4$ the Ni(IV) complex of (CO)$_4^{-4}$ which is entirely incredible. However, the fractional charge of the iron ($-$II) is expected to be *less* negative than -2; it is also interesting to note that no indication is found of a low-lying 4s-like orbital. It may be remembered that CO, like PF$_3$, is one of the few Lewis bases which are hardly Brønsted bases at all. It might be argued that in the complexes having really strong π-back-bonding, such as Cr(CO)$_5^{--}$ and Fe(CO)$_4^{--}$ (Ti(CO)$_6^{--}$ has not yet been reported) the ligands act as Lewis acids rather than as Lewis bases. The 3d^5-system $(S = 1/2)$ V(CO)$_6$ has no tendency to dimerize. The dark blue solid may show collective effects like the transitions to a conduction band in crystalline $TiCl_3$ whereas the pale green colour of the vapour (cf. the spectrum given by HAAS and SHELINE, 1966) *must* correspond to a weak band in the red, which is probably due to transitions between the three orbitals of the lower sub-shell separated in energy to some extent by the Jahn-Teller effect. Many carbonyls tend to dimerize or form carbonyl-bridged oligomers, or to form cluster compounds to be discussed in Chapter 10.

At a certain time, it was believed that the complexes of phosphines R_3P and arsines R_3As have strong π-back-bonding effects to supposedly low-lying empty 3d- or 4d-orbitals. This hypothesis is much less popular today; most of the absorption spectra and other physical measurements can be explained by predominant σ-bonding like in the amines. However, the colourless phosphorus (III) halides are known to form complexes such as $Ni(PCl_3)_4$ and $Ni(PBr_3)_4$ and PF_3 is even more effective than CO to stabilize low oxidation states. KRUCK (1967) has published an extensive review of PF_3 complexes such as $Cr(PF_3)_6$, $Fe(PF_3)_4^{--}$, $Fe(PF_3)_5$, $Co(PF_3)_4^-$, $Ru(PF_3)_4^{--}$, $Rh(PF_3)_4^-$, $Pd(PF_3)_4$, $Os(PF_3)_4^{--}$ and $Ir(PF_3)_4^-$ and mixed hydride complexes such as $Co(PF_3)_4H$, $Ru(PF_3)_4H_2$ and $Ir(PF_3)_4H$. Most of these complexes are colourless, and others lemon-yellow; since many more than one PF_3 ligand occur, it is not a satisfactory solution to assume PF_3^{--}, and we have to accept the unusually low oxidation states of the central atoms such as $Co(-I)$ and $Ir(-I)$. The interesting question is now: which are the orbitals to which the π-back-bonding takes place? It must be recognized that CO has the first excited levels at much higher energy than of R_3P and I^- to be discussed below, and that π-back-bonding cannot be excluded entirely of spectroscopic reasons though it may be that R_3P and PF_3 have effects of continuum orbitals rather than of definite 3d-orbitals. However, it is an important difference that the effects of chemical bonding in CO are very strong and produce a considerable mixing of l-values, in this case 2s and 2p in the orbitals of symmetry type σ. Hence, it is easier to understand that highly excited π-anti-bonding orbitals of CO still can influence the chemical bonding with a central atom such as $Fe(-II)$ or $Co(-I)$. On the other hand, phosphines and related compounds having longer internuclear distances and a variety of other characteristics are expected to have smaller non-diagonal elements of energy between high-lying, unoccupied orbitals of the ligand and d-orbitals of the central atom. It is not quite easy to understand the stability of PF_3 complexes of d-group elements; but the non-positive oxidation states seem entirely well-defined.

On the other hand, certain diatomic ligands such as NO, O_2 and N_2 pose far more serious problems regarding the oxidation states than do CO, CN^- and PF_3. One charateristic feature of the suspect ligands is the tendency to form mixed complexes with innocent ligands (though frequently of low electronegativity) and only one suspect ligand. The simplest hypothesis is that the series C_2^{--}, CN^-, CO, N_2, NO^+ is isoelectronic and occur in complexes. The number of cation ligands is low, and NO^+ is certainly the simplest. It seems to occur in mixed cyanide-complexes such as the violet $Mn^I(CN)_5NO^{-3}$, the red nitroprusside $Fe^{II}(CN)_5$ NO^{-2}, orange $[Fe^{-II}(CO)_3NO]^-$ red-brown $Co^{-I}(CO)_3NO$ and Ir^{-I} $[P(C_6H_5)_3]_3NO$ (cf. MALATESTA, 1967) and if anything, the low oxidation states are stabilized even more than in the unsubstituted complexes, though $Ir^{III}Br_5NO^-$ is also known. The green solid $Mn(CO)(NO)_3$ reported by BARRACLOUGH and LEWIS (1960) might be the only d-group $M(-III)$.

However, the truth regarding nitrosyl complexes must be very complicated. The (bent, not linear) molecules colourless FNO, yellow ClNO and brown BrNO obviously have electron transfer bands where the halide reduces the central atom from N(III) to N(II). Though the typical $M(NO)^I$ complexes seem to have the triatomic group linear, this does not necessarily always be the case. Actually, one has the impression that there exist weakly bound NO complexes such as the "brown ring test" for nitrate in concentrated sulphuric acid:

$$Fe(H_2O)_6^{++} + NO - Fe(H_2O)_5NO^{++} + H_2O . \qquad (9.11)$$

The complex has $S = {}^3/_2$ which might be taken as an anti-ferromagnetic coupling between $S = 2$ of a central iron (II) and $S = {}^1/_2$ of the ligand (cf. JØRGENSEN, 1963). The dividing line between loose NO adducts and $(NO)^I$ complexes might be the presence of σ-anti-bonding electrons in the loose adducts. Thus, VAN VOORST and HEMMERICH (1966) discuss the electron spin resonance spectra of $Fe(CN)_5NO^{-3}$ and $Fe(CN)_5NOH^{--}$ giving evidence for the lowest empty π orbitals of NO (containing one electron in the free molecule) having the energy situated between the lower sub-shell of the octahedral complex, accommodating at most six electrons, and the two σ-anti-bonding orbitals. Said in other words, the two latter complexes are Fe(II) and not Fe(I). This is also true for (9.11).

In certain cases, the bonding of NO to a central atom can be so strong as to remind one about HIEBER's pseudo-atoms. Thus, the group $(RuNO)^{+3}$ occurs in a very large number of complexes, the oldest example being the purple $Ru(NO)Cl_5^{-2}$. CHARONNAT prepared a large number of trans-$Ru(NO)X_4Y$ and there is evidence that Y situated on the one side of the axis continuing through the RuNO nuclei has anomalous long $Y - Ru$ distance and exchanges readily, showing clear-cut trans-effect. There is no objection to distribute the spectroscopic oxidation states Ru(II) and

$(NO)^I$ inside the group. GRIFFITH (1967) has written an excellent book about the chemistry of Ru, Rh, Os and Ir. However, when one comes to $Pt(NO)Cl_5^{-2}$ there is the objection against Pt(II) that this oxidation state rarely forms octahedral complexes, and it is possible that it involves Pt(IV) and NO^-, or that NO is not at all innocent. JOHNSON and MCCLE- VERTY (1966) published a review on d-group nitrosyl complexes; we are only going to comment on a few more recent observations.

One of the most enigmatic questions in nitrosyl complex chemistry is the black and the brick-red isomers of $Co(NH_3)_5NO^{+2}$. HALL and TAGGART (1965) and DALE and HODGKIN (1965) resolved crystal struc- tures containing the black cation which is monomeric. Apparently, it is a cobalt (III) complex of NO^- like diamagnetic $Co(CN)_5NO^{-3}$. The black chloride shows a weak paramagnetism (ODELL et al., 1965) might be interpreted as an anti-ferromagnetic coupling between low-spin Co(II) and neutral NO for which BOSTRUP (1963) also gives chemical arguments. The black colour and lack of non-destructive solvents prevent spectro- scopic argumentation. There is a growing suspicion (RAYNOR, 1966; MERCER et al., 1967; GANS, 1967) that the red isomer is dimeric contain- ing a *cis*-hyponitrite bridge

$$\left[(NH_3)_5\,Co^{III} \diagdown \underset{O}{\diagup} N\!=\!N \diagdown \underset{O}{\diagup} Co^{III}(NH_3)_5 \right]^{+4}$$

in agreement with the absorption spectrum.

In certain cases, the central atom is bound to a point between N and O, i.e. there exists an irregular triangle MNO. Thus, the complex Co $dtc_2(NO)$ has Co above the plane of four sulphur atoms whereas Fe dtc_2 (NO) has an almost linear grouping FeNO. On the other hand, RAE (1967) found that the line NO forms an angle $137°$ with the axis of the square pyramid of Fe $mnt_2(NO)$. The two bidentate ligands are $dtc^- = (C_2H_5)_2$ NCS_2^- and $mnt^{--} = (NC)_2C_2S_2^{--}$. DOMENICANO et al. (1966) described the dark red compounds $Ru\,dtc_3$ and $Ru\,dtc_3NO$ where the former contains octahedral $Ru(III)S_6$ and the second a NO group *cis*- to the sulphur atom from a unidentate ligand. Apparently, $Ru(II)NS_5$ contains NO^I.

NO^- is isoelectronic with O_2 and would have $S = 1$ if there is no other atom attached in a non-linear way. Several stable molecules R_2NO with $S = 1/2$ are known. The oldest case is FREMY's violet anion $ON(SO_3)_2^{--}$ (having an orange, diamagnetic dimer) and the brown liquid $[(CH_3)_3C]_2$ NO described by HOFFMANN and HENDERSON (1961). The M.O. descrip- tion of such species and their visible absorption was discussed by JØRGEN- SEN (1963, p. 129).

Though NO is thermodynamically highly instable relative to N_2 and O_2, there is no safe evidence for dimerization. It may be pointed out that the ground term is $^2\Pi$ consisting of two adjacent levels with $\Omega = 1/2$ and $\Omega = 3/2$. Since the former level is exclusively populated at low temperature and, somewhat accidentally, has the g-factor zero, the molecule looses its paramagnetism by cooling.

Many of the same intricate problems as for NO occur for the O_2 complexes where the isolated ligand is known in four different steps of reduction, viz. $O_2^+(S = 1/2)$, $O_2(S = 1)$, $O_2^-(S = 1/2)$ and $O_2^{--}(S = 0)$ having one, two, three and four electrons in the two degenerate π-anti-bonding orbitals which are used for π-back-bonding in linear MCO. One would expect the peroxides O_2^{--} to show the simplest behaviour, but we have already the problem to determine whether the ligand is bidentate or unidentate. Thus, the red $Cr^V(O_2)_4^{-3}$ and orange $Nb^V(O_2)_4^{-3}$ are undoubted $3d^1$ and $4d^0$-systems; but the peroxide has one oxygen atom at shorter distance from the central atom than the other oxygen atom; and one has the impression that the chemical bond goes from M to a point between the two oxygen atoms, much in the same way as in certain NO complexes. Hence, it is an open question whether $N = 4$ or 8. One would have expected that the internuclear distance $O - O$ would increase regularly from O_2^+ to O_2^{--} as the number of π-anti-bonding electrons increases from 1 to 4. However, the Madelung potential stabilizes the anions and make the superoxides O_2^- and peroxides O_2^{--} contract perceptibly in the solid state. Thus, the crystal structure of $Cr(O_2)_2(NH_3)_3$ was implied to involve O_2^- and Cr(II) though there is no doubt that it contains Cr(IV) and contracted peroxo ligands (cf. JØRGENSEN, 1963). There is no clear-cut case known of a monomeric superoxo complex, like the general case that ligands having positive S are relatively rare, probably because they show strong anti-ferromagnetic coupling. There are a few interesting dimeric complexes. WERNER oxidized the brown diamagnetic $(NH_3)_5CoO_2Co(NH_3)_5^{+4}$ which is a definite Co(III) peroxo complex to dark green $(NH_3)_5CoO_2$ $Co(NH_3)_5^{+5}$. The electron paramagnetic resonance spectrum of this cation shows that it contains two equivalent cobalt atoms, and that the density of uncompensated spin has a certain value close to the cobalt nuclei as indicated by the nuclear hyperfine-structure. However, it can be discussed whether it is a partly delocalized superoxo complex of Co(III) or whether it is a resonance between Co(III) and Co(IV) in analogy to H_2^+. The stereochemistry of these complexes have also been rather much discussed; the crystal structures by VANNERBERG (1965) show that CoO_2Co does not have the $O - O$ bond exactly perpendicular on the line $Co - Co$ as once proposed by VLČEK; but the angles are not extremely different from this suggestion. Also HOOH in gaseous state is strongly bent.

FALLAB (1967) has reviewed the reversible (frequently dark brown) and irreversible reaction products of O_2 with cobalt (II) complexes of amines or amino-acids. BAYER and SCHRETZMANN (1967) have reviewed the more general problem of oxygen adducts. Hemoglobin with O_2 is known to be diamagnetic; this may be a low-spin iron (II) complex back-bonding to O_2 in much the same way as to olefins in ZEISE's anion; or it may be a low-spin Fe(III) antiferromagnetically coupled to O_2^-. VASKA (1964) studied the adducts of a variety of neutral molecules to the quad-ratic iridium (I) complex $Ir(CO)Cl[P(C_6H_5)_3]_2$. By the way, this complex can be formed from C_2H_5OH, $P(C_6H_5)_3$ and iridium chlorides by heating; the CO is abstracted from an ethanol molecule. The adduct of this com-plex with H_2 is an octahedral Ir(III) complex having two hydride ligands. It can be discussed whether the O_2 adduct is a five-coordinate Ir(I) with olefine O_2 or a distorted octahedral Ir(III) peroxo complex containing the chromophore $Ir(III)CO_2P_2Cl$.

VASKA also prepared adducts with SO_2. Cases are known, such as octahedral $Ru(NH_3)_5SO_2^{++}$, which are not more surprising than sulfite complexes bound with one σ-bonding lone-pair such as $Co^{III}(NH_3)_5SO_3^+$. However, it would not seem that Ir(I) just accepts a fifth ligand, neutral SO_2. On the other hand, it is doubtful whether the Lewis basicity of the central atom goes to the point where the ligand can be described as sulphur (II) SO_2^{--}.

N_2 is very unreactive at room temperature. Certain organisms can reduce nitrogen to ammonia, possibly via an enzyme containing molyb-denum (V). But the only well-defined complexes of N_2 known at present are formed by cobalt, ruthenium and osmium, whereas carbonyl analo-gues such as $Ni(N_2)_4$ have not yet been reported. ALLEN and SENOFF (1965) isolated salts of yellow $Ru(NH_3)_5N_2^{++}$ and ALLEN and STEVENS (1967) of pale yellow $Os(NH_3)_5N_2^{++}$. They seem to be octahedral d^6-systems $(S=0)$ with linear MNN groups comparable to the isoelectronic MCO. The nitrogen molecule in these compounds was derived from hydrazine H_2NNH_2, but ALLEN and BOTTOMLEY (1968) demonstrated that a solution of ruthenium (II) in aqueous ammonia is able *selectively* from air to fix N_2 in preference to O_2. HARRISON, WEISSBERGER and TAUBE (1968) studied the ultraviolet spectra of $Ru(NH_3)_5N_2^{++}$ having a strong band ($\varepsilon = 16000$) at 45.2 kK and a dimer $(NH_3)_5RuN_2Ru(NH_3)_5^{+4}$ with an even stronger band ($\varepsilon = 47000$ per 2 Ru) at 38.1 kK. The high intensity of these bands might suggest Ru(III) complexes of N_2^- (perhaps anti-ferromagnetic) but the writer believes in TAUBE's explanation involving inverted electron transfer from Ru(II) to the otherwise very high-lying empty M. O. of N_2. These intense bands explain also the diffi-culty of obtaining the spectrum of $Ru(NH_3)_6^{++}$ whereas $Ruen_3^{++}$ can be studied with appropriate precautions (SCHMIDTKE and GARTHOFF, 1966).

SACCO and ROSSI (1967) prepared yellow $Co^{III}H_3L_3$ ($L=P(C_6H_5)_3$) from $CoX_2L_2 + BH_4^-$ in ethanol. This compound reacts with N_2 forming

$CoHL_3N_2$ of which ENEMARK et al. (1968) studied the crystal structure which is a trigonal bi-pyramid with N_2 and H^- on the axis; the distances $N - N$ and $N - Co(I)$ are 1.16 and 1.80 Å. It is orange-red.

PEARSON (1963, 1966) proposed a division of Lewis acids (anti-bases) into *hard acids* such as Be^{+2}, Al^{+3}, La^{+3}, Th^{+4} and *soft acids* such as Cu^+, Pd^{+2}, Ag^+, Hg^{+2}, Tl^{+3} and a division of Lewis bases into *hard bases* such as F^-, SO_4^{--}, $RCOO^-$ and *soft bases* such as I^-, RS^-, R_3P and H^-. Hard acids react preferentially with hard bases and soft acids react preferentially with soft bases. There has been very much discussion whether PEARSON's concepts are new and whether they are useful. It is quite clear that they are not entirely new; the alchemists' ideas of idealized mercury reacting with idealized sulphur are rather similar to the reaction between soft acids and soft bases; and GOLDSCHMIDT's geochemical ideas about lithophilic and chalcophilic elements correspond to hard and soft reactants. Anybody familiar with inorganic qualitative analysis using sulphide recognizes characteristic features of PEARSON's classification. Among the opponents of PEARSON is FAJANS (1967) arguing that the concepts are meaningless and that the quanticule theory is better (we return to this question in Chapter 11) and WILLIAMS and HALE (1966) (cf. also the excellent text-books by PHILLIPS and WILLIAMS, 1965 and 1966) who argue that electrostatic attraction determines the hard-hard interactions, and that the soft-soft interactions are specific from case to case (as is Lewis acidity as contrasted to Brønsted acidity in a definite solvent) and correspond essentially to covalent bonding.

Several authors have attempted to introduce quantitative scales of softness. Professor R. F. HUDSON who now is at the University of Kent in Canterbury held a symposium on the subject at the Cyanamid European Research Institute in Geneva in 1965. Among the arguments used were kinetics of carbon (IV) and platinum (II) compounds. A second symposium was organized by Professor M. J. FRAZER at Northern Polytechnic, Holloway, London in 1966. We may for a moment return to the origin of a comparable treatment of complex formation constants. Already ABEGG and later J. BJERRUM (1950) emphasized that the halides F^-, Cl^-, Br^-, I^- constitute a set where certain metallic elements (such as Ag and Hg) prefer I^- and other (such as Al and Fe) prefer F^- with the three other halides in the same universal order. AHRLAND, CHATT and DAVIES (1958) compared the available complex formation constants and classified the central ions according to (a) character, i.e. PEARSON's hard LEWIS acids, or (b) character i.e. PEARSON's soft LEWIS acids. However, PEARSON expressed many original ideas. Thus, he pointed out that "good" metals as defined in Chapter 8 are *ipso facto* soft, e.g. Na, Al and Cu. Another interesting point is that any π-back-bonding central atom is soft; hence the formation of carbonyl or ethylene complexes is a sufficient, though not necessary, reason to classify a central atom in a given oxidation state as soft. It may be remarked that until recently, carbonyl complexes were not known of oxidation states higher than M(II) such as $Fe(CN)_5CO^{-3}$ and $Fe(CO)_4I_2$, but then MALATESTA prepared $Ir(CO)Br_5^{-2}$ and $Ir(CO)_2I_4^-$; and $Ru(CO)Cl_5^{-2}$ is also known (cf. the review by CALDERAZZO, 1967). Quite generally, the soft central atoms seem to fall into several categories, whereas the behaviour of the hard central atoms seems to be determined by one parameter, such as the square of the ionic charge divided by the ionic radius. It had previously been remarked (JØRGENSEN, 1963) that the s^2 systems such as Pb(II) and Bi(III) show (b) character toward I^- and RS^- but not toward CN^- and amines. This can be explained by the situation that ligands with one lone-pair tend to make the s-orbital more σ-anti-bonding. PEARSON (1963) suggested that a given element is the softer the lower the oxidation state. Especially the propensity to form unmixed carbonyls seems to be connected with low oxidation states; it must be remembered that the mixed halide-carbonyl complexes of M(II) and M(III) may have pure σ-bonding of CO without any π-back-bonding. JØRGENSEN (1964b) suggested that the softness is not necessarily a monotonic function of the oxidation state. Though the softness obviously decreases tremendously from $Mn(-I)$ over Mn(I) to Mn(II), the minimum may be reached close to Mn(III), and Mn(VII) may be moderately soft though not as much as Mn(I). One may imagine two different reasons for this non-monotonic variation; the softness may be a monotonically decreasing function of the fractional charge δ, and with a given set of ligands, δ may have a maximum at Mn(II); or the softness may be connected with adjacent orbitals having energies not far from that of the partly filled shell. These adjacent orbitals would be the manganese 4s orbital in Mn(I) and the ligand orbitals in Mn(VII). JØRGENSEN (1966) elaborated this idea further and considered the non-innocent ligands as an extreme case of softness. If the hard species has the groundstate well isolated from any excited states to which the transitions have large oscillator strength, whereas the soft species has low-lying excited states to which the transitions have high probability, a certain connection with *electric polarizability* α would be obtained The metallic materials fall at one extreme in these two classifications. However, a closer analysis (JØRGENSEN, 1966; SALZMANN and JØRGENSEN, 1968; JØRGENSEN, 1968b) shows that there is not an absolute correlation between α and softness; Cs(I) has higher α than Ag(I) and Ba(II) higher α than Cd(II) though in both cases, the d^{10} systems are far softer.

The electric polarizability has not so great an interest for chemistry as one might have expected. FAJANS (1967) discussed the case of diatomic molecules containing a small ion (say Li^+ or F^-) and a highly polarizable ion (say I^- or Cs^+) giving rise to a smaller polarity of LiI and CsF than of, say, NaF. However, in most cases, this is a rather minor effect, and what FAJANS calls *deformability* could not be connected with multipole moments of lower order than 16 in cubic crystals such as AgCl and AgBr and is rather caused by spherically symmetric modifications of the radial functions to be discussed below in connection with continuum back-bonding. However, there are quite interesting facts about the α values (SALZMANN and JØRGENSEN, 1968; JØRGENSEN, 1968b). In an isoelectronic series, α does not necessarily decrease monotonically with increasing oxidation states; the values are e.g. $Se(-II) > Br(-I) > Kr > Rb(I) > Sr(II) \sim Y(III) \sim Zr(IV) < Nb(V)$ and rather different from the values for gaseous species $Br^- \gg Kr > Rb^+ > \ldots$ It is also certain that Fe(III) has higher α than Fe(II) or the isoelectronic Mn(II). FAJANS legitimately pointed out that in compounds, α are not strictly additive. However, the extent to which an approximate rule of additivity is obeyed is surprising. It may be remembered that the validity of additive rules is not changed if α of each species having the ionic charge z is changed by Cz, where C is an arbitrary constant, since we can only determine α via the refractive indices for macroscopically neutral materials. The zero-points can be fixed by looking for the negligible α-value expected for Li(I), Mg(II) or Al(III). Similar problems occur in the electric conductances of individual ions in solution, or in the determination of ionic radii.

AHRLAND (1967 and 1968) compared changes of free energy ΔG and of enthalpy ΔH by complex formation and concluded that in aqueous solution, the typical hard-hard complexes are formed endothermically (because of e.g. F^- forming strong hydrogen bonds to the solvent which are broken by the formation of complexes such as $Al(H_2O)_5F^{+2}$ or $Fe(H_2O)_5F^{+2}$) and the reaction only proceeds because of the tremendous increase of entropy by releasing the solvent molecules ordered around the reactants $Al(H_2O)_6^{+3}$ and F^-; whereas the typical soft-soft complexes are formed exothermically (e.g. Hg^{++} reacting with I^- or CN^-) and ΔG and ΔH are roughly similar because the changes of entropy are far less important. SO_4^- and $RCOO^-$ behave in a way analogous to F^- whereas OH^- does not behave as a typical hard ligand. AHRLAND (1968) proposed as softness parameter the energy needed for removing the z electrons from a neutral atom M to form the aqua ion of M^{+z} in solution. As discussed in connection with eq. (4.40) the writer had previously introduced this quantity in connection with an entirely different problem. KLOPMAN (1968) treated the softness series at the Northern Polytechnic conference and proposed a somewhat more complicated calculation elaborated on KLOPMAN's theory for chemical bonding in heteronuclear molecules. In both treatments, the influence of the solvent is very important for determining the softness parameter, though it is only included as the reciprocal dielectric constant neglecting chemical differences.

The ligands having high α or having low electronegativity x are under most circumstances the softest. One might argue that the soft central atoms as a whole have low ionization energies; however, this is not an universal rule; Tl(III) is much softer by chemical standards than one would have expected according to this regularity. Originally, AHRLAND (1966) proposed that central atoms simultaneously ought to have a large number of d-electrons and high α in order to be soft. This would suggest that back-bonding to the ligands is of decisive importance for the soft-soft interactions. Though backbonding to empty 5d-orbitals in I^- has particularly bad press among spectroscopists because the first excitations in ionic iodides in crystals or aqueous solution do not occur before 44 kK and are essentially $5p \to 6s$ transitions, whereas the $5p \to 5d$ excitations can be detected above 50 kK (cf. eq. (6.6) and the review by JØRGENSEN, 1967 b) it is still conceivable that RS^-, R_3P and R_3As have moderately important back-bonding to empty d-orbitals though no very clear-cut positive evidence is available. However, there is an aspect of the soft ligands with which chemists specialized in aqueous solutions are not familiar, viz. that the hydride anion H^- is also soft. One would not normally speak about low-lying d-orbitals in this case, and it cannot even be argued that the ionization energy is particularly low. Actually, Ru(II), Rh(III), Re(VII), Ir(III) and Pt(II) hydride complexes tend to have their excited levels at *higher* energy than analogous chlorides, and $H(-I)$ cannot at all be compared with gaseous H^- having the very low ionization energy 6 kK. By the same token, carbanions R_3C^- are very soft, whereas carbonium ions R_3C^+ are moderately hard. In many ways, the methyl complexes MCH_3 are closely comparable to hydride MH. JØRGENSEN (1964) discussed the *symbiosis*, the tendency of soft ligands to flock together. It is well-known from preparative chemistry that ligands such as H^-, CO, CH_3^-, CN^- and R_3P tend toward mutual stabilization. Thus, $Co(NH_3)_5F^{++}$ is far more stable than $Co(NH_3)_5I^{++}$ whereas the isoelectronic $Co^{III}(CN)_5X^{-3}$ and $Mn^I(CO)_5X$ are far more stable for X = I, H or CH_3 and are not yet prepared for X = F. This symbiosis is the opposite of that found experimentally by J. BJERRUM (1950) for ligands of comparable softness such as NH_3 and H_2O, where the mixed complexes usually are slightly more stable than expected from statistical considerations.

The back-bonding does perhaps not take place to empty d-orbitals but rather to the continuum of orbitals having positive one-electron energy (JØRGENSEN, 1967 c). Energy-wise, the effects may be rather small, and the positions of the lowest excited levels are not modified in a way one can recognize. Ligands having low ionization energies and hence low x can be invaded by electrons of central atoms having moderate to low ionization energy. Another way of saying the same thing is that the electronic density of the central atom is slightly distorted by the presence of soft ligands. It is clear that the qualitative variation will be similar to that of electric polarizability α of which a fair to large proportion originates in excitations to the continuum (SALZMANN and JØRGENSEN, 1968). However, the chemical effect is different because it seems to be of shorter range of excitation energy; it would be one more breakdown (with moderate effects, energy-wise) of the L.C.A.O. description involving only a few basis orbitals of each atom.

Another such effect which may be connected with rather subtle aspects of polarizability are the *hypersensitive pseudoquadrupolar transitions* in 4f-group complexes. JØRGENSEN and JUDD (1964) pointed out that most 4f-group M(III) have one or two transitions from the ground-level to excited J-levels of which the intensity varies much more with the choice of ligands than the other absorption bands. The hypersensitive transitions have selection rules (they go from J to $(J-2)$ and in the case of Russell-Saunders coupling from L to $(L-2)$ keeping S constant) as if they were electric quadrupole radiation. However, this is nearly impossible in a strict sense, though the numerical calculations (CARNALL, FIELDS and WYBOURNE, 1965; CARNALL, FIELDS and RAJNAK, kindly communicated preprint) show that if the intensity expressions are parametrized, perfect agreement is obtained with experiment. These authors also explained the rather surprising fact that most transitions to the last J-level of excited terms having the same S as the groundstate are relatively weak. The physical origin of the hypersensitive transitions seems to be some failure of the description of the molecules interacting with the electromagnetic field of the light waves as some continuous dielectric medium. From a chemical point of view, it is striking that ligands presenting "resonance between equivalent structures" such as acetylacetonates and carboxylate complexes have very strong hypersensitive transitions. However, the planar, triangular molecules $NdBr_3$ and NdI_3 in gaseous state (GRUEN and DeKOCK, 1966) and octahedral 4f-group MX_6^{-3} (RYAN and JØRGENSEN, 1966) also show strong hypersensitive transitions. They have been studied for nitrate complexes by GALLAGHER (1964) and ABRAHAMER and MARCUS (1967) and for multidentate carboxylates by CHOPPIN, HENRIE and BUIJS (1966) and for β-diketonates by KARRAKER (1967) among other authors. JUDD (1966) pointed out that certain low symmetries allow linear terms in the ligand field, what had not previously been recognized, and that these low-symmetry components might be the origin of the pseudoquadrupole transitions. BLASSE and BRIL (1967) studied the luminescence of Eu(III) in crystals such as garnets and found intense $^5D_0 \rightarrow {}^7F_2$ transitions, whereas $^5D_0 \rightarrow {}^7F_1$ normally is stronger (the transition $^5D_0 \rightarrow {}^7F_0$ between states which are spherically symmetric to a very high approximation is extremely weak). Similar hypersensitive luminescence is well known from Eu(III) β-diketonate complexes. However, in the writer's opinion, the hypersensitivity is more a chemical effect and it is not based on low symmetry of the chromophore alone. This is particularly true for the 4f-group hexahalides which are known to possess the symmetry O_h to a high precision.

MISUMI, KIDA and ISOBE (1968) studied the circular dichroism of 4f-group complexes of the optically active, but colourless, ligand levo-propylenediaminetetraacetate. The strong discontinuities at those internal $4f^q$-transitions which have magnetic dipole intensities are technically a Pfeiffer effect induced by the adjacent, optically active ligand. The Faraday effect, i.e. circular dichroism (or less specifically, rotation) induced by external magnetic fields was studied for lanthanide solutions by BRIAT et al. (1966). The broad background of rotation is mainly connected with the 4f → 5d transitions in the ultra-violet as discussed by BERGER et al. (1964). However, RUBINSTEIN and BERGER (1965) found that Yb(III) has the opposite sign of that of most lanthanides. It is interesting to speculate that those compounds having the unusual sign of the Verdet constant measuring the Faraday rotation at wavenumbers below the first strong absorption band (viz. NO, O_2, tetrahedral complexes such as $TiCl_4$ and MnO_4^- and octahedral complexes such as $Fe(CN)_6^{-3}$ and $IrCl_6^-$) all have a transition where the number of angular node-planes *"l" decreases* by one unit ("f" → 3d; 4f → 5d with the exception of possible electron transfer transitions "g" → 4f) whereas most organic molecules having the usual sign of the Verdet constant may have Rydberg orbitals with an additional node-plane.

Bibliography

ABRAHAMER, I., and Y. MARCUS: Inorg. Chem. 6, 2103 (1967).
ABRAHAMS, S. C., A. P. GINSBERG, and K. KNOX: Inorg. Chem. 3, 558 (1964).
AHRLAND, S., J. CHATT, and N. R. DAVIES: Quart. Rev. (London) 12, 265 (1958).
— Struct. Bonding 1, 207 (1966).
— Helv. Chim. Acta 50, 306 (1967).
— Struct. Bonding 5, 118 (1968).
ALBANO, V., P. L. BELLON, and V. SCATTURIN: Chem. Commun. 507 (1966).
ALLEN, A. D., and C. V. SENOFF: Chem. Commun. 621 (1965).
—, and J. R. STEVENS: Chem. Commun. 1147 (1967).
—, and F. BOTTOMLEY: Can. J. Chem. 46, 468 (1968).
ALMENNINGEN, A., O. BASTIANSEN, and A. HAALAND: J. Chem. Phys. 40, 3434 (1964).
—, G. GUNDERSEN, and A. HAALAND: Acta Chem. Scand. 22, 328 (1968).
BADDLEY, W. H.: J. Am. Chem. Soc. 88, 4545 (1966).
BAENZINGER, N. C., and R. E. RUNDLE: Acta Cryst. 2, 258 (1949).
BAIRD, M. C., J. T. MAGUE, J. A. OSBORN, and G. WILKINSON: J. Chem. Soc. (A) 1347 (1967).
—, G. HARTWELL, and G. WILKINSON: J. Chem. Soc. (A) 2037 (1967).

BARRACLOUGH, C. G., and J. LEWIS: J. Chem. Soc. 4842 (1960).
BASOLO, F., and R. G. PEARSON: Advan. Inorg. Chem. Radiochem. 3, 1 (1961).
— — Progr. Inorg. Chem. 4, 381 (1962).
BAYER, E., and P. SCHRETZMANN: Struct. Bonding 2, 181 (1967).
BERGER, S. B., C. B. RUBINSTEIN, C. R. KURKJIAN, and A. W. TREPTOW: Phys. Rev. 133, A 723 (1964).
BIRD, P. H., and M. R. CHURCHILL: Chem. Commun. 403 (1967).
BJERRUM, J.: Chem. Rev. 46, 381 (1950).
BLASSE, G., and A. BRIL: J. Chem. Phys. 47, 5442 (1967).
BOSTRUP, O.: Acta Chem. Scand. 17, 1029 (1963).
BOWDEN, F. L., and A. B. P. LEVER: Organometall. Chem. Rev. 3, 227 (1968).
BRIAT, B., M. BILLARDON, J. BADOZ, and J. LORIERS: Anal. Chim. Acta 34, 465 (1966).
CALDERAZZO, F.: Halogen Chem. 3, 383 (1967). London: Academic Press.
CARNALL, W. T., P. R. FIELDS, and B. G. WYBOURNE: J. Chem. Phys. 42, 3797 (1965).
— —, and K. RAJNAK: private communication.
CHATT, J., L. A. DUNCANSON, and L. M. VENANZI: J. Chem. Soc. 4456 (1955).
CHOPPIN, G. R., D. E. HENRIE, and K. BUIJS: Inorg. Chem. 5, 1743 (1966).
CLARK, H. C., J. D. COTTON, and J. H. TSAI: Inorg. Chem. 5, 1582 (1966).
COTTON, F. A., and G. WILKINSON: Advanced Inorganic Chemistry, 2. Ed. (1. Ed. 1962). New York: Interscience 1966.
CRAMER, R., and G. W. PARSHALL: J. Am. Chem. Soc. 87, 1392 (1965).
DALE, D., and D. HODGKIN: J. Chem. Soc. 1364 (1965).
DENNING, R. G., F. R. HARTLEY, and L. M. VENANZI: J. Chem. Soc. (A) 1322 (1967).
DHAR, D. N.: Chem. Rev. 67, 611 (1967).
DODD, D., and M. D. JOHNSON: J. Chem. Soc. (A) 34 (1968).
DOMENICANO, A., A. VACIAGO, L. ZAMBONELLI, P. L. LOADER, and L. M. VENANZI: Chem. Commun. 476 (1966).
ENEMARK, J. H., B. R. DAVIS, J. A. McGINNETY, and J. A. IBERS: Chem. Commun. 96 (1968).
FAJANS, K.: Struct. Bonding 3, 88 (1967).
FALLAB, S.: Chimia 21, 538 (1967).
GALLAGHER, P. K.: J. Chem. Phys. 41, 3061 (1964).
GANS, P.: J. Chem. Soc. (A) 943 (1967).
GREEN, M., R. B. L. OSBORN, A. J. REST, and F. G. A. STONE: Chem. Commun. 502 (1966).
GRIFFITH, W. P.: The Chemistry of the Rarer Platinum Metals (Os, Ru, Ir and Rh). London: Interscience 1967.
GRUEN, D. M., and C. W. DeKOCK: J. Chem. Phys. 45, 455 (1966).
HAALAND, A., and J. E. NILSSON: Chem. Commun. 88 (1968).
HAAS, H., and R. K. SHELINE: J. Am. Chem. Soc. 88, 3219 (1966).
HALL, D., and A. A. TAGGART: J. Chem. Soc. 1359 (1965).
HARRISON, D. F., E. WEISSBERGER, and H. TAUBE: Science 159, 320 (1968).
HOFFMANN, A. K., and A. T. HENDERSON: J. Am. Chem. Soc. 83, 4671 (1961).
JØRGENSEN, C. K.: Absorption Spectra and Chemical Bonding in Complexes. Oxford: Pergamon Press 1962. U. S. distributor: Addison-Wesley.
— Inorganic Complexes. London: Academic Press 1963.
— Mol. Phys. 7, 417 (1964a).
— Inorg. Chem. 3, 1201 (1964b).
—, and B. R. JUDD: Mol. Phys. 8, 281 (1964).
— Struct. Bonding 1, 234 (1966).

— Advances in Chemistry Series **62**, 161 (1967a). Washington D. C.: American Chemical Society.
— Halogen Chem. **1**, 265 (1967b). London: Academic Press.
— Struct. Bonding **3**, 106 (1967c).
— Chemical Applications of Spectroscopy (Ed. B. G. WYBOURNE). New York: Interscience 1968a.
— Rev. Chim. Minerale (Paris) in press (1968b).
JOHNSON, B. F. G., and J. A. McCLEVERTY: Progr. Inorg. Chem. **7**, 277 (1966).
JOHNSON, M. P., and D. F. SHRIVER: J. Am. Chem. Soc. **88**, 301 (1966).
JONES, G. D., and R. A. SATTEN: Phys. Rev. **147**, 566 (1966).
JUDD, B. R.: J. Chem. Phys. **44**, 839 (1966).
KARRAKER, D. G.: Inorg. Chem. **6**, 1863 (1967).
KLOPMAN, G.: J. Am. Chem. Soc. **86**, 1463 and 4550 (1964).
— J. Am. Chem. Soc. **90**, 223 (1968).
KRISHNAMURTHY, R., W. B. SCHAAP, and J. R. PERUMAREDDI: Inorg. Chem. **6**, 1338 (1967).
KRUCK, T., u. M. NOACK: Chem. Ber. **97**, 1693 (1964).
— Angew. Chem. **79**, 27 (1967).
L'EPLATTENIER, F., and F. CALDERAZZO: Inorg. Chem. **6**, 2092 (1967).
MALATESTA, L.: Progr. Inorg. Chem. **1**, 284 (1959).
— Helv. Chim. Acta Fasciculus extraordinarius Alfred Werner, p. 147, (1967).
MANSMANN, M.: Z. Krist. **122**, 375 (1965).
MASSEY, A. G., A. J. PARK, and F. G. A. STONE: J. Am. Chem. Soc. **85**, 2021 (1963).
McGINNETY, J. A., and J. A. IBERS: Chem. Commun. 235 (1968).
MERCER, E. E., W. A. McALLISTER, and J. R. DURIG: Inorg. Chem. **6**, 1816 (1967).
MISUMI, S., S. KIDA, and T. ISOBE: Spectrochim. Acta **24 A**, 271 (1968).
NIEUWPOORT, W. C.: Philips Progress Rep. **20**, Suppl. no. 6 (1965).
NÖTH, H., u. G. SCHMID: Angew. Chem. **75**, 861 (1961).
NYMAN, C. J., C. E. WYMORE, and G. WILKINSON: J. Chem. Soc. (A) 561 (1968).
ODELL, A. L., R. W. OLLIFF, and A. A. TAGGART: J. Chem. Soc. 6024 (1965).
OLAH, G. A., and R. H. SCHLOSBERG: J. Am. Chem. Soc. **90**, 2726 (1968).
ORGEL, L. E.: J. Inorg. Nucl. Chem. **2**, 137 (1956).
OSBORN, J. A., A. R. POWELL, and G. WILKINSON: Chem. Commun. 461 (1966).
—, F. H. JARDINE, J. F. YOUNG, and G. WILKINSON: J. Chem. Soc. (A) 1711 (1966).
PANATTONI, C., G. BOMBIERI, U. BELLUCO, and W. H. BRADDLEY: J. Am. Chem. Soc. **90**, 798 (1968).
PARSHALL, G. W.: J. Am. Chem. Soc. **86**, 361 (1964).
PEARSON, R. G.: J. Am. Chem. Soc. **85**, 3533 (1963).
— Science **151**, 172 (1966).
PHILLIPS, C. S. G., and R. J. P. WILLIAMS: Inorganic Chemistry, Vol. I and II. Oxford: Clarendon Press 1965 and 1966.
PREUSS, H.: private communication.
RAE, A. I. M.: Chem. Commun. 1245 (1967).
RAYNOR, J. B.: J. Chem. Soc. (A) 997 (1966).
ROBERTSON, R. E., and H. M. McCONNELL: J. Phys. Chem. **64**, 70 (1960).
ROBIETTE, A. G., G. M. SHELDRICK, and R. N. F. SIMPSON: Chem. Commun, 506 (1968).
ROSSMANITH, K.: Monatsh. Chem. **95**, 1424 (1964).
ROUNDHILL, D. M., and G. WILKINSON: J. Chem. Soc. (A) 506 (1968).
RUBINSTEIN, C. B., and S. B. BERGER: J. Appl. Phys. **36**, 3951 (1965).
RYAN, J. L., and C. K. JØRGENSEN: J. Phys. Chem. **70**, 2845 (1966).
SACCO, A., and M. ROSSI: Chem. Commun. 316 (1967).

SALZMANN, J. J., and P. MOSIMANN: Helv. Chim. Acta **50**, 1831 (1967).
—, and C. K. JØRGENSEN: Helv. Chim. Acta **51**, 1276 (1968).
SCHMIDTKE, H. H., and D. GARTHOFF: Helv. Chim. Acta **49**, 2039 (1966).
SCOTT, D. R., and R. S. BECKER: J. Organometall. Chem. **4**, 409 (1965).
— J. Organometall. Chem. **6**, 429 (1966).
SCOTT, R. N., D. F. SHRIVER, and L. VASKA: J. Am. Chem. Soc. **90**, 1079 (1968).
SHRIVER, D. F.: J. Am. Chem. Soc. **85**, 3509 (1963).
— Struct. Bonding **1**, 32 (1966).
STAMMREICH, H., K. KAWAI, Y. TAVARES, P. KRUMHOLZ, J. BEHMOIRAS, and
    S. BRIL: J. Chem. Phys. **32**, 1482 (1960).
STROHMEIER, W.: Struct. Bonding **5**, 96 (1968).
THROUGHTON, P. G. H., and A. C. SKAPSKI: Chem. Commun. 575 (1968).
TSAI, J. H., J. J. FLYNN, and F. B. BOER: Chem. Commun. 702 (1967).
VAN VOORST, J. D. W., and P. HEMMERICH: J. Chem. Phys. **41**, 3914 (1966).
VANNERBERG, N. G.: Acta Cryst. **18**, 449 (1965).
VASKA, L.: Science **145**, 920 (1964).
WALSH, A. D.: J. Chem. Soc. 2266 (1953).
WARREN, L. F., and M. F. HAWTHORNE: J. Am. Chem. Soc. **89**, 470 (1967).
WILKINSON, G., and F. A. COTTON: Progr. Inorg. Chem. **1**, 1 (1959).
WILLIAMS, R. J. P., and J. D. HALE: Struct. and Bonding **1**, 249 (1966).
ZEISS, H., P. J. WHEATLEY, and H. J. S. WINKLER: Benzenoid-metal Complexes.
    New York: Ronald Press 1966.

10. Homopolar Bonds and Catenation

Organic chemistry presents the most conspicuous and numerous examples of the stability of a "backbone" system of catenated atoms of the same element. The whole concept of oxidation states is the least adaptable to this situation. Nobody would expect an essential difference between the chemical bonding in methane CH_4 and ethane C_2H_6. This means that two of the eighteen electrons of the ethane molecule must be so firmly occupied by forming a single carbon-carbon σ-bond that both carbon atoms, obviously being equivalent, still can be considered in some sense to be carbon (IV) like in methane. This situation cannot be incorporated in the idea of spectroscopic oxidation states, and the difficulties inherent in the homonuclear $C-C$ bond are the same as discussed for the case of H_2. Actually, the situation can be made rather similar to the heteronuclear interhalogens. There is a sense in which CH_3CF_3 is approaching, though perhaps only to a small extent, the asymmetry of a carbanion CF_3^- being bound to a carbonium ion CH_3^+. This argument would seem far more artificial in the case of CH_3CCl_3 or CH_3CBr_3.

Seen from this point of view, it is not entirely unreasonable that the spectroscopic oxidation state of CH_3X can be a function of whether X has a markedly higher electronegativity x than CH_3 or a markedly lower x, the cases of roughly identical x remaining irremediably indeterminate. Thus, CH_3F and CH_3Cl are definitely the trihydrido complex of carbon (IV) adding a fourth ligand, $F(-I)$ or $Cl(-I)$, whereas a long series of organometallic compounds such as $Zn(CH_3)_2$, $Hg(CH_3)_2$, $Hg(CH_3)X$, $Pb(CH_3)_4$, $Pb(CH_3)_2X_2$, cis-quadratic $Au(CH_3)_2(H_2O)_2^+$, $Mn(CO)_5CH_3$, $Co(salen)CH_3$ (see eq. 7.14) and $Cr(CH_3)_6^{-3}$ undoubtedly are complexes of CH_3^- (the trihydrido complex of carbon (II) or the triprotonated adduct of carbon $(-IV)$ as one prefers; this choice seems rather arbitrary) of the central atoms having well-defined oxidation states Zn(II), Hg(II), Pb(IV), Au(III), Mn(I), Co(III) and Cr(III). If simple aryl groups such as C_6H_5 are considered (and not heavily conjugated systems such as $C(C_6H_5)_3$ which are able to exist as individual molecules with $S=1/2$) comparable conclusions can be drawn. In the cations $N(CH_3)_4^+$ or $P(C_6H_5)_4^+$ the assumption of carbonium ligands produces the oxidation states $N(-III)$ and $P(-III)$ whereas carbanion ligands correspond to the more reasonable N(V) and P(V). Consequently, the quaternization reaction

$$N^{III}(CH_3)_3 + CH_3I = [N^V(CH_3)_4]^+ + I^- \qquad (10.1)$$

involves an oxidation of N(III) to N(V) whereas both $B(C_6H_5)_3$ and $B(C_6H_5)_4^-$ contain boron (III).

It can be discussed whether CH_3I is a carbonium iodide ($-I$) or an iodine (I) methide complex. If linear $(CH_3)_2I^-$ had been reported, it would have been considered as the iodine (I) di-carbanion complex. On the other hand, $(C_6H_5)_2I^+$ is known to be a bent molecule in close analogy to ICl_2^+ and they can best be considered as iodine (III) complexes. The same problems are posed by the sulphonium cations R_3S^+ and the corresponding R_3Se^+ and R_3Te^+. Whereas oxonium cations R_3O^+ undoubtedly are carbonium adducts of oxygen ($-II$), the R_3S^+ represent a choice between carbonium adducts of sulphur ($-II$) or carbanion complexes of sulphur (IV). The tendency toward the second alternative becomes stronger in direction of tellurium (IV) where pyramidal complexes such as $TeCl_3^+$ also are known. We are discussing the interesting stereochemistry of tellurium below, but we may already mention the surprising fact that $(CH_3)_2TeI_2$ exists in two modifications, one being a *cis*-quadratic monomeric compound and the other being $Te(CH_3)_3^+$ $Te(CH_3)I_4^-$, the anion being square-pyramidal with the methyl group perpendicular on the Te(IV)I_4 plane (EINSTEIN, TROTTER and WILLISTON, 1967). It is quite common for s^2-systems to form pyramidal complexes, such as SO_3^{--}, ClO_3^-, BrO_3^-, $SnCl_3^-$, IO_3^- and XeO_3.

The catenation in sulphur-containing compounds does not produce as stable homopolar bonds as $C-C$, but in many ways, the indifferent prolongation of a chain with a number of sulphur atoms is perhaps even more surprising than the addition of CH_2-links to a n-alkane chain. Thus, dithionate $O_3SSO_3^{--}$ is an extraordinarily unreactive anion used for the precipitation of many complex cations. The $S-S$ bond is an exception to PAULING's rule against strong bonding between adjacent atoms having the same sign of the fractional charge, as is also F_5SSF_5. These two compounds might have had oxide or fluoride bridges (in analogy to Nb_2Cl_{10} or Ru_4F_{20}) and are quite surprising. Trithionate $O_3SSSO_3^{--}$ and the subsequent polythionates $(O_3S)S_{n-2}(SO_3)^{--}$ have a central chain of sulphur atoms which admittedly are chemically reactive, but which illustrate the most archaic ideas of each sulphur atom just being bivalent. We return in Chapter 11 to the question about what a classification can be made of such compounds, but we have to realize that though the most stable modification of sulphur is cyclic S_8, less stable cyclic modifications are S_6, S_{10}, S_{12} and nitrogen-substituted forms such as S_7NH. The anion S_2^{--} is closely comparable with O_2^{--} but tends to prolongate to S_3^{--}, S_4^{--}, S_5^{--},... which tend to form mixtures difficult to separate. Foss (1960, 1967) wrote two reviews on the stereo-chemistry of catenated sulphur, selenium and tellurium compounds. Brown et al. (1968) reported yellow, quadratic Se_4^{++}; N. J. Bjerrum recently found deep purple Te_4^{++}.

Certain low-spin p^q-systems have a regular stereochemistry where one does not need to take mixing of l-values into account; in other cases, the M.O. must have mixed p and s (or other even parity) character. Thus, the linear p^2-systems XMX formed by Cl(I), Br(I), Kr(II), I(I) and Xe(II) are readily understood; the two lone-pairs $p\pi c$ and $p\pi s$ form an equatorial belt of electronic density, and the empty orbital $p\sigma$ is is σ-anti-bonding to the two ligands. Again, the quadratic p^4-systems such as BrF_4^-, ICl_4^- and XeF_4 have their lone-pair $p\sigma$ perpendicular on the plane of MX_4 in close analogy to the quadratic low-spin d^8-systems having $d\sigma$ perpendicular on the plane (Jørgensen, 1967). On the other hand, the pyramidal s^2-systems mentioned above must mix s and $p\sigma$ character along the trigonal axis much in the same way as for NH_3, PF_3 and $AsCl_3$. One would have expected that tellurium (II) complexes would behave like iodine (III) and xenon (IV) and form quadratic complexes. This is also true for $Te(thiourea)_4^{++}$ but Foss and collaborators (1967) found that a number of cis-$Te(II)S_2X_2$ (where S originates in sulphur-containing ligands such as thiourea $(NH_2)_2CS$) have very short $Te-S$ and long $Te-X$ distances. Though $trans$-$Te(II)S_2X_2$ exist under certain conditions (Foss, 1967; Hendra and Jovic, 1968) the analogy to TeH_2 and $Te(CH_3)_2$ is striking. What is more surprising is that Husebye (1966) found that $Te(S_2P(OCH_3)_2)_2$ has two short $Te-S$ distances in cis-position and not at all the rectangular chromophore $Ni(II)S_4$ charac-acterizing low-spin $Ni(S_2P(OR)_2)_2$. The writer (1968) is going to discuss sulphur-containing ligands further. Dr. S. Husebye was so kind as to inform the writer about the fact that the diethyldiselenophosphinates $M(Se_2P(C_2H_5)_2)_2$ with $M = Se$ and Te again have two short distances $M-Se$ with the angle SeMSe close to $90°$. Of course, one may insist that these two molecules are cis-quadratic selenium (II) and tellurium (II) complexes; but one cannot avoid the question how the central selenium atom knows that the oxidation states are $P(V)Se(-II)Se(II)Se(-II)P(V)$. An even more pernicious question is the following: when RSe^- are oxidi-zed to diselenide RSeSeR, what is the difference between this catenated situation and PSeSeSeP in Husebye's complex? Also ClSSCl is a rather difficult compound to assign oxidation states, and even hydrazine NH_2NH_2 and its derivatives pose certain problems. Bent I_3^+ can be construed as a di-iodo cis-complex of iodine (III); but I_5^- is also L-shaped and looks suspiciously like the adduct of two I_2 molecules end-on to one

iodide anion (cf. the review by WIEBENGA, HAVINGA and BOSWIJK, 1961).
Catenated ligands are known to occur; S_5^{--} occur in $Pt(S_5)_3^{--}$ having a
normal $Pt(IV)S_6$ octahedral chromophore (JONES and KATZ, 1967). Red
to purple $cp_2 NbS_2X$ probably are Nb(V) complexes of S_2^{--} though
TREICHEL and WERBER (1968) felt that the S—S distance was too short
and similar to S_2 (cf. the discussion of peroxo complexes).

Returning to organic compounds, the central, secondary carbon
atom in propane $CH_3CH_2CH_3$ undoubtedly is slightly different from the
terminal primary carbon atoms. We have quaternary carbon atoms in
diamond, the simplest case being in neopentane $C(CH_3)_4$. Sometimes
the secondary carbon atoms can be sufficiently activated to achieve
observable Brønsted acidity of the protons attached; both acetylacetone
and esters of malonic acid loose one proton from the central CH_2 link in
strong bases; and the reaction is even rapid, as compared to the slow
de-protonation of "pseudo-acids" such as CH_3NO_2 to $CH_2NO_2^-$. However,
most of these differences are so small that they cannot be readily incorporated in a description of carbon (II) lone-pairs coordinated to carbon
(IV) centers as one would need for a consistent assignment of oxidation
states. We are not going either to discuss the subject of aromatic compounds which is fascinating, but again only can be brought in connection
with our main topic by extremely artificial extrapolations.

Whereas $V(CO)_6$ does not dimerize, manganese forms a dimeric
carbonyl $(OC)_5MnMn(CO)_5$ as do Tc and Re. One might imagine various
types of chemical bonding in this dimer. The least chemical alternative
would be antiferromagnetic coupling (the material is macroscopically
diamagnetic) between two square-pyramidal entities $Mn(CO)_5$ each
having $S=1/2$. However, the Mn—Mn distance 2.92 Å is sufficiently
short to represent a decent chemical bond, (the dissociation energy is
7 kK), and the choice is really between two other alternatives: either
each Mn has the lower sub-shell (xy, xz, yz) filled with six electrons, and
the seventh electron from each of the two manganese atoms fills a bonding
M.O. mainly constructed from the two $d\sigma$ orbitals directed along the
OCMnMnCO axis; or the two manganese atoms have the conditional
oxidation state Mn[I] (cf. the discussion of pyrites and related crystals in
Chapter 8) and are connected by a σ-bonding orbital not very d-like. It
may be remembered that such an orbital may be constructed from an
appropriate linear combination of $3d\sigma$ and 4s (with modified radial
function) nearly annihilating the equatorial belt of $3d\sigma$; but this is by
no means the only available possibility. The yellow colour of the dimer
is not much deeper than of a variety of $Mn(CO)_5X$ (GRAY et al., 1963)
and one has the feeling that the $Mn(CO)_5$ moiety is a kind of pseudo-
halogen, as also confirmed by the replacement of two halides in many
organometallic compounds by $Fe(CO)_4$ (cis-octahedral) or of one halide

with $Mn(CO)_5$ or $Co(CO)_4$. It is definite that the filled lower sub-shell provide the conditional oxidation states Mn[I] and Re[I] to the dimeric carbonyls; it would probably not be suitable to ascribe them spectroscopic oxidation states Mn(0) and Re(0), though it is not known at present whether the σ-bonding orbital connecting the two M[I] is predominantly $d\sigma$ or not. Certain halide-carbonyls have normal halide bridges such as $(OC)_4MnBr_2Mn(CO)_4$ containing cis-$Mn(I)C_4Br_2$.

A variety of dimers are isoelectronic with $Mn_2(CO)_{10}$, thus $Cr_2(CO)_{10}^{--}$ and heteronuclear species such as $CrMo(CO)_{10}^{--}$ which may be compared with $MnRe(CO)_{10}$. What is perhaps more surprising is that the square-pyramidal species green $Co(CN)_5^{-3}$ and blue $Co(CNR)_5^{+2}$ under certain conditions dimerize to ADAMSON's purple, diamagnetic $Co_2(CN)_{10}^{-6}$ and to COTTON's red $Co_2(CNR)_{10}^{+4}$ hence representing a dimerization of low-spin Co(II) to a situation involving Co[III]. Normally, such species rather form MCNM' bridges (SHRIVER, 1966).

An even more unexpected complex is yellow $(OC)_5CrHCr(CO)_5^-$ prepared by HANDY et al. (1966). As one extreme, one may think of a protonation of $Cr_2(CO)_{10}^{--}$. However, we saw in Chapter 9 that hydride $(-I)$ is capable of forming bridges between two atoms (or six in crystalline LiH) and one may also consider this complex as two $Cr(CO)_5$ groups connected with a central $H(-I)$, and hence essentially containing Cr[O]. This opinion is supported by the compound $(CO)_5ReHRe(CO)_4Mn(CO)_5$ studied by CHURCHILL and BAU (1967) which can be dissected in $Re(CO)_5^+$ bridged by $H(-I)$ to a residue $ReMn(CO)_9$ BEHRENS and SCHWAB (1964) report the red $(OC)_5CrICr(CO)_5$ and dark blue $Cr(CO)_5I$ both having $S = 1/2$. BEHRENS and HERRMANN (1966) prepared the yellow diamagnetic analog $(OC)_5CrICr(CO)_5^-$ to the hydride-bridged complex. Another type of bridging ligands is SO_2 and $SnCl_2$ studied by VLČEK and BASOLO (1966)

$$(NC)_5CoSCo(CN)_5^{-6} \quad\quad (NC)_5CoSnCo(CN)_5^{-6}.$$

with O above and O below the S, and Cl above and Cl below the Sn.

The dark red adduct of SO_2 has the extraordinarily high $\varepsilon = 50\,000$ of a band at 22.4 kK. It is conceivable, but by no means certain that cobalt has the oxidation state Co(III). One might also argue the presence of central atoms S(VI) and Sn(IV) having anions $Co(CN)_5^{-4}$ occupying one coordination position. One difficulty for the interpretation of these species is that SO_2 and $SnCl_2$ normally are considered to have only one lone-pair, as has hydride, by the way.

Most compounds containing niobium and tantalum in apparent low oxidation states are cluster compounds of the type Nb_6 or Ta_6 discussed below. It is so much more striking that the low-spin d^6-systems $Nb(CO)_6^-$

and $Ta(CO)_6^-$ are known; they show no tendency toward oxidation to $M(CO)_6$. KEBLYS and DUBECK (1964) prepared the red, volatile crystals of $C_2H_5HgTa(CO)_6$ obviously containing Hg(II) whereas one may argue whether a σ-bonding electron pair connects this atom with Ta[I].

The dimeric cobalt carbonyl exists in two forms, a $M-M$ bridged $(OC)_4CoCo(CO)_4$ and one bridged with two carbonyl groups $(OC)_3Co(CO)_2$ $Co(CO)_3$ showing a certain similarity with ketones R_2CO. The two isomers are so close to have the same free energy that they occur in thermal equilibrium in solution (NOACK, 1964). The former isomer is a pseudo-halogenic coupling with a σ-bond of two Co(I) as also known from $(OC)_4CoHgCo(CO)_4$ and $Cl_3SiCo(CO)_4$ (cf. the discussion preceeding eq. 9.8). If one insists on the oxidation numbers of the constituents to add up to the external charge in protonic units, these compounds are in a way Hg^{II} and Si^{IV} complexes of $Co^{-I}(CO)_4$. This is essentially the question of whether the σ-bond is considered heteropolar, being the anion $Co(CO)_4^-$ polarized to some extent, or whether the two additional electrons are assumed to constitute a homopolar σ-bond (like in ethane) between Hg[II] and Co[I]. If the dicobalt-octacarbonyl isomer with two CO bridges really were comparable to ketones, it would be a case of σ-back-bonding to form CO^{--} ligands connecting two Co(II) entities. This is not a very probable description because we know no cases of $Co(CO)_3X_2$. The simultaneous existence of a weak $Co-Co$ bond (though the distance is only 2.52 Å) completing the distorted hexa-coordination of two Co[III] is not very probable either. Metallo-organic chemistry involving more complicated ligands presents a few confusing cases of simultaneous σ- and π-bonding by different parts of the organic ligand; we are not going to discuss this situation further here.

Dimeric acetates and other carboxylates seem to occur in two categories. The red, diamagnetic Cr^{II} acetate and adducts with a variety of neutral ligands L (HERZOG and OBERENDER, 1963) and the corresponding green Rh^{II} acetate (JOHNSON, HUNT and NEUMANN, 1963) form dimers $L_2M_2(RCOO)_4$ where each M atom is surrounded by an octahedron consisting of one oxygen from each of the four bridging, bidentate carboxylate groups and, perpendicular to the plane of these four oxygen atoms, one atom from L and the other M atom. This category is strictly comparable with dimanganese-decacarbonyl; two Cr[III] (their $S=3/2$ must be strongly antiferromagnetically coupled, as is also true for the "basic rhodo ion" $(NH_3)_5CrOCr(NH_3)_5^{+4}$ according to SCHÄFFER, 1958 and WILMARTH, GRAFF and GUSTIN, 1956) or two Rh[III] $(S=0)$ are connected with a σ-orbital either consisting of the positive linear combination of the two $d\sigma$-orbitals or having some other constitution. One cannot readily ascribe the spectroscopic oxidation states Cr(II) and Rh(II) to these materials. On the other hand, it is now recognized (JØRGENSEN,

1963; DUBICKI and MARTIN, 1966 also reporting the narrow band at 30.7 kK of chromium acetate) that the dimeric Cu(II) carboxylates though having the same stereochemistry have only a weak anti-ferromagnetic coupling between two square-pyramidal $Cu(II)O_4L$ entities. The reason for this profound difference is that the two Cu(II) central atoms have $d\sigma$ filled, and not half-occupied as would be the case for Cr(II) and Rh(II), and one would need a reorganization of the electronic structure to form a Cu—Cu σ-bond (the weak bonding present is technically δ-bonding between the two half-filled $d\delta c$ orbitals). The dimeric copper (II) carboxylates are essentially kept together by the bidentate nature of the ligands, and may be compared with the complexes $OBe_4(O_2CR)_6$ (which is also known for Co(II) in the case of the pivalate where $R = C(CH_3)_3$) where a central atom oxygen $(-II)$ is surrounded by four Be(II) in a regular tetrahedron having six bidentate carboxylates along its sides so each $Be(II)O_4$ is four-coordinate. In some cases of central $O(-II)$ bound to three transition group ions, the anti-ferromagnetic coupling can be very strong; thus, $OM_3(RCO_2)_6L_3^+$ are known of M = Cr(III) and Fe(III) and discussed by JØRGENSEN and ORGEL (1961). LECOQ DE BOISBAUDRAN discovered blue $OIr_3(SO_4)_6(H_2O)_3^{-3}$ having the mixed oxidation state one Ir(III) and two Ir(IV). The dark green $NIr_3(SO_4)_6(H_2O)_3^{-4}$ and $NIr_3(SO_4)_6(NH_3)_3^{-4}$ and red-brown $NIr_3(SO_4)_6(OH)_3^{-7}$ were studied by DELÉPINE over a period of 64 years and were finally shown (DELÉPINE, 1959 and JØRGENSEN, 1959) to be reversibly reduced by two electrons to the pale lemon-yellow iridium (III) complex $NIr_3(SO_4)_6(H_2O)_3^{-6}$. The oxidized forms are diamagnetic; they do not seem to be electronically ordered but are lacking two electrons in a combination of the lower sub-shells which is anti-bonding to the $p\sigma$ orbital perpendicular to the NIr_3 plane. The central nitrogen $(-III)$ is known to form other planar complexes such as $N(Si(CH_3)_3)_3$. From a chemical point of view, it is interesting that DELÉPINE's dark green species is formed by deprotonation of NH_4^+ in boiling concentrated sulphuric acid (!) in the presence of $IrCl_6^{-3}$.

COTTON (1966) discussed the dimeric $Tc_2X_8^{-2}$ and $Re_2X_8^{-2}$ (which can be reduced to $Re_2X_8^{-3}$) involving two quadratic MX_4 groups connected with a perpendicular M—M bond shorter than in the metallic elements Tc and Re. COTTON argues that one σ orbital, two π orbitals and one δ orbital (i.e. $d\sigma$, $d\pi c$, $d\pi s$ and $d\delta c$ of one M combined with δds of the other having the MX_4 plane turned 45°) are involved in the M—M bond. If eight electrons really were employed entirely in the M—M bond region, the complexes contain no additional 5d-like electrons and might have the conditional oxidation state Re[VII].

The general subject of cluster compounds such as $Nb_6Cl_{12}^{++}$, $Nb_6Br_{12}^{++}$, $Ta_6Cl_{12}^{++}$, $Ta_6Br_{12}^{++}$, $Mo_6Cl_8L_6^{++}$ and $Mo_6Cl_8X_6^{-4}$ has been reviewed by

Cotton (1966), Kepert and Vrieze (1967) and Bulkin and Rundell (1967). These compounds contain a central octahedron of six M atoms connected with a circumscribed cubic structure of twelve halide ligands (on the two-fold axes of the octahedron) or a cube of eight halides (on the three-fold axes of the octahedron). In both cases, the immediate neighbours of M are a square of four X; but each X is shared by two M in the former case and by three in the latter case. These complexes contain a central cavity which must contain considerable electronic density. In a certain sense, the clusters may be considered as minute chunks of the metallic elements having the ligands chemi-adsorbed on the surface in a regular pattern. As seen from the reviews, there has been made several attempts to M.O. calculations on clusters, and it is possible finding assumptions to allow for a closed-shell structure though with low-lying excited levels. The observed, intense absorption bands in the visible are remarkably independent of X and have very small band-widths corresponding to the rigidity and high mass of the M_6 octahedron. The treatment in terms of $M-M$ bonding and anti-bonding d-like orbitals constitute a close analogy to the linear combination of "π", i.e. $2p\sigma$, orbitals in large, planar, aromatic hydrocarbons such as anthracene, tetracene or coronene.

Like these hydrocarbons, certain cluster compounds can be reduced or oxidized in one-electron steps, say to $Ta_6X_{12}^{+3}$ and $Ta_6X_{12}^{+4}$. There is crystallographic evidence (Burbank, 1966) that some of these cations no longer are regular cubic but contains tetragonally elongated Ta_6. This may either be taken as evidence for Jahn-Teller effect (i.e. that some electrons are removed from a set of degenerate delocalized M.O.) or as evidence for partly electronic ordering. The cluster compounds have produced other surprises. Thus, the carbonyl $Rh_6(CO)_{16}$ contains a central octahedron Rh_6 not having all eight faces equivalent. Actually, this black but diamagnetic material has the other surprising property of CO bridges to three different Rh atoms on an octahedral face (4 cases) whereas R_3CO is not known from organic chemistry though R_3NO is. Albano, Chini and Scatturin (1968) recently studied the isomorphous $Co_6(CO)_{16}$ and reduction products such as $Co_6(CO)_{15}^{-}$ and $Co_6(CO)_{14}^{-4}$. Apparently, a few of the CO ligands can be replaced by electron pairs.

The compounds $CRu_6(CO)_{17}$ (Johnson et al., 1967) and the analogous $CRu_6(CO)_{14}[C_6H_3(CH_3)_3]$ were recently shown by Mason and Robinson (1968) to contain a central carbon atom in the middle of the octahedron Ru_6. Hence, this is an interstitial carbide of our metaphorical chunk of metal. There are a few other cases known of carbon atoms with high coordination numbers in carbonyls; thus $C[Fe(CO)_3]_5$ contains a square-pyramidal coordination of the central carbon atom by five iron atoms which is not readily explained (BRAYE et al., 1962). It is obvious that time is not yet ripe for classifying the oxidation states of such compounds. MARONI and SPIRO (1968) studied the Raman spectrum of $Bi_6(OH)_{12}^{+6}$ (the most common hydrolysis product of bismuth (III) in acidic aqueous solution) and pointed out that the stereochemistry of this ion is very similar to $Ta_6X_{12}^{+2}$ and one may argue that the six lone-pairs of Bi(III) fill the central cavity. However, this complex is colourless; the excited states involving additional node-planes of this central electronic distribution may have high energy, if the radius of the cavity is sufficiently small. Both solid (BRODERSEN et al., 1965) and gaseous $Pt^{II}Cl_2$ is known to contain comparable Pt_6Cl_{12} (SCHÄFER et al., 1967, also detecting gaseous Pd_6Cl_{12}) whereas solid $PdCl_2$ contains squares $Pd(II)Cl_4$ sharing two corners, though the cluster modification was prepared by the latter authors. WCl_3 was shown by SIEPMAN et al. (1967) to be $(W_6Cl_{12})Cl_6$.

The most numerous class of cluster compounds involves central triangles M_3 as we previously met in $Os_3(CO)_{12}$. COTTON and collaborators demonstrated that red $Re^{III}Cl_4^-$ is trimeric $Re_3Cl_{12}^{-3}$ and prepared a large number of other $Re_3Cl_xL_{12-x}^{+9-x}$ and $Re_3Br_xL_{12-x}^{+9-x}$ having very characteristic absorption spectra (for a review, see COTTON, 1966). If six electrons are reserved for the bonding in the central triangle, we are left with Re[V] systems each containing two 5d-like electrons. It is not certain that a similar argument can be used for showing that $Os_3(CO)_{12}$ contains the d^6-systems Os[II] which are distorted from the symmetry O_h. BRADFORD and NYHOLM (1968) report linear Os_3 chains in $X[Os(CO)_4]_3X$.

Other recent examples of clusters are the phosphine complexes of the diamagnetic Au_6^{+2} studied by CARIATI et al. (1967) and Bi_5^{+3} and Bi_8^{++} investigated by CORBETT (1968). DRENTH et al. (1964) measured ultraviolet spectra of the dimeric $(C_6H_5)_3MM(C_6H_5)_3$ (M = Sn and Pb) and the tetrahedral $M[M'(C_6H_5)_3]_4$ with M = Ge, Sn, Pb and M' = Sn and Pb, but M' not being lighter than M. According to HARADA et al. (1968), there is no evidence for 3d-orbitals in the spectrum of SiH_4 having the first shoulder at 63 kK. $Si(CH_3)_4$ has a Rydberg-like band at 60 kK comparable to that at 62 kK of neo-pentane. LUCKEN (1969) recently reviewed the evidence for and against the importance of 3d-orbitals for the chemical bonding in sulphur compounds.

Cluster complexes of elements having distinctly different electronegativity such as IAg_3^{+2} and $TeAg_8^{+6}$ have been reviewed by BERGERHOFF (1964). It is very interesting that the central atom iodine (− I) and tellurium (− II) are able to coordinate as many silver (I) ligands; the complexes are prepared by dissolving the otherwise very insoluble AgI or Ag_2Te in a strong solution of $AgNO_3$ and are colourless or pale yellow; there seems to be no difficulty for defining spectroscopic oxidation states in these cases. Also PAg_6^{+3} and $AsAg_6^{+3}$ have been reported.

A subject closely related to the cluster-chemistry is the relative abundance of fragments produced by the mass-spectrography of complicated molecules which serve to determine the molecular weight, and in the case of high-resolution mass-spectrographs also the number of atoms of each element. However, this investigation is just at its beginning at the moment. Colloid gold particles have also a rather characteristic absorption spectrum of the aqueous suspension (DOREMUS, 1964) and the empirical experience of glass-makers how to colour silicate glasses with finely dispersed copper, selenium and gold producing bright red samples (WEYL, 1959) is interesting. There are many unresolved problems of related character in solution chemistry. Thus, it is known that elemental selenium dissolves in concentrated sulphuric acid with green colour, and elemental tellurium with an intense purple colour.

Bibliography

ALBANO, V., P. CHINI, and V. SCATTURIN: Chem. Commun. 163 (1968).

BEHRENS, H., u. R. SCHWAB: Z. Naturforsch. 19 b, 768 (1964).

—, u. D. HERRMANN: Z. Naturforsch. 21 b, 1234 (1966).

BERGERHOFF, G.: Angew. Chem. 76, 697 (1964).

BRADFORD, C. W., and R. S. NYHOLM: Chem. Commun. 867 (1968).

BRAYE, E. H., L. F. DAHL, W. HÜBEL, and D. L. WAMPLER: J. Am. Chem. Soc. 84, 4633 (1962).

BRODERSEN, K., G. THIELE, u. H. G. SCHNERING: Z. Anorg. Allgem. Chem. 337, 120 (1965).

BROWN, I. D., D. B. CRUMP, R. J. GILLESPIE, and D. P. SANTRY: Chem. Commun. 853 (1968).

BULKIN, B. J., and C. A. RUNDELL: Coordin. Chem. Rev. 2, 371 (1967).

BURBANK, R. D.: Inorg. Chem. 5, 1491 (1966).

CARIATI, F., L. NALDINI, G. SIMONETTA, and L. MALATESTA: Inorg. Chim. Acta 1, 315 (1967).

CHURCHILL, M. R., and R. BAU: Inorg. Chem. 6, 2086 (1967).

CORBETT, J. D.: Inorg. Chem. 7, 198 (1968).

COTTON, F. A.: Quart. Rev. (London) 20, 389 (1966).

DELÉPINE, M.: Ann. Chim. (Paris) [13] 4, 1115 (1959).

DOREMUS, R. H.: J. Chem. Phys. 40, 2389 (1964).

DRENTH, W., M. J. JANSSEN, G. J. M. VAN DER KERK, and J. A. VLIEGENTHART: J. Organometall. Chem. 2, 265 (1964).

DUBICKI, L., and R. L. MARTIN: Inorg. Chem. 5, 2203 (1966).

EINSTEIN, F., J. TROTTER, and C. WILLISTON: J. Chem. Soc. (A) 2018 (1967).

FOSS, O.: Advan. Inorg. Chem. Radiochem. 2, 237 (1960).

— Selected Topics in Structure Chemistry, p. 145. Universitetsforlaget, Oslo, 1967.

GRAY, H. B., E. BILLIG, A. WOJCICKI, and M. FARONA: Can. J. Chem. 41, 1281 (1963).

HANDY, L. B., P. M. TREICHEL, L. F. DAHL, and R. G. HAYTER: J. Am. Chem. Soc. 88, 366 (1966).

HARADA, Y., J. N. MURRELL, and H. H. SHEENA: Chem. Phys. Letters 1, 595 (1968).

HENDRA, P. J., and Z. JOVIC: J. Chem. Soc. (A) 911 (1968).

HERZOG, S., and H. OBERENDER: Z. Chem. 3, 67 (1963).
HUSEBYE, S.: Acta Chem. Scand. 20, 24 and 2007 (1966).
JØRGENSEN, C. K.: Acta Chem. Scand. 13, 196 (1959).
—, and L. E. ORGEL: Mol. Phys. 4, 215 (1961).
— Inorganic Complexes. London: Academic Press 1963.
— Halogen Chem. 1, 265 (1967). London: Academic Press.
— Inorg. Chim. Acta Rev. 2 (1968).
JOHNSON, B. F. G., R. D. JOHNSTON, and J. LEWIS: Chem. Commun. 1057 (1967).
JOHNSON, S. A., H. R. HUNT, and H. M. NEUMANN: Inorg. Chem. 2, 960 (1963).
JONES, P. E., and L. KATZ: Chem. Commun. 842 (1967).
KEBLYS, K. A., and M. DUBECK: Inorg. Chem. 3, 1646 (1964).
KEPERT, D. L., and K. VRIEZE: Halogen Chem. 3, 1 (1967). London: Academic Press.
LUCKEN, E. A. C.: Struct. Bonding, in press.
MARONI, V. A., and T. G. SPIRO: Inorg. Chem. 7, 183 (1968).
MASON, R., and W. R. ROBINSON: Chem. Commun. 468 (1968).
NOACK, K.: Helv. Chim. Acta 47, 1064 and 1555 (1964).
SCHÄFER, H., U. WIESE, K. RINKE u. K. BRENDEL: Angew. Chem. 79, 244 (1967).
SCHÄFFER, C. E.: J. Inorg. Nucl. Chem. 8, 149 (1958).
SHRIVER, D. F.: Struct. Bonding 1, 32 (1966).
SIEPMAN, R., H. G. VON SCHNERING u. H. SCHÄFER: Angew. Chem. 79, 650 (1967).
TREICHEL, P. M., and G. P. WERBER: J. Am. Chem. Soc. 90, 1753 (1968).
VLČEK, A. A., and F. BASOLO: Inorg. Chem. 5, 156 (1966).
WEYL, W. A.: Coloured Glasses. London: Dawson's of Pall Mall 1959.
WIEBENGA, E. H., E. E. HAVINGA, and K. H. BOSWIJK: Advan. Inorg. Chem. Radiochem. 3, 133 (1961).
WILMARTH, W. K., H. GRAFF, and S. T. GUSTIN: J. Am. Chem. Soc. 78, 2683 (1956).

11. Quanticule Oxidation States

It might be argued that the nine first chapters of this book are an expression of an extreme opinion in the inorganic coordination chemist, viz. that most chemical bonds are intrinsically heteropolar and always, at least as a Gedankenexperiment, can be split heterolytically in a coordinating part containing at least one lone-pair; and an electron-pair acceptor. In Chapter 10, it was realized that in an entirely homopolar situation, the concept of spectroscopic oxidation states is reduced to an absurdity; it would be necessary to classify H_2 as the adduct of the electrophile H^+ with the nucleophile H^- and C_2H_6 as the adduct of CH_3^+ with CH_3^-. It may also be argued that the usual valence-bond description makes exactly the opposite mistake; it considers all chemical bonds as homopolar to the first approximation, and it looks in the hybridization theory for the six orbitals responsable for the six bonds formed in SiF_6^{--}, PCl_6^-, gaseous SF_6 or solid MnO, BaTe, LaN or TmP. However, the two extreme attitudes: that one can always find heteronuclear characteristics of a bond between two different elements; and that the atomic constituents are close to neutral, are not on an equal footing from the view-point of applied group theory. Of whatever is the reason, the heterolytic extreme predicts the symmetry types of the groundstate and the low-lying excited states correctly in a large majority, though not in all, d- and f-group complexes. On the other hand, H_2, N_2, P_4, diamond etc. are all in the situation of a fairly isolated, non-degenerate groundstate though the degeneracy number e of two hydrogen atoms is four, of two nitrogen atoms $e = 400$ if we restrict ourselves to the electron configuration p^3, and of four phosphorus atoms $e = 20^4 = 160000$. The number of resonance structures is very large, and much larger than expected by most chemists, and has no resemblance whatsoever with the lowest excited states. However, it is reasonable to ask about what one should do in the cases the spectroscopic oxidation states are unreasonable or impossible to define. We already mentioned the idea in Chapter 9 to reserve two electrons for a σ-bond between two manganese atoms having approximately closed-shell structure with six d-like electrons in the lower sub-shell and hence the conditional oxidation state Mn[I] in $(OC)_5MnMn(CO)_5$. FAJANS' proposal (1959, 1961) of *quanticules* is a very interesting suggestion in this connection.

It is a pity that FAJANS' ideas did not find more understanding among the chemists at a time, between 1930 and 1940, where they would have

constituted a most useful counterweight bringing balance in the rather
extreme valence-bond descriptions prevailing at that time. This does not
mean that the writer agrees with certain of FAJANS' opinions; but he
recognizes that these ideas were most important and interesting precur-
sors of the preponderant electron configurations and hence, indirectly,
the spectroscopic oxidation states. According to FAJANS, a quanticule
is one or more (most frequently an even number) electrons quantized
with respect to one *or more* nuclei. The quanticules quantized with
respect to one nucleus are our conventional nl-shells. Following a suggestion by LEWIS, FAJANS considers the closed-shell atomic cores isoelectronic with the noble gases (or with certain other structures, such as the
28 electrons of Cu^+ or 68 electrons of Hf^{+4}) as natural building blocks
of molecules which are not modified significantly by chemical bonding
(except that certain borderline cases, such as Cu^+, Xe or Yb^{++} can be
oxidized under some circumstances). The interesting part of the theory
for the chemists are the quanticules belonging to two nuclei, such as the
e_2^{--} combining the two protons in H_2, or the two CH_3 fragments in C_2H_6,
or the six electrons e_6^{-6} occurring in benzene C_6H_6 as the loosest bound
quanticule. The stereochemical consequencies of loose lone-pairs and
of ligands can be comparable; thus, NH_3 is N^{+5} with three pairs connecting protons and one quanticule lone-pair, whereas NH_4^+ is N^{+5} with four
equivalent $(e_2^{-2})H^+$ or again, SO_3^{--} is S^{+6} with three O^{--} and a lone-pair,
as contrasted to SO_4^{--} being S^{+6} with four O^{--}. It is seen that the two-
electron quanticules essentially do the work of classical valence-lines,
but that heteronuclear compounds *either are ionic or not*. FAJANS also
elaborated a few special theses, such as N_2 not being $N^{+3}(e_6^{-6})N^{+3}$ but
$N^{+5}(e_{10}^{-10})N^{+5}$ where the ten-electron quanticule is isoelectronic with
neon; and acetylene being $H^+(C^{+4}(e_{10}^{-10})C^{+4})H^+$ abbreviated $H^+CC^{-2}H^+$
in analogy to the explosive dibromo-acetylene $Br^+CC^{-2}Br^+$ whereas
dibromoethane has two-electron quanticules connecting both H^+ and
Br^+ tightly to the carbon skeleton.

   Seen from the point of view of modern quantum mechanics, it is an
interesting question what one means exactly with the statement about
a set of electrons being quantized with respect to a set of nuclei. This is
by no means a trivial problem. Two helium atoms at large mutual distance undoubtedly have two electrons belonging to each atom. The writer
feels that FAJANS has invented the spectroscopic oxidation state; but the
writer cannot agree in the sharp division between ionic and covalent
compounds. The small discontinuity that LiH, BeH and BH seem to be
polarized hydrides and CH, NH, OH and FH are connected with a
quanticule (FAJANS, 1964) is much more readily explained in M.O.
theory by the relative number of bonding and non-bonding electrons.
It is very sympathetic that FAJANS emphasized the unreasonable aspects

of resonance between numerous structures; but this cannot be used to support the statement which once was taken as dogmatic truth in Russia, that a definite molecule has only one chemical formula (cf. the review about the question of how to write chemical species by RASCH, 1966). The writer, of course, agrees that benzene is benzene and not some time-average of two Kekulè structures. But though the preponderant configuration of the helium atom in its groundstate *is* $1s^2$, we also know that the total wavefunction contains more than one configuration, though 99.2% of the squared amplitudes can be ascribed to $1s^2$. Unfortunately, this problem is much more serious than 0.8% in atoms heavier than helium (JØRGENSEN, 1962) and one cannot decide arbitrarily that a monatomic entity has Ψ corresponding to a well-defined configuration, though in nearly all cases, it has a preponderant electron configuration. The other aspect of FAJANS' quanticule theory which cannot be seriously supported by M. O. theory is the absolute choice between polarized, ionic compounds (cf. the discussion of electric polarizabilities p. 227) and covalent molecules connected with diatomic or polyatomic quanticules. In the series of crystalline KF, CaO, ScN and TiC, there must be a gradual transition from more ionic to less ionic compounds.

We write the atomic constituent lacking all electrons participating in diatomic or polyatomic quanticules in curly brackets, such as $N\{III\}$ in NH_3 and $N\{V\}$ in NH_4^+ and call this quantity which takes lone-pair electrons into account in the atomic population the *quanticule oxidation state*. Hence, the quanticule oxidation state of diamond is $C\{IV\}$ and of all the carbon atoms and hydrogen atoms in neopentane, cyclohexane and all other alkanes $C\{IV\}$ and $H\{I\}$. The obvious disadvantage of this treatment is that no distinction is made between 2s and 2p electrons of carbon though we know that the difference by a factor between 1.5 and 2 in the ionization energy is conserved in molecules according to photoelectron spectroscopy. An interesting aspect of the diamond question is the isotypic and isoelectronic compounds CuBr, ZnSe, GaAs and their limit, Ge. If one pair of electrons is reserved to each chemical bond, the quanticule oxidation states of the constituents are $Cu\{I\}Br\{VII\}$, $Zn\{II\}Se\{VI\}$, $Ga\{III\}As\{V\}$ and $Ge\{IV\}$ and if one admits that the electron pairs are asymmetrically distributed in such a way that the major part of their density occurs close to the atom having the highest electronegativity, it becomes conceivable that the fractional charges are rather small and in certain cases, say InSb, may be nearly vanishing. On the other hand, the spectroscopic oxidation states $Cu(I)Br(-I)$; $Zn(II)Se(-II)$ and $Ga(III)As(-III)$ extrapolate to an absurdity in the case of germanium.

In the cases where colourless or at most yellow compounds are formed containing M — M bonds, it may still be reasonable to reserve an electron pair to the bond and to talk about Mn{I}, Hg{II}, Ga{III} and Rh{III} in $Mn_2(CO)_{10}$, Hg_2^{++}, GaS and $Rh_2(CH_3COO)_4$, $2H_2O$. In these cases, the values are the same as one would ascribe conditional oxidation states Mn[I],... However, the additional electrons may be doing something else than forming σ-bonding electron pairs; e.g. they may ensure electric conductivity of LaS which contains La[III] but only La{III} in the extended sense of the conduction band being a mega-atomic quanticule; or the excited configuration $1s^22s^22p^53p$ of neon contains Ne[I] in a quite well-defined sense.

We finally define *distributed quanticule oxidation states* in sharp brackets M⟨II⟩,... as the values one would obtain by sharing each two-electron quanticule evenly between the two atoms bound. Thus, C⟨0⟩ in diamond and the alkanes, N⟨0⟩ in NH_3 and NF_3, N⟨I⟩ in NH_4^+ and $N(CH_3)_4^+$ etc. Thus, the sum of distributed quanticule oxidation states is exactly the charge of the polyatomic entity in protonic units, whereas the sum of quanticule oxidation states C{IV}, N{III}... is lacking the bonding electron-pairs in this respect. The concept of distributed quanticule oxidation states is closely related to *formal charges* in organic chemistry. When it was realized that betaïnes and tetra-alkylammonium cations possess one unit of positive charge on the nitrogen atom which cannot be removed by simple deprotonation by OH^- as it can from NH_4^+, it became useful to say that nitrogen carries the formal charge $+1$ in these compounds and 0 in NH_3 and normal amines. In this sense, the reaction (10.1) is really a redox reaction, N⟨0⟩ and I⟨0⟩ changing to N⟨I⟩ and I⟨−I⟩. The idea of distributed quanticule oxidation states (which might be called "formal charges" for brevity; but there is still a nuance of difference — cf. thiophene) has rather serious drawbacks. In the isoelectronic series BH_4^-, CH_4 and NH_4^+ the central atoms get all the responsability for the ionic charge B⟨−I⟩ and N⟨I⟩ which is not in agreement with our ideas of electronegativity; the important feature was, of course, the equal sharing of the bonding electron-pairs. If the four bonds in SO_4^{--} are single bonds, we have S⟨II⟩ and O⟨−I⟩. If the classical picture of $(O=)_2S(-OH)_2$ for sulphuric acid can be extended to the bond-order $1\frac{1}{2}$ for the S — O bonds in sulphate (like the C — O bonds in carboxylates), i.e. six pairs of electrons being employed by the chemical bonding, we have S⟨0⟩ and O⟨$-\frac{1}{2}$⟩. The whole treatment looks rather confusing for MnO_4^- having Mn⟨III⟩ and O⟨−I⟩ if four bonds occur and Mn⟨0⟩ and O⟨$-\frac{1}{4}$⟩ if seven bonds occur. The occurrence of O⟨−I⟩ is another expression for the concept of semi-polar bonds with certainty (it was said) present in R_3NO and very probably in R_3PO and R_2SO.

The consequences for zincblende structures are also rather serious. Elemental Ge⟨0⟩ dissociates (by moving protons from one half of the nuclei to the other half) to $Ga\langle-I\rangle As\langle I\rangle$; $Zn\langle-II\rangle Se\langle II\rangle$ and $Cu\langle-III\rangle$ $Br\langle III\rangle$ showing the opposite trend of the electronegativity differences. This is more acceptable in CO which is isoelectronic with N_2. In order to conserve its very strong triple bond, it has to transform to $C\langle-I\rangle$ and $O\langle I\rangle$ isoelectronic with N. The very small dipole moment observed for CO is then frequently explained as a cancellation of this effect by the intrinsic higher electronegativity of oxygen than of carbon. Usually, the formal charges are not discussed for coordination numbers higher than 4. It seems to be empty words to have $Ti\langle-II\rangle$ and $C\langle II\rangle$ in TiC and $Sc\langle-III\rangle$ and $N\langle III\rangle$ in ScN.

Actually, the concept of quanticule oxidation states $M\{I\}$... seems better founded than of $M\langle I\rangle$... though it may be argued that it is a complicated way of translating classical valence-structures. However, there are various advantages; it is saying less than the classical picture if the bonds are just removing available electrons without statements being made about their spatial distribution; and this is no negligible advantage in cases like SO_4^{--} and MnO_4^{-}. A choice has to be made as to whether only σ-bonds involving exactly two electrons are accepted, or whether multiple bonds are recognized as well. It is quite obvious that we meet never the detailed evidence for Er(III) in pink compounds, nor Er[III] in erbium metal when considering quanticule oxidation states; but it is among the few available lines of thought when one attempts to define oxidation states in almost homopolar compounds.

From the point of view of *formal oxidation numbers*, the most important quality is the additivity to the external charge in protonic units which assures their conservation in chemical reactions. With exception of $M\langle I\rangle$... we actually have no absolute guarantee that the other oxidation *states*, viz. the spectroscopic M(I), the conditional M[I] and the quanticule $M\{I\}$ follow the additivity rule. This can already be seen from the very fact that chemical reactions can result in products where some or all of these oxidation states cannot be determined. Usually, the spectroscopic oxidation states are only considered valid if they fulfil the additivity requirement. However, one may imagine situations where only one constituent has an obvious spectroscopic oxidation state, and that the two or more other constituents cannot be exhaustively discussed. Thus, the preparation of $[Co(NH_3)_6]^{+3}[I_3]_3$ which definitely is a Co(III) complex does not by itself resolve the problem of the oxidation states in I_3^{-}. The conditional and quanticule oxidation states M[I] and $M\{I\}$ are not at all adapted to the additivity and conservation rule, since the number of chemical bonds readily may change by a chemical reaction. Further on, we have the problem discussed in Chapter 8 that two adjacent atoms may have such a distance that it is not clear whether they form a chemical bond or not. One interesting aspect of this problem is why some molecules much more readily dimerize to form a homonuclear bond than other, isoelectronic species. Though CO_2^- can be generated by X-ray irradiation, it is far more stable as the dimer oxalate, whereas $2NO_2$ (red-brown) have nearly the same energy as the colourless N_2O_4. Chlorine (IV) oxide ClO_2 shows no tendency toward association though the isoelectronic SO_2^- occurs as a minor constituent of aqueous solutions of dithionite $S_2O_4^{--}$. On the other hand, dithionate $S_2O_6^{--}$ and S_2F_{10} show no tendency to dissociation to SO_3^- and SF_5. According to GREEN and LINNETT (1960) one of the conditions for simple molecules having $S = 1/2$ not to dimerize is that the electronegativity is not too different for the constituent atoms, as is true for NO, NO_2, ClO_2 and for O_2^+, O_2^-, the ozonide anion O_3^- and the blue I_2^+ (GILLESPIE and MILNE, 1966).

The so-called *hypofluorites* pose an interesting problem (cf. a review by HOFFMAN, 1964). When one tries to prepare trifluoromethanol CF_3OH (of which only salts are known, such as $K^+CF_3O^-$) CF_3OF is obtained. If these atoms all have oxidation numbers, at least two alternatives have to be taken seriously, either O^0F^{-I} or $O^{-II}F^I$ for the terminal atoms. From a purely spectroscopic point of view, one might also argue a certain similarity with oxygen (II) fluoride OF_2 and considering the CF_3 group as a pseudohalogen having its own dimer, hexafluoroethane. The kinetically very stable "teflon", polymerized $-CF_2-$ is by itself a striking exception to the rule of adjacent charges. It is not easy to choose definitely between the three alternatives of oxidation numbers. By the same token, O_2NOF, F_5SOF, F_5SOCF_3, F_5SONF_2, F_5SOSF_5, F_5SOOSF_5, and O_3ClOF (as well as O_3ClF) are known. What is perhaps more surprising is that DUDLEY and CADY (1963) studied the equilibrium between the brown gas $(S = 1/2)$ O_2SOF and its dimer $O_2FSOOSFO_2$ being a colourless peroxo complex. CZARNOWSKI et al. (1968) discussed the related problem of F_2SO_3 which can be considered as the sulphur (VI) complex of two oxide, one fluoride and one OF^- ligand. Quite generally, sulphur (VI) forms unexpected compounds; thus, COHEN and MACDIARMID (1965) prepared SF_5Cl, SF_5Br and NSF_3. The latter, tetrahedral nitride complex can be compared with $NV^VCl_3^+$ (STRÄHLE and DEHNICKE, 1965) made from $ClNVCl_3$. It is, of course, not excluded that one might oxidize the oxygen of SO_4^{--} to SO_4^- or SO_4 without changing the oxidation state S(VI) of the central atom. However, in actual practice, such species seem rapidly to dimerize to peroxodisulphate $[O_3S(O_2^-)SO_3]^{--}$ or re-arrange to the anion of CARO's acid $[O_3S(O_2^-)]^{--}$. It is sometimes possible to prepare new isomers, such as the chlorine (II) peroxide ClOO which was compared with OClO by EACHUS et al. (1967).

Various unstable catenated oxygen fluorides exist, such as FOOF, FO_3F and FO_4F (STRENG, 1963). In all of these cases, the assignment of oxidation states meets serious difficulties. On the other hand, F_3^- does not seem to exist though F_2^- has been detected as a colour center by irradiation of crystals and though ClF_4^- and ClF_5 have been prepared in recent years (cf. JØRGENSEN, 1967). The triatomic molecule Cl_3 was detected by NELSON and PIMENTEL (1967).

When discussing catenation, we have all possible, and a few impossible, cases. Thus, the colourless compound (COATES and PARKIN, 1963) $(CH_3)_3AuP(CH_3)_3$ has a formal similarity with hexamethyl-ethane, but it is quite clear that it is a coordination compound of Au(III) with $P(CH_3)_3$ It is already less clear what to say in cases where the two middle atoms are less different, such as $(CH_3)_3SiC(CH_3)_3$. This poses the question how closely similar two atoms have to be before we consider them bound by an essentially homopolar bond. In solution, we have further difficulties. Thus, the colourless $Au(CH_3)_2(H_2O)_2^+$ studied by MILES, GLASS and TOBIAS (1966) is undoubtedly a *cis*-quadratic gold (III) complex, but because of the *trans*-effect, the two water molecules are loosely bound (whereas the CH_3^- ligands show no tendency to get protonated to CH_4 by the solvent) and one might readily run into problems such as discussed above in connection with tellurium (II). The complexes reported by BEHRENS and MÜLLER (1965) olive-green $Cr(CN)_6^{-6}$, brick-red $Ni[P (C_6H_5)_3]_4$ and yellow $Ni[As(C_6H_5)_3]_4$ can be classified as coordination compounds of Cr(0) and Ni(0) but we are at the same time approaching the topic of cluster-compounds, as we also were in the case of $Bi_6(OH)_{12}^{16}$ and $Sn[Pb(C_6H_5)_3]_4$.

The tendency toward catenation even strongly influences the behaviour of elemental sulphur (MEYER, 1965); for spectra of $S_nO_6^{--}$ and S_n^{--}, see GOLDING (1960). A compound such as realgar As_4S_4 has the four sulphur atoms in a plane, two As above and two As below the plane (SOMMER and BECKE-GOEHRING, 1965). The heteroatomic chains and cycles formed by nitrogen and sulphur were reviewed by BECKE-GOEHRING (1959). For a discussion of the M.O. treatment of $N_4S_4^{-z}$ ($z = 0,1,2,3, 4$) having co-planar N_4 see TURNER and MORTIMER (1966). WEISS and NEUBERT (1966) describe the planar Co, Ni and Pd complexes of two ligands SSNS (connected with the two terminal S atoms to the central atom M) and of two ligands HNSNS (connected to M with the NH group and the terminal S atom). It is obvious that such ligands defy a classification in terms of oxidation states though it may be argued that the characteristic behaviour of Ni(II) and Pd(II) suggest one negative charge SSNS$^-$ and HNSNS$^-$ which can be construed to mean the substitution of sulphur atoms by N$^-$ and by NH in catenated S_4. We mentioned above the platinum (IV) complex $Pt(S_5)_3^{--}$.

The reader may feel at this point that the prudent and cautious attitude would be to think in terms of oxidation numbers being a convenient tool of classification of compounds and reactions, but without any deep physical significance. It must be noted, in all fairness, that we have gone way out spot troubles; the large majority of compounds encountered in inorganic chemistry do not show the pathological behaviour of some of the queerest examples we have inspected. This is also the reason why the chemistry of halogens and xenon, though highly interesting, has not been so much discussed here (cf. the review by JØRGENSEN, 1967) simply because the problems of defining oxidation states are not so hard with the ironical exception of NEIL BARTLETT's first xenon compound*), so possibly being $Xe^+PtF_6^-$. The lanthanides also are undramatic in the sense that they never produce serious doubts about the spectroscopic or at least the conditional oxidation state; but they serve to explain our techniques for gaining the same confidence in partly filled d-shells as we have in detecting a partly filled 4f-shell.

However, there seems to be a kind of inverse variation of the physical content of various possible definitions of oxidation states and the number of compounds to which these definitions can be applied. If we were only interested in monatomic entities in gaseous state, the formal oxidation number is strictly defined, and there is no need of introducing the name spectroscopic oxidation state for the same quantity. In the absence of the extremely rare situation that two different partly filled shells have nearly the same average radii, the conditional oxidation state M[I]... is also always well-defined. But chemistry is interesting because of the interactions between the atomic constituents. We finish in Chapter 12 with a discussion of the classificatory, the taxological, aspects of quantum chemistry which are intimately connected with *oxidation states* as contrasted to oxidation numbers as assisting book-keeping of chemical reactions.

Bibliography

BECKE-GOEHRING, M.: Progr. Inorg. Chem. 1, 207 (1959).

BEHRENS, H., u. A. MÜLLER: Z. Anorg. Allgem. Chem. 341, 124 (1965).

COATES, G. E., and C. PARKIN: J. Chem. Soc. 421 (1963).

COHEN, B., and A. G. McDIARMID: Inorg. Chem. 4, 1782 (1965).

CZARNOWSKI, J., E. CASTELLANO u. H. J. SCHUMACHER: Z. Physik. Chem. 57, 249 (1968).

DUDLEY, F. B., and G. H. CADY: J. Am. Chem. Soc. 85, 3375 (1963).

EACHUS, R. S., P. R. EDWARDS, S. SUBRAMANIAN, and M. C. R. SYMONS: Chem. Commun. 1036 (1967).

FAJANS, K.: Chimia 13, 349 (1959). English version is available from Ulrich's Book Store, 547 East University Avenue, Ann Arbor, Michigan 48104.

— Kwantykulowa teoria wiazania chemicznego. Warszawa: Panstwowe Wydawnictwa Techniczne, 1961.

---

*) However, SLADKY et al. (1968) identified three components $Xe_2F_3^+MF_6^-$; $XeF^+MF_6^-$ and $XeF^+M_2F_{11}^-$ of such compounds all containing xenon(II).

— J. Chem. Phys. **40**, 1773 and **41**, 4005 (1964).

GILLESPIE, R. J., and J. B. MILNE: Inorg. Chem. **5**, 1577 (1966).

GOLDING, R. M.: J. Chem. Phys. **33**, 1666 (1960).

GREEN, M., and J. W. LINNETT: J. Chem. Soc. 4959 (1960).

HOFFMAN, C. J.: Chem. Rev. **64**, 91 (1964).

JØRGENSEN, C. K.: Orbitals in Atoms and Molecules. London: Academic Press 1962.

— Halogen Chem. **1**, 265 (1967). London: Academic Press.

MEYER, B. (editor): Elemental Sulfur-Chemistry and Physics. New York: Interscience 1965.

MILES, M. G., G. E. GLASS, and R. S. TOBIAS: J. Am. Chem. Soc. **88**, 5738 (1966).

NELSON, L. Y., and G. C. PIMENTEL: J. Chem. Phys. **47**, 3671 (1967).

RASCH, G.: Z. Chem. **6**, 297 (1966).

SLADKY, F. O., P. A. BULLINER, N. BARTLETT, B. G. DE BOER, and A. ZALKIN: Chem. Commun. 1048 (1968).

SOMMER, K., u. M. BECKE-GOEHRING: Z. Anorg. Allgem. Chem. **339**, 182 (1965).

STRÄHLE, J., u. K. DEHNICKE: Z. Anorg. Allgem. Chem. **338**, 287 (1965).

STRENG, A. G.: Chem. Rev. **63**, 607 (1963).

TURNER, A. G., and F. S. MORTIMER: Inorg. Chem. **5**, 906 (1966).

WEISS, J., u. H. S. NEUBERT: Z. Naturforsch. **21 b**, 286 (1966).

12. Taxological Quantum Chemistry

Obviously, quantum mechanics has been of some help to chemists. This would have been true even in 1927 when Burrau wrote his paper on the solutions of Schrödinger's equation for the one-electron system H_2^+. Even before the simplest problems of chemical bonding could be tackled, it was already an admirable result to have rationalized the energy levels of atoms and monatomic positive ions. Cynical people may argue that the atomic spectroscopists already had taken care of this classification relating nl-shell configurations to parity and the quantum number J, and in the case of Russell-Saunders coupling, also L and S. However, it is an oversimplification when amateur pragmatic philosophers try to convince us that a new theory is only useful if it predicts new results besides correlating the old, known facts. Even if a new theory only connects previously recognized results, it does not really matter if it is beautiful enough. The situation is rarely that of a crucial experiment so dear to text-book writers. In typical cases, when one performs hundred new experiments, the new theory may explain ten of them successfully and be indifferently compatible (and this can be a very elastic concept) with ninety of them. One most not forget that even in the most decisive type of one crucial experiment, such as that of Michelson, there are still a bunch of parallel theories compatible, e. g. the Lorentz contraction acting on rigid bodies, or Einstein's theory of special relativity. However, the concept of special relativity is entirely different from the revolution introduced by quantum mechanics. One might imagine Blaise Pascal or Sonja Kowalewska as children inventing the special theory of relativity as an intellectually satisfactory game, whereas the necessity of quantum mechanics was a direct blow against us from reality.

It cannot be denied that chemistry conceivably already was reduced to applied mathematics by Schrödinger. However, there are only the simplest molecules such as LiH or Li_2 which have been strictly treated, and nearly all of the theoretical results consist of the comparison af highly approximative calculations with experimental data such as absorption spectra, dipole moments, force constants and internuclear distances (mentioned in the order of generally increasing disagreement in typical cases). The quantum chemists performing large calculations frequently console themselves and attempt to procure a well-deserved euphoric feeling in their students by referring to the *variation principle*. This is an extraordinarily bad excuse. In atoms having atomic numbers higher than 10, the first ionization energy is lower than the correlation energy of the best Hartree-Fock functions. Consequently, the eigenvalue corresponding to this otherwise excellent approximation is situated in the continuum above the first ionization energy, and an infinity of excited states of the same symmetry type as the groundstate have lower energy than the approximate wave-function. Actually, we are only obtaining necessarily better agreement with the groundstate if the energies of the set of ameliorating approximations are situated in the interval between the real groundstate and the first excited state of the same symmetry type.

The interesting conclusions of quantum chemistry and applied group theory which are accumulating are that

1. the atomic constituents of compounds retain spherical symmetry to a certain extent,

2. that excitations of inner shells and other filled orbitals can be recognized,

3. that the discrete one-electron energies frequently have a strong ressemblance with those corresponding to definite spectroscopic oxidation states of the constituent atoms, and

4. that the low-lying excited levels most frequently are correctly classified by zero-electron (in the case of partly filled shells) and one-electron excitations.

We are shortly discussing each of these points. It is obvious that (1) only applies to strongly bound electrons and not to, say, the conduction electrons of a metal. The most clear-cut evidence for ionization of inner shells (and penultimate orbitals in general) comes from X-ray and photoelectron spectroscopy. It is worth once more to emphasize the technicality that since the one-electron energy of such orbitals is highly negative, the excited states lacking one such electron are situated high up in the continuum surrounded by an over-all dense distribution of other states, which may be described by the simultaneous excitation or ionization of two or more looser bound electrons. However, the X-ray spectra in absorption or emission are extremely sharp relative to the total energy of the photon, though their band-width may be in the order of energy differences of chemical interest, say 10 kK. For the chemist, it is even more important that the partly filled shell, or the loosest bound, fully occupied orbitals frequently, but not always, behave as if it has well-defined l-value in the strict sense of separability of the one-electron function in the product of a radial function and a characteristic angular function according to eq. (3.1) or at least in the looser sense of having "l" angular node-planes. In the case of a partly filled 4f or 5f shell, there is no serious doubt that the central atom has excited levels of the same type as a corresponding gaseous ion having a core closed-shell configuration (54 or 86 electrons) to which $4f^q$ or $5f^q$ are added. In the case of partly filled 3d, 4d or 5d shells, there occurs a certain delocalization, and only the major part of the electronic density residing on the central atom retains $l = 2$, whereas the minor part delocalized on the ligands at most is characterized by the two ("l" $= 2$) angular node-planes (or one equivalent node-cone). In the case of p-shells, two categories seem to occur. When the complex has a centre of inversion, it seems to be possible to talk about low-spin linear p^2-systems or low-spin quadratic p^4-systems. Also the closed -shell ligands N(-III), O(-II), F(-I), S(-II), Cl(-I), Se(-II), Br(-I) and I(-I) belong to this category, having two p π orbitals; and one σ orbital which may contain some admixture of s character in the absence of a centre of inversion. The high-spin p^2-system Bi(I) seems to have approximate spherical symmetry. On the other hand, the apparent p^q-systems lacking a centre of inversion such as ("q" $= 2$) O(II) in OF_2, Te(II) in cis-Te(II) S_2X_2, I(III) in ICl_2^+ or Cl(III) in ClF_3 may present strong mixing of s- and p-character in the lone-pairs. This is also true for certain apparent s^2-systems such as pyramidal MX_3 formed by C(II) (if CH_3^- is considered as a hydrido complex), N(III), P(III), S(IV), Cl(V), As(III), Se(IV), Br(V), Sn(II) (in $SnCl_3^-$), Sb(III), Te(IV) (in $TeCl_3^+$), I(V), Xe(VI) (in XeO_3) and Pb(II) (in $Pb(OH)_3^-$). Other s^2-systems are known with a centre of inversion, at least as a time-average result, such as octahedral complexes of Sb(III), Te(IV), I(V), Xe(VI), Tl(I), Pb(II) and Bi(III). On the other hand, one of the six ligands tends to be replaced by the lone-pair (hence having extensive s-p mixing) and square-pyramidal TeF_5^-, IF_5 and XeF_5^+ are formed. Bent dioxo-complexes of N(III), S(IV) and Se(IV) containing one lone-pair and of Cl(III) containing two form a striking contrast to the dioxo-complexes possessing linear symmetry with centre of inversion $D_{\infty h}$ formed by carbon(IV), nitrogen(V) and uranium(VI) containing no lone-pairs. Table 12.1 gives the common oxidation numbers and total spin quantum number S of complexes formed by elements having higher atomic number than calcium. In all cases, the preponderant configuration of the corresponding gaseous ion is indicated (not taking into account the s-p mixing in complexes without centre of inversion) with the exception of d-group central atoms in low oxidation states where the complexes definitely do not fill the subsequent s-orbital. The notation d^0 is used for certain transition ions rather than the equivalent s^2p^6. Ce(IV) and U(VI) are usually called f^0 because of the low-lying electron transfer excited levels. In Chapter 6, it was discussed how excitations such as $4f \rightarrow 5d$, $5f \rightarrow 6d$, $5s \rightarrow 5p$, $6s \rightarrow 6p$ and to a certain approximation $3d \rightarrow 4s$ and in closed-shell ligands e.g. $2p \rightarrow 3s$ in oxide and $5p \rightarrow 6s$ and $5p \rightarrow 5d$ in iodide can be recognized. Hence, in many cases, orbitals belonging to a definite l-shell can be detected via the visible and ultra-violet absorption spectra.

Once the delocalized nature of most M.O. is accepted it is clear why spectroscopic oxidation states cannot *always* be defined. We may discuss a simplified example: one σ-bonding orbital in a diatomic chromophore XY. When the diagonal elements of one-electron energy E_X and E_Y of the atomic orbital of X and of Y are varied, we may think of their values as the two straight lines representing the asymptotes of a hyperbola, whereas the bonding orbital has an eigenvalue of one-electron energy corresponding to the lower branch of the hyperbola and the anti-bonding orbital to the corresponding upper branch. (cf. eq. 3.43). The minimum distance between the two branches is twice the effective non-diagonal element and occurs where the two straight-line asymptotes cross. The classical valence description of essentially neutral atoms X and Y forming a homopolar bond is equivalent to consider, as a first approximation, the horizontal tangent to the lower eigen-value at the crossing point $E_X = E_Y$ of the diagonal elements. The spectroscopic oxidation states $X(I) Y(-I)$ valid for the branch close to the left-hand part of one asymptote, and $X(-I) Y(I)$ valid for the other extreme $E_X \ll E_Y$ might be taken as a series of three approximations valid in each interval:

$E_X \gg E_Y$	$E_X = E_Y$	$E_X \ll E_Y$
$\delta \to 1$	$\delta = 0$	$\delta \to -1$
$X(I) Y(-I)$		$X(-I) Y(I)$

and it might be argued that the spectroscopic oxidation states simply are more or less satisfactory approximations to the fractional charges δ on the two atoms $X^{+\delta} Y^{-\delta}$. However, this would be underestimating the utility of spectroscopic oxidation states. All the time, *one* eigen-value is found as the lower branch of the hyperbola, well separated from the next eigen-value. Hence, a classification of one-electron energies based on the admittedly crude approximation $X^+ Y^-$ *or*, for that matter, $X^- Y^+$, would produce the correct prediction that one isolated eigen-value occurs. This property becomes very useful in polyatomic chromophores such as MX_N. If the relative order of one-electron energies agrees with a definite set of spectroscopic oxidation states, say $M(IV)$ with a partly filled d-shell admittedly delocalized to some extent and the closed-shell ligands $X(-I)$, it does not matter that the fractional charges δ may be $+1.6$ on M and -0.6 on each X. If the diagonal elements of energy form a discrete pattern, the perturbation treatment taking non-diagonal elements into account does not change the *number* of interacting states; and in our usual cases of innocent ligands, it will frequently be possible to recognize the discrete pattern of eigen-values as a distorted version of the pattern of diagonal elements. Thus, in octahedral chromophores MX_6, the lower sub-shell consisting of three orbitals and the upper, σ- anti-bonding sub-shell of two orbitals will be recognized in d-group cases, though the M.O. may have expanded radial functions, definitely have additional radial nodes between M and X, and quite generally are deformed to some extent.

It may be noted in Table 12.1 that ruthenium and osmium are the two only elements varying from zero to ten d-like electrons. This would also be true for manganese if $Mn(NO)_3(CO)$ is accepted as $Mn(-III)$. The opinion that a given element at most can change its oxidation state by eight units is derived from the p-groups.

Spectroscopic studies allow the identification of oxidation states, e.g. the electron transfer bands of the rhodium (IV) complex $RhCl_6^-$ and the narrow-line luminescence of the first spin-forbidden transition of $Mn(IV)O_6$ which would not have been readily detected by conventional analytical technique.

This concept can be extended to chromophores having lower symmetry, but being approximately octahedral. Thus, NYHOLM and VRIEZE (1965) reacted yellow $Rh(III)$ As_3 Cl_2H complexes with $Hg X_2$ to form complexes containing the ligand $HgCl^-$ or $HgBr^-$ and the chromophore $Rh(III)$ As_3Cl_2Hg. The orange colour and spectra of these complexes demonstrate that the position of HgX^- in the spectrochemical series must be comparable with NH_3. Actually, one may either argue that mercury(0) has the lone-pair consisting of a mixture of 6s and 6p-character; or that the mercury atom has the conditional oxidation state Hg[II] and a σ-bonding lone-pair of unspecified composition forming the RhHg bond. The situation is comparable with the red, trigonal-bipyramidal platinum(II) complex $Pt(SnCl_3)_5^{-3}$ of the ligand $SnCl_3^-$ (CRAMER et al., 1963). These authors also prepared a yellow quadratic complex $PtCl_2$-$(SnCl_3)_2^-$. The former complex may either be described as a chloro complex of the cluster $PtSn_5^{+12}$ or as a Pt(II) complex of the Sn(II) ligand having a lone-pair with mixed $5s-5p$ character. In this connection, it is worth remembering that the spectroscopic oxidation state of one atom in a complex may be completely definite, whereas it is uncertain for some of the other atoms. Thus, $Co(NH_3)_6^{+3}$ contains low-spin $3d^6$ Co(III); but the consequences would not be very different whether the ammonia ligands are considered as nitrogen($-$III) with three protons or as nitrogen(III) hydride. Again, the ligand CN^- may be considered as a carbon(II) nitride(-III) and CH_3^- as a carbon(II) trihydride, or one may completely refrain from determining spectroscopic oxidation states of the individual atoms of the polyatomic ligands; but this does not prevent $Cr(CN)_6^{-3}$ and $Cr(CH_3)_6^{-3}$ from containing clearly recognized Cr(III).

It is perhaps worth emphasizing that $Co(NH_3)_6^{+3}$ and $Co(CN)_6^{-3}$ have the total symmetry $^1A_{1g}$ in O_h. It is frequently argued that a totally symmetric Slater determinant has the property that if the fully occupied

Table 12.1. *Spectroscopic oxidation states, preponderant electron configurations and total spin quantum numbers S known for compounds (not containing M-M bonds) of elements having the atomic number Z above 20. The list is not exhaustive. In the cases where an unusual oxidation state only occurs with definite types of ligands, their formulae are given in parentheses. Page numbers are indicated for main references to many oxidation states*

Z	Element	Oxidation state	Config.	S	Page numbers
Z = 21	Scandium	Sc(III)	$3d^0$	$S = 0$	66
Z = 22	Titanium	Ti(IV)	$3d^0$	$S = 0$	59, 175, 194 and 198
		Ti(III)	$3d^1$	$S = 1/2$	164 and 199
		Ti(II)	$3d^2$	rare	134 and 165
Z = 23	Vanadium	V(V)	$3d^0$	$S = 0$	250
		V(IV)	$3d^1$	$S = 1/2$	157 and 194
		V(III)	$3d^2$	$S = 1$	87
		V(II)	$3d^3$	$S = 3/2$	88, 98, 164, 196 and 217
		V(I)	$3d^4$	rare (dip)	164 and 218
		V(0)	$3d^5$	$S = 1/2$ (CO)	164 and 219
		V(−I)	$3d^6$	$S = 0$ (CO)	164 and 218
Z = 24	Chromium	Cr(VI)	$3d^0$	$S = 0$ (common)	196
		Cr(V)	$3d^1$	$S = 1/2$	223
		Cr(IV)	$3d^2$	$S = 1$ (rare)	157, 192 and 223
		Cr(III)	$3d^3$	$S = 3/2$ (common)	73, 85, 88, 98, 104, 106, 170, 181 and 211
		Cr(II)	$3d^4$	$S = 2$ and 1	97, 163 and 239
		Cr(I)	$3d^5$	$S = 1/2$ (rare)	164, 218 and 238
		Cr(0)	$3d^6$	$S = 0$ (CO, PF_3)	164, 218 and 251
		Cr(−I)	$3d^8$	$S = 0$ (CO)	219
Z = 25	Manganese	Mn(VII)	$3d^0$	$S = 0$	196 and 226
		Mn(VI)	$3d^1$	$S = 1/2$	—
		Mn(V)	$3d^2$	$S = 1$ (rare)	97
		Mn(IV)	$3d^3$	$S = 3/2$	88, 157 and 258
		Mn(III)	$3d^4$	$S = 2$ and 1	87, 97 and 198
		Mn(II)	$3d^5$	$S = 5/2$ (common) and $1/2$ (rare)	86, 93, 97, 194, 198 and 217
		Mn(I)	$3d^6$	$S = 0$ (CO, PF_3)	85, 135, 207, 215 and 237
		Mn(−I)	$3d^8$	$S = 0$ (CO)	214

Table 12.1 (continued)

$Z = 26$	Iron	Fe(VI)	$3d^2$	$S = 1$ (O^{--})	—
		Fe(IV)	$3d^4$	$S = 2$ and 1, rare	157, 169 and 192
		Fe(III)	$3d^5$	$S = {}^5/_2$ (common) and ${}^1/_2$	93, 159, 168, 192 and 197
		Fe(II)	$3d^6$	$S = 2$ (common) and 0	85, 97, 127, 163, 170, 197 and 221
		Fe(0)	$3d^8$	$S = 0$ (CO, PF$_3$)	165, 208 and 220
		Fe($-$II)	$3d^{10}$	$S = 0$ (CO, PF$_3$)	208 and 220
$Z = 27$	Cobalt	Co(V)	$3d^4$	$S = 2$ (O^{--})	—
		Co(IV)	$3d^5$	$S = {}^1/_2$ (F^-)	193
		Co(III)	$3d^6$	$S = 2$ (rare, F^-), 1 (rare), 0 (common)	85, 104, 167, 169, 193, 217 and 223
		Co(II)	$3d^7$	$S = {}^3/_2$ (common) and ${}^1/_2$	97, 144, 167, 199, 217 and 238
		Co(I)	$3d^8$	$S = 0$	167, 208, 219 and 223
		Co($-$I)	$3d^{10}$	$S = 0$ (CO, PF$_3$)	208 and 219
$Z = 28$	Nickel	Ni(IV)	$3d^6$	$S = 0$	66, 104, 169, 193 and 217
		Ni(III)	$3d^7$	$S = {}^3/_2$ (rare, NO_3^-) and ${}^1/_2$	156, 169 and 217
		Ni(II)	$3d^8$	$S = 1$ (common) and 0	86, 96, 105, 155, 160, 163 and 215
		Ni(0)	$3d^{10}$	$S = 0$ (CO, PF$_3$)	219 and 251
$Z = 29$	Copper	Cu(III)	$3d^8$	$S = 1$ (F^-) and 0	66
		Cu(II)	$3d^9$	$S = {}^1/_2$ (common)	152, 195 and 239
		Cu(I)	$3d^{10}$	$S = 0$	59, 128 and 163
$Z = 30$	Zinc	Zn(II)	$3d^{10}$	$S = 0$	132
$Z = 31$	Gallium	Ga(III)	$3d^{10}$	$S = 0$	—
		Ga(I)	$4s^2$	$S = 0$ (rare)	165
$Z = 32$	Germanium	Ge(IV)	$3d^{10}$	$S = 0$	215
		Ge(II)	$4s^2$	$S = 0$ (rare)	—
$Z = 33$	Arsenic	As(V)	$3d^{10}$	$S = 0$	—
		As(III)	$4s^2$	$S = 0$	—
		As($-$III)	$4s^2 4p^6$	$S = 0$	187, 242 and 247

Table 12.1 (continued)

Z	Element	Configuration	Species	S	References
Z = 34	Selenium	$3d^{10}$	Se(VI)	$S = 0$ (O—, F⁻)	—
		$4s^2$	Se(IV)	$S = 0$	130
		$4s^24p^6$	Se(−II)	$S = 0$	188, 191, 194 and 247
Z = 35	Bromine	$3d^{10}$	Br(VII)	$S = 0$ (rare, O—)	—
		$4s^2$	Br(V)	$S = 0$	235
		$4s^24p^2$	Br(III)	$S = 0$ (rare)	236
		$4s^24p^4$	Br(I)	$S = 0$	236
		$4s^24p^6$	Br(−I)	$S = 0$	130, 147 and 174
Z = 36	Krypton	$4s^24p^4$	Kr(II)	$S = 0$ (F⁻)	236
Z = 37	Rubidium	$4s^24p^6$	Rb(I)	$S = 0$	—
Z = 38	Strontium	$4s^24p^6$	Sr(II)	$S = 0$	—
Z = 39	Yttrium	$4d^0$	Y(III)	$S = 0$	—
Z = 40	Zirconium	$4d^0$	Zr(IV)	$S = 0$	175 and 210
		$4d^1$	Zr(III)	(rare)	199
Z = 41	Niobium	$4d^0$	Nb(V)	$S = 0$ (common)	86, 175 and 223
		$4d^1$	Nb(IV)	$S = 1/2$	—
		$4d^6$	Nb(−I)	$S = 0$ (CO)	164 and 238
Z = 42	Molybdenum	$4d^0$	Mo(VI)	$S = 0$ (common)	104, 146 and 155
		$4d^1$	Mo(V)	$S = 1/2$	199
		$4d^2$	Mo(IV)	$S = 1$ (rare) and 0	—
		$4d^3$	Mo(III)	$S = 3/2$	89 and 104
		$4d^4$	Mo(II)	(rare)	—
		$4d^6$	Mo(0)	$S = 0$ (CO, PF₃)	166
Z = 43	Technetium	$4d^0$	Tc(VII)	$S = 0$	207
		$4d^1$	Tc(VI)	$S = 1/2$ (F⁻)	—
		$4d^2$	Tc(V)	$S = 1$ and 0	—
		$4d^3$	Tc(IV)	$S = 3/2$	89, 104 and 146

Table 12.1 (continued)

		$4d^4$	Tc(III)	$S = 1$ (rare)	—
		$4d^6$	Tc(I)	$S = 0$ (CO, CN$^-$)	—
$Z = 44$	Ruthenium	$4d^0$	Ru(VIII)	$S = 0$ (O^{--})	65
		$4d^1$	Ru(VII)	$S = 1/2$ (O^{--})	65
		$4d^2$	Ru(VI)	$S = 1$ and 0	65 and 87
		$4d^3$	Ru(V)	$S = 3/2$ (F$^-$)	89
		$4d^4$	Ru(IV)	$S = 1$	146 and 192
		$4d^5$	Ru(III)	$S = 1/2$ (common)	146, 193, 222 and 226
		$4d^6$	Ru(II)	$S = 0$	61, 104, 193, 196, 208, 217, 221 and 224
		$4d^8$	Ru(0)	$S = 0$ (CO, PF$_3$)	208
		$4d^{10}$	Ru(−II)	$S = 0$ (PF$_3$)	221
$Z = 45$	Rhodium	$4d^3$	Rh(VI)	$S = 3/2$ (rare, F$^-$)	—
		$4d^4$	Rh(V)	$S = 1$ (rare)	146 and 258
		$4d^5$	Rh(IV)	$S = 1/2$ (rare)	104, 146, 193, 214 and 258
		$4d^6$	Rh(III)	$S = 0$ (common)	156 and 239
		$4d^7$	Rh(II)	$S = 1/2$ (rare)	209
		$4d^8$	Rh(I)	$S = 0$	220
		$4d^{10}$	Rh(−I)	$S = 0$ (PF$_3$)	
$Z = 46$	Palladium	$4d^6$	Pd(IV)	$S = 0$	105 and 146
		$4d^8$	Pd(II)	$S = 1$ (F$^-$) and 0 (common)	155, 163, 168, 175, 195 and 242
		$4d^{10}$	Pd(0)	$S = 0$ (CN$^-$, PF$_3$)	213 and 220
$Z = 47$	Silver	$4d^8$	Ag(III)	$S = 0$ (rare)	66
		$4d^9$	Ag(II)	$S = 1/2$	59 and 154
		$4d^{10}$	Ag(I)	$S = 0$ (common)	53, 67, 128, 196 and 242
$Z = 48$	Cadmium	$4d^{10}$	Cd(II)	$S = 0$	—
$Z = 49$	Indium	$4d^{10}$	In(III)	$S = 0$	—
		$5s^2$	In(I)	$S = 0$ (rare)	195

Table 12.1 (continued)

Z	Element	Ion	Config	S	Values
Z = 50	Tin	Sn(IV)	$4d^{10}$	$S = 0$	—
		Sn(II)	$5s^2$	$S = 0$	129, 219, 238 and 258
Z = 51	Antimony	Sb(V)	$4d^{10}$	$S = 0$	86 and 197
		Sb(III)	$5s^2$	$S = 0$	129 and 197
		Sb(−III)	$5s^25p^6$	$S = 0$ (rare)	187
Z = 52	Tellurium	Te(VI)	$4d^{10}$	$S = 0$	66 and 86
		Te(IV)	$5s^2$	$S = 0$	130 and 235
		Te(II)	$5s^25p^2$	$S = 0$	236
		Te(−II)	$5s^25p^6$	$S = 0$	188, 194, 206 and 242
Z = 53	Iodine	I(VII)	$4d^{10}$	$S = 0$	66
		I(V)	$5s^2$	$S = 0$	130
		I(III)	$5s^25p^2$	$S = 0$	114 and 236
		I(I)	$5s^25p^4$	$S = 0$	114 and 236
		I(−I)	$5s^25p^6$	$S = 0$ (common)	131, 147, 228 and 242
Z = 54	Xenon	Xe(VIII)	$4d^{10}$	$S = 0$ (O−−)	—
		Xe(VI)	$5s^2$	$S = 0$ (O−−, F−)	130 and 235
		Xe(IV)	$5s^25p^2$	$S = 0$ (F−)	114 and 236
		Xe(II)	$5s^25p^4$	$S = 0$ (F−, Cl−)	114, 236 and 252
Z = 55	Caesium	Cs(I)	$5s^25p^6$	$S = 0$	131
Z = 56	Barium	Ba(II)	$5s^25p^6$	$S = 0$	—
Z = 57	Lanthanum	La(III)	$5s^25p^6$	$S = 0$	—
Z = 58	Cerium	Ce(IV)	$5s^25p^6$	$S = 0$	61 and 148
		Ce(III)	$4f^1$	$S = 1/2$	61, 120 and 190
Z = 59	Praseodymium	Pr(IV)	$4f^1$	$S = 1/2$ (O−−, F−)	61 and 197
		Pr(III)	$4f^2$	$S = 1$ (common)	74, 76, 122 and 211
Z = 60	Neodymium	Nd(IV)	$4f^2$	$S = 1$ (rare, F−)	61 and 148
		Nd(III)	$4f^3$	$S = 3/2$ (common)	77, 195, 211 and 229
		Nd(II)	$4f^4$	$S = 2$ (rare)	122 and 188

Table 12.1 (continued)

$Z = 61$	Promethium	$4f^4$	Pm(III)	$S = 2$	—
$Z = 62$	Samarium	$4f^5$	Sm(III)	$S = {}^5/_2$ (common)	63
		$4f^6$	Sm(II)	$S = 3$ $(J = 0)$	123 and 188
$Z = 63$	Europium	$4f^6$	Eu(III)	$S = 3$ $(J = 0)$	63, 73, 172, 197 and 229
		$4f^7$	Eu(II)	$S = {}^7/_2$	123, 128, 181, 188 and 210
$Z = 64$	Gadolinium	$4f^7$	Gd(III)	$S = {}^7/_2$	72, 179 and 188
$Z = 65$	Terbium	$4f^7$	Tb(IV)	$S = {}^7/_2$ (O^{--}, F$^-$)	61 and 197
		$4f^8$	Tb(III)	$S = 3$	122
$Z = 66$	Dysprosium	$4f^8$	Dy(IV)	$S = 3$ (rare, F$^-$)	61 and 148
		$4f^9$	Dy(III)	$S = {}^5/_2$ (common)	—
		$4f^{10}$	Dy(II)	$S = 2$ (rare, Cl$^-$)	122 and 188
$Z = 67$	Holmium	$4f^{10}$	Ho(III)	$S = 2$	210
$Z = 68$	Erbium	$4f^{11}$	Er(III)	$S = {}^3/_2$	74, 78, 81, 179 and 210
$Z = 69$	Thulium	$4f^{12}$	Tm(III)	$S = 1$	63
		$4f^{13}$	Tm(II)	$S = {}^1/_2$ (rare)	63, 122 and 188
$Z = 70$	Ytterbium	$4f^{13}$	Yb(III)	$S = {}^1/_2$ (common)	63, 172 and 217
		$4f^{14}$	Yb(II)	$S = 0$	63, 124 and 188
$Z = 71$	Lutetium	$4f^{14}$	Lu(III)	$S = 0$	—
$Z = 72$	Hafnium	$4f^{14}$	Hf(IV)	$S = 0$	—
$Z = 73$	Tantalum	$5d^0$	Ta(V)	$S = 0$ (common)	86, 146 and 219
		$5d^1$	Ta(IV)	$S = {}^1/_2$	—
		$5d^6$	Ta($-$I)	$S = 0$ (CO)	239
$Z = 74$	Tungsten	$5d^0$	W(VI)	$S = 0$ (common)	86, 146, 198 and 219
		$5d^1$	W(V)	$S = {}^1/_2$	199
		$5d^2$	W(IV)	$S = 1$ and 0	219
		$5d^3$	W(III)		—
		$5d^6$	W(0)	$S = 0$ (CO, PF$_3$)	—

Table 12.1 (continued)

Z	Element	Config	Species	Spin	References
Z = 75	Rhenium	$5d^0$	Re(VII)	$S = 0$ (common)	207
		$5d^1$	Re(VI)	$S = 1/2$	157 and 199
		$5d^2$	Re(V)	$S = 1$ (rare) and 0	219
		$5d^3$	Re(IV)	$S = 3/2$	89 and 196
		$5d^6$	Re(I)	$S = 0$ (CO, CN$^-$, PF$_3$)	238
		$5d^8$	Re($-$I)	$S = 0$ (CO, PF$_3$)	215
Z = 76	Osmium	$5d^0$	Os(VIII)	$S = 0$ (O$^-$)	—
		$5d^1$	Os(VII)	$S = 1/2$ (rare)	—
		$5d^2$	Os(VI)	$S = 1$ (F$^-$) and 0	87 and 146
		$5d^3$	Os(V)	$S = 3/2$ (F$^-$)	89
		$5d^4$	Os(IV)	$S = 1$	87, 146, 174, 193 and 196
		$5d^5$	Os(III)	$S = 1/2$	146 and 193
		$5d^6$	Os(II)	$S = 0$	193, 208, 217 and 224
		$5d^8$	Os(0)	$S = 0$	208
		$5d^{10}$	Os($-$II)	$S = 0$ (PF$_3$)	220
Z = 77	Iridium	$5d^3$	Ir(VI)	$S = 3/2$ (F$^-$)	89 and 146
		$5d^4$	Ir(V)	$S = 1$ (F$^-$)	—
		$5d^5$	Ir(IV)	$S = 1/2$	150, 173, 193 and 196
		$5d^6$	Ir(III)	$S = 0$ (common)	105, 150, 193, 208, 213, 217, 226 and 240
		$5d^8$	Ir(I)	$S = 0$	213, 220 and 224
		$5d^{10}$	Ir($-$I)	$S = 0$ (PF$_3$)	220
Z = 78	Platinum	$5d^4$	Pt(VI)	$S = 1$ (F$^-$)	87 and 146
		$5d^5$	Pt(V)	$S = 1/2$ (F$^-$)	146 and 252
		$5d^6$	Pt(IV)	$S = 0$	105, 132, 146, 151, 193 and 237
		$5d^8$	Pt(II)	$S = 0$	129, 168, 200, 207, 213, 242 and 258
		$5d^{10}$	Pt(0)	$S = 0$ (PF$_3$)	213
Z = 79	Gold	$5d^7$	Au(IV)	$S = 1/2$ (rare, F$^-$)	—
		$5d^8$	Au(III)	$S = 0$	152, 234 and 251

Table 12.1 (continued)

Z	Element	Config	Species	S	References
		$5d^9$	Au(II)	$S = 1/2$ (rare)	154
		$5d^{10}$	Au(I)	$S = 0$	218
		$6s^2$	Au(−I)	$S = 0$ (rare)	195
$Z = 80$	Mercury	$5d^{10}$	Hg(II)	$S = 0$	218
$Z = 81$	Thallium	$5d^{10}$	Tl(III)	$S = 0$	165 and 199
		$6s^2$	Tl(I)	$S = 0$	129, 137, 196 and 216
$Z = 82$	Lead	$5d^{10}$	Pb(IV)	$S = 0$	234
		$6s^2$	Pb(II)	$S = 0$	130, 137, 212 and 219
$Z = 83$	Bismuth	$5d^{10}$	Bi(V)	$S = 0$ (rare)	—
		$6s^2$	Bi(III)	$S = 0$	130 and 242
		$6s^2 6p^2$	Bi(I)	$S = 1$ ($J = 0$) (rare)	115
		$6s^2 6p^6$	Bi(−III)	$S = 0$ (rare)	187
$Z = 84$	Polonium	$6s^2$	Po(IV)		130
		$6s^2 6p^2$	Po(II)		—
$Z = 88$	Radium	$6s^2 6p^6$	Ra(II)	$S = 0$	—
$Z = 89$	Actinium	$6s^2 6p^6$	Ac(III)	$S = 0$	—
$Z = 90$	Thorium	$6s^2 6p^6$	Th(IV)	$S = 0$	62 and 191
$Z = 91$	Protactinium	$6s^2 6p^6$	Pa(V)	$S = 0$	—
		$5f^1$	Pa(IV)	$S = 1/2$	81 and 125
$Z = 92$	Uranium	$6s^2 6p^6$	U(VI)	$S = 0$	148, 191 and 196
		$5f^1$	U(V)	$S = 1/2$ (rare)	81, 125 and 148
		$5f^2$	U(IV)	$S = 1$	81, 125, 148, 194, 198, 210 and 217
		$5f^3$	U(III)	$S = 3/2$	62 and 125
$Z = 93$	Neptunium	$6s^2 6p^6$	Np(VII)	$S = 0$ (rare, O—)	65
		$5f^1$	Np(VI)	$S = 1/2$	81 and 148
		$5f^2$	Np(V)	$S = 1$	81

Table 12.1 (continued)

Z	Element	Config	Species	S	
		$5f^3$	Np(IV)	$S=3/2$	125, 148 and 210
		$5f^4$	Np(II)	$S=2$	62 and 83
Z=94	Plutonium	$5f^2$	Pu(VI)	$S=1$	81 and 148
		$5f^3$	Pu(V)	$S=3/2$	–
		$5f^4$	Pu(IV)	$S=2$	82, 125 and 148
		$5f^5$	Pu(III)	$S=5/2$	62 and 125
Z=95	Americium	$5f^3$	Am(VI)	$S=3/2$ (O⁻⁻)	–
		$5f^4$	Am(V)	$S=2$ (O⁻⁻)	–
		$5f^5$	Am(IV)	$S=5/2$ (O⁻⁻, F⁻)	
		$5f^6$	Am(III)	$S=3$ $(J=0)$	83 and 125
Z=96	Curium	$5f^6$	Cm(IV)	$S=3$ $(J=0)$ (O⁻⁻, F⁻)	65 and 83
		$5f^7$	Cm(II)	$S=7/2$	–
Z=97	Berkelium	$5f^7$	Bk(IV)	$S=7/2$	65
		$5f^8$	Bk(III)	$S=3$	–
Z=98	Californium	$5f^9$	Cf(III)	$S=5/2$	–
		$5f^{10}$	Cf(II)	?	126
Z=99	Einsteinium	$5f^{10}$	Es(III)	$S=2$	126
Z=100	Fermium	$5f^{11}$	Fm(III)	$S=3/2$	126
Z=101	Mendelevium	$5f^{12}$	Md(III)	$S=1$	126
		$5f^{13}$	Md(II)	$S=1/2$	126

orbitals are re-arranged in linear combinations according to an arbitrary, unitary transformation, the many-electron wavefunction Ψ is not changed at all. This argument is valid for the Slater determinant; but the trouble is that the actual Ψ even for systems having the total symmetry A_{1g} or A_1 and $S = 0$ is not exactly a Slater determinant. However, the physical origin of the individuality of inner shells and penultimate orbitals in general is based on comparison with one-electron excitations or ionization. This is the difficulty for the description of diamond or CH_4 in terms of four *equivalent orbitals* situated close to each carbon atom. If one only considers the groundstate without allowing optical excitation, the actual electronic density of systems having $S = 0$ can be distributed in a rather arbitrary way on orbitals which are not eigen-functions of a one-electron operator. This remark is not by itself devastating for the concept of equivalent orbitals (related to the two-electron quanticules in Chapter 11) but one has to be rather cautious (cf. JØRGENSEN, 1962).

The word "taxonomy" means the classification of species as done in zoology or botany. When we use the word *taxology* about quantum chemistry, we are trying to draw the attention to the meta-theoretical, the higher-type aspect in BERTRAND RUSSELL's sense, propensity of preponderant electron configurations suitably chosen to *classify correctly* the symmetry types of the groundstate and the lowest excited levels. In the case of one-electron excitation, this classification of excited levels is closely related to the general existence of one-electron functions. However, in the specific case of partly filled d or f shells, we have zero-electron excitations in the sense of the $(4l+2)!/[q!(4l+2-q)!]$ individual (though partly degenerate) states of the configuration l^q having differing energy according to the symmetry of the chromophore and according to various coupling schemes. The interesting point is now that the numerous excited levels predicted frequently all or nearly all are observed. No criterion is safer for the presence of a definite spectroscopic oxidation state than the observation of such a complicated manifold of excited levels with the predicted symmetry types.

The paradoxical situation is that this classification works even though we know that Ψ of many-electron systems do not correspond to well-defined configurations. Further on, it is possible to determine spin-pairing energy parameters D and other parameters of interelectronic repulsion which satisfactorily describe the distribution of the excited levels of d^q or f^q configurations. The whole theory of such configurations is a masquerade played by Nature; it is *as if* the preponderant configurations are taxologically valid.

Frankly, we must confess that we do not yet know a good explanation why the taxological aspects of quantum chemistry are so successful. We may know in a few centuries. It is to be hoped that our species has not at that time reverted to the stone-age; or is extinct. But if people still devote their efforts to try to understand chemistry, there is hope. The empirical facts of chemistry have been fashioned into many theoretical descriptions with time; most of these descriptions have been abandoned again; but there remain always some pleasant memories. Thus, the idea of elements or principles have oscillated between extremes. If elements are indestructible and invariant prototypes of matter which can be exposed in a nice collection in a cupboard, we are close to the opinion prevailing at the end of the 19. Century. If principles occur everywhere, the world looked the simplest in 1930, when it apparently consisted only of electrons and protons, whereas today, elementary particles are nearly as numerous as BERZELIUS' elements. Obviously, the phlogistonists protected a part of the truth; if CROOKES' cathode-ray tube had been invented a little earlier, it would have been evident that phlogiston is electrons.

By the same token, positive valency and (negative or positive) oxidation numbers constitute two extremes between which chemistry has oscillated since BERZELIUS. The present book admittedly takes the side of the oxidation numbers and attempts to find heteronuclear aspects in compounds where only a small separation of fractional charges of the individual atoms occur. But chemistry is not alone; the consideration of the rainbow and the construction of spectroscopic equipment coming down along the ages through work of DESCARTES, NEWTON, FRAUNHO-FER, BUNSEN and KIRCHHOFF finally joined the chemical facts and produced the new concept of spectroscopic oxidation states. Since the properties of matter surrounding us are essentially dependent on the electronic structure alone, the combination of chemistry and spectroscopy is of outmost importance for our preliminary and provisional, but ever progressing understanding of the complicated and fascinating world.

Bibliography

CRAMER, R. D., E. L. JENNER, R. V. LINDSAY, and U. G. STOLBERG: J. Am. Chem. Soc. 85, 1691 (1963).

JØRGENSEN, C. K.: Orbitals in Atoms and Molecules. London: Academic Press 1962.

NYHOLM, R. S., and K. VRIEZE: J. Chem. Soc. 5331 (1965).

Author Index

References of the type (Y, see X) are citations of publications by X and Y, or by X, Z, Y, . . . References are not given to the writer, who is cited on the majority of the pages. In the alphabetic order, ä is represented by ae; ö and ø by oe; and ü by ue. Slightly more unusually, å is represented by a.

Subject Index

Only a few individual elements of special interest are given, because Table 12.1 on pp. 259—267 gives references to the specific oxidation states of elements having the atomic number above 20. A few concepts related to the names of two physicists, such as Hartree-Fock wavefunctions, and Russell-Saunders coupling, and in a few cases to one, such as the Landé parameter and Laporte's rule, are more readily found in the author index.

Type-setting and printing: Meister Druck, Kassel